ENVIRONMENTAL ISSUES AND GREEN BUILDINGS

ENVIRONMENTAL ISSUES AND GREEN BUILDINGS

MANSOUR SEPEHR, PH.D., PE

Mansour Sepehr, Ph.D., PE
Environmental Issues and Green Buildings

Published by Spines Publishing Platform
ISBN: 979-8-90001-378-7

Spines

CONTENTS

To my wife, Parvin Sepehr and my children

Samantha, Sohalia, Marjan and Mehrdad (Mike)

ABOUT THE AUTHOR

Dr. Sepehr is a California registered professional engineer (PE). He holds a doctoral degree in Civil and Environmental Engineering from Utah State University. He is the founder and President of SOMA Environmental Engineering, Inc. located in the Bay Area, California. During the last 30 years he has performed extensive environmental studies at various hazardous waste sites impacted by toxic chemicals. Dr. Sepehr has applied different remediation technologies for the removal of toxic chemicals from prospective construction sites aims to mitigate human health risks and minimize environmental impacts.

Since 2018, Dr. Sepehr was retained by California State University, East Bay Campus Department of Construction Management to teach undergraduate and graduate courses in environmental issues, sustainability, and concept of green buildings as well as courses in engineering and construction management.

INTRODUCTION

The term green building or high-performance building refers to the quality and characteristics of the actual structure created using the principals and methodologies of sustainable construction. Green buildings can be defined as human-health supporting facilities designed and built in a resource efficient manner using ecologically based principles. In fact, a truly sustainable green building which completely utilizes renewable energy systems, closed material loops, creating a healthy living environment and is in full integration with surrounding landscape is "non-existent". However, over the years since the development of green building idea incremental improvements are being made to fulfill its ideal concept.

The emergence of green building or sustainable construction is the response of construction industry to the rapid negative changes in earth's environment and its ecosystem. Three major changes are motivating construction industry to develop an ethical response:

1. Accelerated destruction of planetary ecosystem including massive destruction of life and biodiversity and fear of global warming,
2. Enormous increase in population and demand for natural resources by both developed and developing countries such as Brazil, Russia, India, and China (BRIC),
3. Green Building movement is coinciding with similar transformation in manufacturing, tourism, agriculture, medicine, which adopted various approaches toward greening their activities.

People like me, coming from the environmental and ecological protection side of the equation, can be driven by idealistic notions and perhaps a bit of utopian thinking. But folks from the construction or engineering side might prefer a bit more hard-headed, show me the money kind of reason for embracing sustainability. Well, the good news is that following the sustainability principle includes triple bottom line. The triple bottom line includes three major principals that should be satisfied by the green building notion. These principles include 1) social, 2) economic, and 3) ecological aspects of the high-performance green building movement. As the global population keeps growing and the emerging huge economies of China and India development to create a consumer base with aspirations of emulating the US and European model of living, demand for resources is skyrocketing – steel, cement, lumber, oil – these and other key commodities are subject to growing competition. Since much of this demand is still met from virgin sources – quarried ores, old-growth forests – and not from recycled ones, this growth in demand is accompanied by accelerated destruction of planetary ecosystems and significant changes to the biogeochemical cycles such as hydrologic and carbon cycles in our planet. The outcome of such disturbances will cause the occurrence of global warming which we are trying to avoid.

The principles of sustainability limit destruction of planetary resource and over-harvesting of natural resources and disappearance of certain materials which are about to vanish from the earth crust. Sustainable construction provides a framework for considering the effects of material selection on human health, and the environment. It has been proven that many human health problems are related to different materials used in our daily lives. A high-performance building uses whole-building design to achieve social, economic, and environmental performance that is substantially better than conventional practice. This approach requires that the design team fully collaborate from the project inception in a process often referred to as integrated design.

This book has been designed not only to consider the pre-construction activities in selecting the construction site, but it also offers a step-by-step procedure in designing the high-performance building that meets the requirements of sustainable construction principles. The design parameters can be utilized to claim credits toward rating of the built environment as outlined in U.S. Green Building Council (USGBC) rating system called "Leadership in Energy, and Environmental Design (LEED)".

The first chapter of this book outlines the methodology in selecting building location which will help the project team and the owner to evaluate:

1. The environmental conditions of building site, by conducting Phase I Environmental Site Assessment (ESA) to evaluate if the past activities at the site, have left chemicals in soil, soil gas, and

groundwater beneath site. If the outcome of ESA indicates the possibility of the presence of toxic chemicals in subsurface, then the additional field investigation should be conducted to evaluate the extent of toxic chemicals in the subsurface, before the commencement of construction activities.

2. If the site is located within the sensitive land protection areas or within the historical or high priority zones, then the location of the future built environment within these zones has a profound impact in achieving the desired green building rating using LEED guidelines. For instance, if the building site is within the wetland area or flood zone, not only would the permitting agency not grant the construction permit, but rating such structure upon construction would not be considered as a green structure.

LEED mandates conducting Phase I ESA before the construction of school projects, otherwise, such structures cannot be rated using a LEED rating system. In another word, conducting phase I ESA is a prerequisite by LEED for rating school projects. We will discuss the LEED Building Rating System for high performance green buildings in Chapter 6.

The second chapter of this book deals with preparation of stormwater pollution prevention plan (SWPPP) which is mandated by California general construction permit requirements for prevention of sediment-laden stormwater discharge into the stormwater collection network during rainy season and disturbance of soils by construction activities. Preparation of SWPPP also is required by LEED Building Rating system and carries one credit for preparation and implementation of SWPPP during construction phase of those projects, see Chapter 6.

The remaining chapters of this book (chapters 3-13) discuss how to design the built environment to reduce the building utility bills by reducing the use of electricity, water, wastewater, and promoting indoor air quality during construction and post-construction periods to enhance the health and well-being of the future building's occupants. In addition, this book discusses how to choose building material with low emissivity, with high recycling content that can be reused for saving energy and preventing the generation of greenhouse gases which are the major reason for climate change of our planet. Reducing waste during construction activities and building occupancy is extremely important to gain credits for achieving high rating using the LEED rating system introduced by Unted State Green Building Council (USGBC) in 1998. These issues will be the major topics that will be discussed in this book.

ONE
PRECONSTRUCTION ACTIVITIES

PROJECT SITE SELECTION:

Selecting the right location for the built environment plays a major role in complying with the required characteristics of high-performance buildings. Even, prior to the advent of green buildings, the construction site had to be free of toxic substances to support the health and safety of the construction crew as well as the future building occupants. For example, in California the lending institutions do not finance the construction projects prior to receiving a Phase I Environmental Site Assessment (ESA) report indicating that there is no recognized environmental condition at the site which may pose human health risk to the construction workers as well as the future occupants. In addition, the construction permitting agencies will not issue the construction permit unless they receive approval from the health agencies indicating that the project site is clean and free of toxic material and will not pose health risk to the public and site occupants in the future.

Selecting the project site for construction of a high-performance building is more critical than the conventional buildings. The selected land/site should not only environmentally meet certain standards set forth by the US Environmental Protection Agency (US, EPA) and or local regulatory agencies, it should also meet certain conditions set forth by United States Green Building Council (USGBC), as part of achieving targeted Leadership in Energy and Environmental Design (LEED) rating (i.e., certified, silver, gold, or platinum) by gaining and accumulating credits for in different categories as specified by LEED rating system, see Chapter 6. In another word, the project site should be located within the downtown areas close to public transportation, shopping centers, and supermarkets for the future inhabitants to avoid driving for running errands daily. Avoiding driving and instead taking public transportation, biking and walking as much as possible will promote human health, enhance air quality, and reduce carbon dioxide emission which will eventually reduce the concentration of greenhouse gases in the atmosphere, a number one reason for causing global warming.

For construction of certain green buildings such as school projects, there is a prerequisite or mandatory requirement for conducting Phase I Environmental Site Assessment (ESA) per ASTM Standard E-1527-21, otherwise those projects cannot be rated per LEED requirements. For other projects, conducting such Phase I ESA qualifies the project to gain certain credits toward rating. If contamination is found, Phase II ESA per ASTM Standard E1903-19 must be conducted. If contaminated, the site must be investigated to evaluate the extent of contamination in soil, soil gas, and groundwater. Once the site is fully characterized, the project team should remove contaminants from different media under the regulatory oversight agencies. The remediation scheme should use the most stringent of federal, state, or local agencies cleanup levels to protect the future site occupants, as well as the environment. If there are state or local equivalents of Phase I and Phase II ESAs, those may be used in lieu of the ASTM Standards.

As it becomes clear, the project site should be studied environmentally to avoid exposing the future inhabitants to toxic chemicals and reducing potential lawsuits and occurrence of sick building syndrome (SBS) within the built environment. As such, prior to any planning for constructing a high-performance building conducting a Phase I ESA is highly recommended. Conducting the Phase I ESA not only is the requirement of many lending institutions for granting construction loans but is also mandated/encouraged by LEED rating system set forth by the USGBC. The following is a brief description of steps required in conducting Phase I ESA per ASTM E1527-21.

PHASE I ENVIRONMENTAL SITE ASSESSMENT

Phase I Environmental Site Assessment (ESA) is conducted in accordance with American Society for Testing and Materials (ASTM) document. This document needs to be followed and is entitled (ATSM E 1527-NN). NN is the version number of the document which is updated every 8 years.

The purpose of this practice is to define good commercial and customary practice in the United States of America for conducting an environmental site assessment of a parcel of commercial real estate with respect to the range of contaminants within the scope of the Comprehensive Environmental Response, Compensation, and Liability Act (CERCLA) and petroleum products.The current version of this document is ASTM E 1527-21. This practice is intended for use on a voluntary basis by parties who wish to assess the environmental condition of commercial real estate by collecting known and reasonably ascertainable information without collecting soil, soil gas and groundwater samples. While use of this practice is intended to constitute all appropriate inquiries for purposes of site assessment, it is not intended that its use be limited to that purpose. This practice is intended primarily as an approach to conducting an inquiry designed to identify recognized environmental conditions in connection with a subject property. The term recognized environmental condition means (1) the presence of hazardous substances or petroleum products in, on, or at the subject property due to a release to the environment; (2) the likely presence of hazardous substances or petroleum products in, on, or at the subject property due to a release or likely release to the environment; or (3) the presence of hazardous substances or petroleum products in, on, or at the subject property under conditions that pose a material threat of a future release to the environment. A de minimis condition is not a recognized environmental condition.

Conducting Phase I ESA using ASTM guidelines is intended to reflect good commercial and customary practice, which is designed to assist the user in developing information about the environmental condition of any property which is going through transaction and the owner/lending institution is planning to invest heavily in developing this property.

The Phase I ESA must be performed by an environmental professional. No practical standard can be designed to eliminate the role of judgment and the value and need for experience in the party performing the inquiry. The professional judgment of an environmental professional is, consequently, vital to the performance of all appropriate inquiries.

No environmental site assessment can wholly eliminate uncertainty regarding the potential for recognized environmental conditions in connection with any site. The performance of this practice is intended to reduce, but not eliminate, uncertainty regarding the potential for recognized environmental conditions in connection with a subject property, and this practice recognizes reasonable limits of time and cost.

All appropriate inquiries do not mean an exhaustive assessment of a property. There is a point at which the cost of information obtained, or the time required to gather it outweighs the usefulness of the information and, in fact, may be a material detriment to the orderly completion of transactions. One of the purposes of this practice is to identify a balance between the competing goals of limiting the costs and time demands inherent in performing an environmental site assessment and the reduction of uncertainty about unknown conditions resulting from additional information.

THE COMPONENTS OF PHASE I ESA

In conducting the Phase I ESA, the following items will be reviewed to evaluate if there are any recognized environmental conditions at the site. These items are only reviewed from the existing documents which have been already filed and can be found in federal, state, city and or local regulatory websites and files. No extra attempt is required by the environmental specialist to ascertain and prepare the answers in connection with the environmental condition of the site. The following is the list of items that need to be reviewed and investigated by the environmental specialist in conducting Phase I ESA:

- Review of City Directory & Permits
- Review of Past Activities Conducted at the Site
- Review of Relevant State and Regulatory Databases and Records
- Historical Information Such as Sanborn Maps and Aerial Photographs
- Inspection of Site and Surrounding Properties
- Interviews with Key Persons; and
- Development of a Report

The following are the descriptions of the above-mentioned items and the source of information to be gathered:

RECORDS REVIEW

The records review includes searching through Federal, State and Local lists which identify problem sites. The approximate minimal search distance may vary depending on the lists; however, it is typically a one-mile radius for state and Federal hazardous waste site lists and a one-quarter mile radius for other lists. The search is performed by a data management computer service. Information about the subject property's physical setting includes:

- Aerial photographs
- City directories
- Radius map
- Sanborn maps
- Tax records and
- Historical topographical maps

The above information can be found in a computerized database provided by Environmental Data Record (EDR) of LIGHTBOX, Inc.

Periodical aerial photographs taken by various government agencies provide valuable information about the past development activities at the site and development of nearby facilities next to the subject property. These aerial photos can be utilized to identify the date of site development and type of the surrounding facilities from as early as 1880s until the current date.

The City-Directory Database reveals the type of activities that were conducted at the site and nearby properties. Most importantly, the city directory shows if any businesses such as dry-cleaning facilities, gas stations, plating shops or manufacturing facilities or auto mechanic shops, or auto body shops have operated at the site and nearby properties. The City-Directory also shows the type of permits that have been issued for the subject property including construction and remodeling permits for the site and neighboring properties. The information helps to search these addresses in governmental databases such as Regional Water Quality Control Board's GeoTracker, or Department of Toxic Substances Control (DTSC)'s ENVIROSTOR Database or the local city or county's websites for further evaluation of the environmental condition of the subject property.

Radius Map includes very important information about the site and surrounding facilities. It indicates if the site exists in any of the hazardous waste site databases, and it also includes the address of the known hazardous waste sites within the one-mile-radius of the site. It gives a narrative about each of the hazardous waste sites and indicates if these sites are still active or have received regulatory closure from the DTSC, RWQCB or the local agencies. It also may include approximate groundwater flow direction, depth to groundwater, elevation of the site and type of prevailing soil around the subject site. In addition, it reveals if the site is located within 100-year flood or 500-year flood zones or located within US wetland areas. It also includes approximate average radon gas concentration in subsurface in the county where the site is located. The latest Radius map database has advanced features that can display Sanborn maps and aerial photographs and historical topographic maps which may assist the environmental engineer to see if the site historically was located within a hazardous waste site or located in a proximity of any potentially hazardous waste site such as a gasoline service station, mechanic shop or dry-cleaning and autobody facility.

Sanborn maps are detailed maps of U.S. cities and towns in the 19th and 20th centuries. Originally published by the Sanborn Map Company (Sanborn), the maps were created to allow fire insurance companies to assess their total liability in urbanized areas of the United States. Since they contain detailed information about properties and individual buildings in approximately 12,000 U.S. cities and towns, Sanborn maps are valuable for documenting changes in the built environment of American cities over many decades.

Sanborn held a monopoly over fire insurance maps almost throughout the 20th century, but the business declined as US insurance companies stopped using maps for underwriting in the 1960s. The last Sanborn fire maps were published in microfilm 1977, but old Sanborn maps remain useful for historical research into urban geography. The license for the maps was acquired by a land data company called Environmental Data Resources (EDR), and EDR was acquired in 2019 by a real estate services company, called LightBox, Inc.

Sanborn Maps is a helpful data package provided by EDR. These maps go back over 100 years and provide extremely good information about the site's history and the surrounding facilities. They also provide the type of activities and structures built around the subject site prior to its development. In addition, it can guide the environmental engineer to figure out when the site was developed, and the type of businesses conducted in the proximity of the site.

It is important to note that Phase I ESA report should recommend the possibility of presence of the lead and asbestos at the site. This will help the property owner to survey the site prior to site demolition/deconstruction activities. It should be noted that no lead or asbestos need to be surveyed if the site was developed after 1979.

Unfortunately, not every property has been mapped in the past, therefore, certain properties may have not been included in Sanborn maps and as was mentioned, after 1977 no Sanborn maps were developed.

Tax records are sometimes important to evaluate the current owner of the property and date the subject property was recorded under the current ownership and help to find out if tax bill is current.

The historical topographical maps may show the historical changes in land topography over the years which can be helpful to understand the historical changes around the subject site.

SITE INSPECTION AND SURROUNDING PROPERTIES

The objective of site reconnaissance of publicly accessible areas is to obtain information indicating the likelihood of identifying recognized environmental conditions in connection with the site.

The environmental engineer should inspect the site and the surrounding facilities to make sure that there is nothing which may indicate on-going hazardous waste storage or their release to the environment. The objective of site reconnaissance of publicly accessible areas is to obtain any information indicating the likelihood of identifying recognized environmental conditions in connection with the site. During the site reconnaissance, the inspector should make notes and take photos and look for the presence of the following features:

- Hazardous substances and/petroleum products,
- Aboveground or underground hazardous substances, or petroleum product storage tanks, hazardous substances, and unidentified substance containers, electrical or mechanical equipment likely to contain fluids,
- Interior stains or corrosion,
- Strong, pungent, or noxious odors,
- Pools of liquids, drains, sumps, and clarifiers pits,
- Ponds and lagoons,
- Stained soil or pavement,
- Stressed vegetation,
- Solid waste disposal or evidence of fill materials, wastewater discharges,
- Wells and transformers, and
- Septic system.

During the site inspection, the site reconnaissance should include a visual inspection for indications of water intrusions or the presence of active mold growth on readily accessible interior and exterior surfaces. Confirmation sampling is not included for Phase I ESA assessment. However, readily accessible areas should be observed for visual or olfactory indications of mold, and for areas of any visible water damage or olfactory indications of mold.

INTERVIEWS WITH KEY PERSONS

As part of Phase I ESA investigation and in accordance with the Landowner Liability Protections (LLPs) offered by the Small Business Liability Relief and Brownfields Revitalization Act of 2001 (the "Brownfields Amendments"), the environmental professional should conduct a written or an oral interview with a person who is the most knowledgeable about the site in order to get additional information, if any, about the history of the site. At a minimum, a written interview in the form of a questionnaire should be prepared and sent to that person. A copy of an example questionnaire is included at the end of this chapter.

ASBESTOS-CONTAINING MATERIAL

Although the asbestos survey is not part of the Phase I ESA requirement, the report should have a narrative to explain if asbestos is suspected to be used in construction of the site. If per its development date the use of asbestos in construction material is justified, then the report should recommend an asbestos survey before demolition/deconstruction of site for remodeling or redevelopment activities. The following is a brief description of building materials that may have contained asbestos.

Asbestos is the name for a group of naturally occurring silicate minerals that can be separated into fibers. The fibers are strong, durable, and resistant to heat and fire. They are also long, thin, and flexible, so they can even be woven into cloth. Because of these qualities, asbestos has been used in thousands of industrial, maritime, automotive, scientific, and building products. During the 20th century, some 30 million tons of asbestos have been used in industrial sites, homes, schools, shipyards, and commercial buildings in the United States. Common asbestos-containing materials (ACMs) include pipe-covering, insulating cement, insulating block, boiler insulation materials, board, fireproofing spray, joint compound, vinyl floor tile, ceiling tile, mastics, roofing products, and duct insulation for Heating, Ventilation and Air Conditioning (HVAC) applications. Inhalation of asbestos fibers can result in deleterious health effects. The DTSC has classified friable, finely divided, and powdered wastes containing greater than one percent (1.0%) asbestos as hazardous waste. Asbestos includes all mineralogical forms of chrysotile, amosite, crocidolite, tremolite, anthophyllite, and actinolite. A friable material is one that can be reduced to powder or dust under hand pressure when dry. The federal government has banned the production of friable asbestos-containing materials (ACMs) since 1979. Asbestos generally becomes a health hazard when individual fibers become airborne and are inhaled by the public. ACMs, therefore, are not considered a health hazard if they are properly maintained and not disturbed. No regulation requires property owners to remove ACMs if they are maintained properly. ACMs must be removed prior to demolition or renovations where their disturbance becomes imminent. Section 25915, Division 20 of the California Health and Safety Code, dated 1988, indicates that owners of public and commercial buildings constructed prior to 1979 must disclose the presence of asbestos to their tenants and workers doing business or work in areas where they could be exposed to asbestos. Additionally, Section 25915 also requires that knowledge of ACMs be disclosed during real estate transactions. According to the EPA, prior to the 1970s many types of building products and insulation materials used in homes contained asbestos.

National Emission Standards for Hazardous Air Pollutants (NESHAP)'s section 61.145, part (c), subpart (1) requires removal of all regulated asbestos-containing material (RACM) from a facility being demolished or renovated before any activity begins that would break up, dislodge, or similarly disturb the material or preclude access to the material for subsequent removal. Table 1-1 summarizes a sample list of typical suspect asbestos-containing materials:

Cement Pipes	Elevator Brake Shoes
Cement Wallboard	HVAC Duct Insulation
Cement Siding	Boiler Insulation
Asphalt Floor Tile	Breaching Insulation
Vinyl Floor Tile	Ductwork Flexible Fabric Connections
Vinyl Sheet Flooring	Cooling Towers
Flooring Backing	Pipe Insulation
Construction Mastics (Floor Tile, Carpet, Ceiling Tile, etc.)	Heating and Electrical Ducts
Acoustic Plaster	Electrical Cloth
Textured Paints/Coatings	Electrical Wiring Insulations
Decorative Plaster	Electrical Cloth
Ceiling Tiles & Lay-in Panels	Chalkboards
Packing Materials (for wall/floor penetrations)	Fire Doors
Spray-Applied Insulation	Roofing Shingles
Blown-in Insulation	Roofing Felt
Fireproofing Materials	Base Flashing
Taping Compounds (thermal)	Thermal Paper Products
Packing Materials (for wall/floor penetrations)	Fire Doors
High Temperature Gaskets	Caulking/Putties
Laboratory Hoods/Tabletops	Adhesives
Laboratory Gloves	Wallboard
Fire Blankets	Joint Compounds
Fire Curtains	Vinyl Wall Covering
Elevator equipment Panels	Spackling Compounds
Note: *The above list does not include every product/material that may contain asbestos. It is intended as a general guide to show which types of materials may contain asbestos.*	

Table 1-1 List of Suspect Asbestos-Containing materials

LEAD-BASED PAINT

As asbestos, the investigation and sampling of the site for determination of the lead contamination and its extent in soils and possibly in groundwater is not the scope of the Phase I ESA. However, the following is a narrative to understand the implication of lead in environment which needs to be addressed as part of reporting and recommendation in Phase I ESA.

Lead-based paint (LBP) or other surface coatings (paint, varnish, stain, or other applied coating) are considered material if they contain an amount of lead equal to, or more than one milligram per square centimeter (1.0 mg/cm^2), more than 5,000 ppm, or more than 0.5 percent by weight. Section 1017 of the Housing and Urban Development (HUD) Guidelines, Residential Lead-Based Paint Hazard Reduction Act of 1992, otherwise known as "Title X, defines a LBP hazard as "any condition that causes exposure to lead that would result in adverse human health effects" resulting from lead-contaminated dust, bare, lead-contaminated soil, and/or lead-contaminated paint that is deteriorated or present on accessible, friction, or impact surfaces. Therefore, under Title X, intact lead-based paint on most walls and ceilings would not be considered a "hazard", although the paint should be maintained and monitored to ensure that it does not deteriorate and become a hazard. It is important to note that LBP may create a lead hazard when being removed. The condition of these materials must be monitored when they are being disturbed. In the event LBP is subject to abrading, sanding, torching and/or cutting during demolition or renovation activities, there may be regulatory issues that must be addressed. Engineering controls and air monitoring shall be implemented during disturbance of chipping or deteriorated LBP to eliminate the risk of public and worker exposure. Prior to any demolition work, a licensed asbestos and lead inspector should conduct a site inspection to determine a necessity for any abatement work.

Abatement projects producing hazardous lead waste must meet land disposal restrictions under the Resource Conservation and Recovery Act (RCRA). Cal/OSHA requires that worker safety protection be implemented to prevent lead exposure above the Permissible Exposure Level (PEL) during construction activities. The PEL for lead is 50 $\mu g/m^3$.

SEARCHING GOVERNMENTAL RECORDS

The next step is to search for government records to find out if the property address is included in government databases due to unauthorized release of chemicals in the past. These sites may include the following regulatory websites:

- WWW.Geotracker.CA.GOV:

GeoTracker is the Water Boards' data management system for sites that impact, or have the potential to impact, water quality in California, with emphasis on groundwater. GeoTracker contains records for properties that require cleanup, such as Leaking Underground Storage Tank (LUST) Sites, Department of Defense Sites, and Cleanup Program Sites. GeoTracker also contains records for various unregulated projects as well as permitted facilities including:

1. Irrigated Lands,
2. Oil and gas production,
3. Operating Permitted USTs, and
4. Land Disposal Sites.

GeoTracker portals retrieve records and view integrated data sets from multiple State Water Board programs and other agencies. Users can view these data through a Google Maps through GIS interface.

- WWW.ENVIROSTOR.CA.GOV:

ENVIROSTOR is the Department of Toxic Substances Control's data management system for tracking the cleanup, permitting, enforcement and investigation efforts at hazardous waste facilities and sites with known contamination or sites where there may be reasons to investigate further. To search for specific cleanup sites or hazardous waste facilities, use the Site/Facility address.

WHO SHOULD PERFORM PHASE I ESA AND WHAT ISSUES BE EXCLUDED?

Any environmental scientist/engineer or geologist, geographer with proficiencies in geologic Information system (GIS) is qualified to conduct Phase I ESA. The Phase I ESA excludes the following items:

- Inspection for asbestos
- Inspection for lead-based paint (LBP)
- Wetlands inspection
- Environmental compliance issues such as storm water
- Sampling of soil, groundwater, or soil gas

REPORTING

Finally, after conducting record search, site reconnaissance and conducting interviews the Phase I ESA report should be prepared and should contain, a copy of related maps, records and photos showing the interior and exterior of the property as well as the descriptions of the nearby hazardous waste sites that may have impacted the site or may impact it in the future along with conclusion and recommendation to the site owner/client.

Once the Phase I ESA is complete, the Environmental Professional will summarize what concerns were identified on the property and make recommendations about what actions, if any, are needed to address these concerns.

By definition, a **recognized environmental condition** (REC) indicates known contamination or the potential for the subsurface to have been impacted by contamination (either from the subject property or possibly from an offsite source(S)).

A **controlled recognized environmental condition** (CREC) identifies that the property has been impacted by contamination which has been investigated and remediated; however, contamination remains and would require additional work if redeveloped.

A **historical recognized environmental condition** (HREC) identifies a release that impacted the subject property, which has been investigated and remediated, meeting unrestricted use criteria.

The report at a minimum should include the following:

1. How many years of historical records were reviewed,
2. From aerial photos and Sanborn maps indicate when the site was developed and elaborate on the neighboring properties and discuss their activities briefly,
3. If the Radius map indicates the site is located within wetland area or within 100-year or 500-year flood zone,

4. Summarize the results of radius search and list the number of hazardous waste sites that are in the higher/lower elevations with respect to the site and indicate their potential impact to the subject property,
5. Summarize the results of local government databases search, such as GeoTracker, ENVIROSTOR and local county, and city website search results,
6. Closest waterbody to the site,
7. Site elevation and soil type,
8. Depth to groundwater and groundwater flow direction,
9. If the site was found in hazardous waste material databases,
10. List the type of businesses conducted at the subject site by reviewing City Directory database,
11. Comment if the site may have a potential lead or asbestos problem, based on its development age, and finally,
12. If you recommend conducting a Phase II Environment Site Assessment (ESA).

If the owner of the project is planning to redevelop the site and start the construction activities, it is important to see if the property is in a flood-prone area. Such information can be learned by conducting Phase I ESA. Clearly, buildings should not be constructed in flood-prone areas, due to the high potential for disasters that result not only in human loss, but also in enormous environmental and resource impacts caused by the cycle of destruction and rebuilding.

This is such a vital matter that the Federal Emergency Management Agency (FEMA) has become deeply involved in issues of flood mapping and insurance. Specifically, in support of the National Flood Insurance Program (NFIP). FEMA has undertaken a massive program of flood hazard identification and mapping to produce Flood Hazard Boundary Map, Flood Insurance Rate Maps, and Flood Boundary and Floodway Maps. EDR Sanborn maps, which is produced first by the fire insurance industry, uses these maps to determine if the site is located within 100-year or 500-year flood zones. The term 100-year of 500-year flood zone is somehow misleading. It is not the flood which will occur every 100 years or 500 years, rather it is the flood elevation that has a one percent chance of being equaled or exceeded each year. Thus the 100-year flood could occur more than once in a relatively short period of time.

To earn points when attempting to certify a green building using Green Globes, the elevation of the building site must be at least 3 feet above the 100-year floodplain.

PHASE II ENVIRONMENTAL SITE ASSESSMENT

When a Phase I ESA identifies a recognized environmental condition (REC) or the potential for impacts to the subsurface at a site, most clients request to evaluate the potential impacts by performing Phase II Environmental Testing. The presence of any type of REC or an environmentally impacted property can greatly reduce its value. Stakeholders want to reduce liability and future cleanup expenses on their investment by conducting a Phase II ESA, in which a subsurface investigation tests soil, soil gas and/or groundwater to identify sources and extent of environmental impacts at the site. The extent of the impact is usually revealed during the follow up work though additional site investigation prior to the removal and remediation activities under regulatory oversight program.

If a buyer or lender doesn't have the full scope and realistic picture of a seemingly limited REC, they could be leaving themselves exposed to hidden risks. Phase II ESA can limit this risk, and in many cases, it can also protect against substantial long-term costs and complex environmental liabilities.

The purpose of a Phase II ESA report is to evaluate the presence or absence of, petroleum products or hazardous substances in the site's subsurface. A trained, licensed, experienced staff of geologists and engineers that possesses expertise in Phase II Environmental project design performs these assessments per the ASTM 1903-19 Standard Guide.

ASTM E1903-19 provides some basic parameters for Phase II Environmental Assessments, but there is a huge amount of Professional judgment that goes into the final report. A lot more goes into the scope of work than just the number of samples taken – for example what type of drilling and sample collection is appropriate for the site and the kinds of contaminants you are looking for. That's why it pays to engage an informed, experienced due diligence consultant who understands what's involved in the assessment and the level of certainty your institution requires.

When designing a Phase II ESA scope, the environmental professional accounts for any areas of concern, chemicals of concern, local geology and/or site access issues as well as local, state, and federal regulations. An accurate, helpful Phase II ESA delivers local knowledge of geologic and regulatory environments and then interprets geological and chemical data to the client so that they are fully informed of their business risk.

Drilling methods used most often by scientists and geologists during Phase II Environmental Testing projects include:

- Direct Push Technology (DPT)
- Hollow Stem Auger
- Hand Auger
- Mud Rotary
- CPT Drilling

WHEN IS A LIMITED PHASE II ESA NEEDED?

There are circumstances in the due diligence process when a potential property owner does not want to invest in a full Phase II ESA. They may instead opt for a limited Phase II sampling, which is conducted to confirm the presence of a pollutant and may be limited by locations sampled, number of samples, media sampled, or a combination. A buyer may conduct a limited Phase II ESA to evaluate the following scenarios to inform their transaction decisions:

1. The identified REC from a Phase I ESA is minor or limited in scope,
2. To confirm a REC that presents more of a risk than the buyer is willing to accept,
3. To identify a REC that requires more discovery (a full-scale Phase II ESA) to confirm the extent of contamination.

If the buyer decides to go further and evaluate the extent of contamination and evaluate the associated cleanup costs, the involvement of the local environmental oversight agency is recommended. Depending upon the nature of contamination and the impacted media (e.g., soil or groundwater) one of the following agencies might be involved:

1. Regional Water Quality Control Board
2. Department of Toxic Substances Control or

3. Local County's Environmental Health Department

Once, the project owner or its consultant contacted one of the above agencies, then based on the nature of contamination and affected media one the above agencies through the Memorandum of Understanding will be involved to direct the responsible party during the site characterization and remediation phase of the impacted property.

A Sample Questionnaire Need to Be Completed by the Most Familiar Person to the Project Site:

USER QUESTIONNAIRE

Site Address:

Respondent Name: ______________________

For Landowner Liability Protections (LLPs) offered by the Small Business Liability Relief and Brownfields Revitalization Act of 2001 (the "Brownfields Amendments"), the user must provide the following information (if available) to the environmental professional. As per the Brownfields Amendments, failure to provide this information could result in a determination that "all appropriate inquiry" is not complete.

Contact Information:

Phone:______________________________

Address:_____________________________

Email:_______________________________

1. Site Building(s) Age, Construction Type, and Current Use

____

____

____

____

____

2. Environmental Cleanup Liens

Are you aware of any environmental cleanup liens, against the property that are filed or recorded under federal, tribal, state or local law?

☐ Yes ☐ No

3. Activity and Land Use Limitations

Are you aware of any AULs, such as engineering controls, land use restrictions or institutional controls that are in place at the site and/or have been filed or recorded in a registry under federal, tribal, state or local law?

☐ Yes ☐ No

4. Specialized Knowledge or Experience

Do you have any specialized knowledge or experience related to the property, nearby properties, or are you involved in the same line of business as the current or former occupants of the property or an adjoining property so that you would have specialized knowledge of the chemicals and processes used by this type of business? Please include any pertinent information. Feel free to attach additional pages, as needed.

☐ Yes ☐ No

____

____

____

____

____

5. Purchase Price vs. Fair Market Value

Does the purchase price paid for this property reasonably reflect the fair market value of the property?

☐ Yes ☐ No

6. Commonly Known or Reasonably Ascertainable Information

Are you aware of commonly known or reasonably ascertainable information about the property that would help the environmental professional to identify conditions indicative of releases or threatened releases?

☐ Yes ☐ No

6a. Do you know the past uses of the property? Please include any pertinent information, as needed.

☐ Yes ☐ No

____

____

____

6b. Do you know of specific chemicals that are present or were once present at the property?

☐ Yes ☐ No

____

____

____

6c. Do you know of any spill(s) or other chemical releases that have taken place at the property?

☐ Yes ☐ No

____

____

____

6d. Do you know of any environmental cleanups that have taken place at the property?

☐ Yes ☐ No

____

____

____

7. Obviousness of the Presence or Likely Presence of Contamination

As the user of this ESA, based on your knowledge and experience related to the property are there any obvious indicators that point to the presence or likely presence of contamination at the property?

☐ Yes ☐ No

____

____

____

Name:____

Signature& Date:__________________________

REFERENCES:

ASTM Standard 1527-21, Practice for Environmental Site Assessment: Phase I Environmental Site Assessment Process

ASTM Standard E-1903-19 Practice for Environmental Site Assessments: Phase II Environmental Site Assessment Process

Comprehensive Environmental Response, Compensation and Liability Act (CERCLA) 1980 Superfund Public Law: P.L.96-510

USGBC- United States Green Building Council, 1998.

TWO
CONSTRUCTION GENERAL PERMIT

BACKGROUND

In 1972, the Federal Water Pollution Control Act (also referred to as the Clean Water Act [CWA]) was amended to provide that the discharge of pollutants to waters of the United States from any point source is unlawful unless the discharge is in compliance with National Pollutant Discharge Elimination System (NPDES) provisions. The 1987 amendments to the CWA added Section 402(p), which establishes a framework for regulating municipal and industrial storm water discharges under the NPDES Program.

On November 16, 1990, the U.S. Environmental Protection Agency (USEPA) published final regulations that established storm water permit application requirements for specified categories of industries. The regulations provide that the dischargers of storm water to waters of the United States from construction projects that encompass five or more acres of soil disturbance are effectively prohibited unless the discharge is in compliance with an NPDES

Permit. Regulations (Phase II Rule) that became final on December 8, 1999, lowered the permitting threshold from five acres to one acre. While federal regulations allow two permitting options for storm water discharges (Individual Permits and General Permits), the State Water Board has elected to adopt only one statewide General Permit at this time that will apply to most storm water discharges associated with construction activity.

On August 19, 1999, the State Water Board reissued the General Construction Storm Water Permit (Water Quality Order 99-08-DWQ). On December 8, 1999, the State Water Board amended Order 99-08-DWQ to apply to sites as small as one acre. The General Permit accompanying this fact sheet regulates storm water runoff from construction sites.

Regulating many storm water discharges under one permit will greatly reduce the administrative burden associated with permitting individual storm water discharges. To obtain coverage under this General Permit, dischargers shall electronically file the Permit Registration Documents (PRDs), which include a Notice of Intent (NOI), Storm Water Pollution Prevention Plan (SWPPP), and other compliance related documents required by this General Permit and mail the appropriate permit fee to the State Water Board. It is expected that as the storm water program develops, the Regional Water Quality Control Boards (Regional Water Boards) may issue General Permits or Individual Permits containing more specific permit provisions. When this occurs, this General Permit will no longer regulate those dischargers.

WHO IS EXEMPT FROM DOING SWPPP?

The following is the list of activities that do not require the preparation of SWPPP:

a. Routine maintenance to maintain original line and grade, hydraulic capacity, or original purpose of the facility,
b. Disturbances to land surfaces solely related to agricultural operations such as disking, harrowing, terracing, leveling, and soil preparation,
c. Discharges of storm water from areas on tribal lands; construction on tribal lands is regulated by a federal permit,
d. Discharges of storm water within the Lake Tahoe Hydrologic Unit. The Lahontan Regional Water Board has adopted its own permit to regulate storm water discharges from construction activity in the Lake Tahoe Hydrologic Unit (Regional Water Board 6SLT). Owners of construction projects in this watershed must apply for the Lahontan Regional Water Board permit rather than the statewide Construction General Permit.
e. Construction projects within the Lahontan region must also comply with the Lahontan Region Project Guideline for Erosion Control (R6T-2005-0007 Section), which can be found at:Lake Tahoe Municipal NPDES Permit | Lahontan Regional Water Quality Control Board
f. Construction activity that disturbs less than one acre of land surface, unless part of a larger common plan of development or the sale of one or more acres of disturbed land surface.

g. Construction activity covered by an individual NPDES Permit for storm water discharges.
h. Landfill construction activity that is subject to the Industrial General Permit.
i. Construction activity that discharges to Combined Sewer Systems.
j. Conveyances that discharge storm water runoff combined with municipal sewage.
k. Discharges of stormwater identified in CWA

EPA'S SMALL CONSTRUCTION EROSIVITY FACTOR

EPA's Stormwater Phase II Final Rule provides an option called "Small Construction Rainfall Erosivity Waiver". This waiver applies to small construction sites between 1 and 5 acres, and allows permitting authorities to waive those sites that do not have adverse water quality impacts. Dischargers eligible for this waiver are exempt from Construction General Permit Coverage. To obtain the waiver, the discharger must certify to the State Water Board that small construction activity will occur only when the rainfall erosivity factor is less than 5 ("R" in the Revised Universal Soil Loss Equation). The rainfall erosivity will be discussed further in this chapter. The period of construction activity begins at an initial earth disturbance date and ends with final stabilization. Where vegetation will be used for final stabilization, the date of installation of a practice that provides interim non-vegetative stabilization can be used for the end of the construction period.

The operator must agree (as a condition waiver eligibility) to periodically inspect and properly maintain the area until the criteria for final stabilization as defined in the General Permit have been met. If use of this interim stabilization eligibility condition was relied on to qualify for the waiver, signature on the waiver with a certification statement constitutes acceptance of and commitment to complete the final stabilization process. The dischargers must submit a waiver certification to the State Board prior to commencing construction activities.

USEPA funded a cooperative agreement with Texas A&M University to develop an online rainfall erosivity calculator. Dischargers can access the calculator from EPA's website at:

http://lew.epa.gov/

Use of the calculator allows the discharger to determine potential eligibility for the rainfall erosivity waiver. It may also be useful in determining the time periods during which construction activity could be waived from permit coverage.

OBTAINING AND TERMINATING PERMIT COVERAGE

The Legally Responsible Person (LRP) must obtain coverage under this General Permit, except in two limited circumstances. First, where the construction of pipelines, utility lines, fiber-optic cables, or other linear underground/overhead projects will occur across several properties, the utility company, municipality, or other public or private company that owns or operates the linear underground/ overground project is responsible for obtaining coverage under the General Permit. Second, where there is a lease of a mineral estate (oil, gas, geothermal, aggregate, precious metals, and/or industrial metals), the lessee is responsible for obtaining coverage under the General Permit. To obtain the coverage, the LRP or other entity described above must file Permit Registration Documents (PRDs) prior to the commencement of construction activity. Failure to obtain coverage under this General Permit for stormwater discharges to waters of the United States is a violation of the CWA, California Water Code, State Water Resources Control Board, General Permit for Stormwater Discharges (September 2009).

To obtain coverage under this General Permit, LRPs must electronically file the PRDs, which includes a notice of Intent (NOI), Stormwater Pollution Prevention Plan (SWPPP), and other documents required by the General Permit, and mail the appropriate permit fee to the State Water Board. It is expected that as the stormwater program develops, the Regional Water Boards may issue General Permit or Individual Permits that contain more specific permit provisions.

RISK DETERMINATION OF THE SITE

To evaluate the overall risk of letting rainfall run-off leaves the project site and discharges into the state receiving waters, the involved risk needs to be evaluated. There are two major requirements related to site planning and risk determination in this General Permit. Therefore, the project's overall risk is broken up into two elements:

1. Project sediment risk (the relative amount of sediment that can be discharged, given the project location details) and
2. Receiving water risk (the risk sediment discharges pose to the receiving waters).

PROJECT SEDIMENT RISK

Project Sediment Risk is determined by multiplying the R, K, and LS factors from the Revised Universal Soil Loss

Equation (RUSLE) to obtain an estimate of project-related bare ground soil loss expressed in tons/acre. The RUSLE equation is as follows:

$$A = (R)(K)(LS)(C)(P)$$

Where:

A =The rate of sheet and rill erosion in tons per acre,
R =Rainfall-runoff erosivity factor
K =Soil erodibility factor
C = Cover factor (erosion Control)
P = Management operation and support practices & sediment control

The C and P factors are given values of 1.0 to simulate bare ground conditions.

There is a map option and a manual calculation option for determining soil loss. For the map option, the R factor for the project is calculated using the online calculator at http://lew.epa.gov/

The product of K and LS are shown on Figure 2-1, the statewide map of K*LS. To determine soil loss in tons per acre, multiply the R factor by the value of K times LS from the map.

Soil loss of less than 15 tons/acre is considered low sediment risk.

Soil loss between 15 and 75 tons/acre is medium sediment risk.

Soil loss over 75 tons/acre is considered high sediment risk.

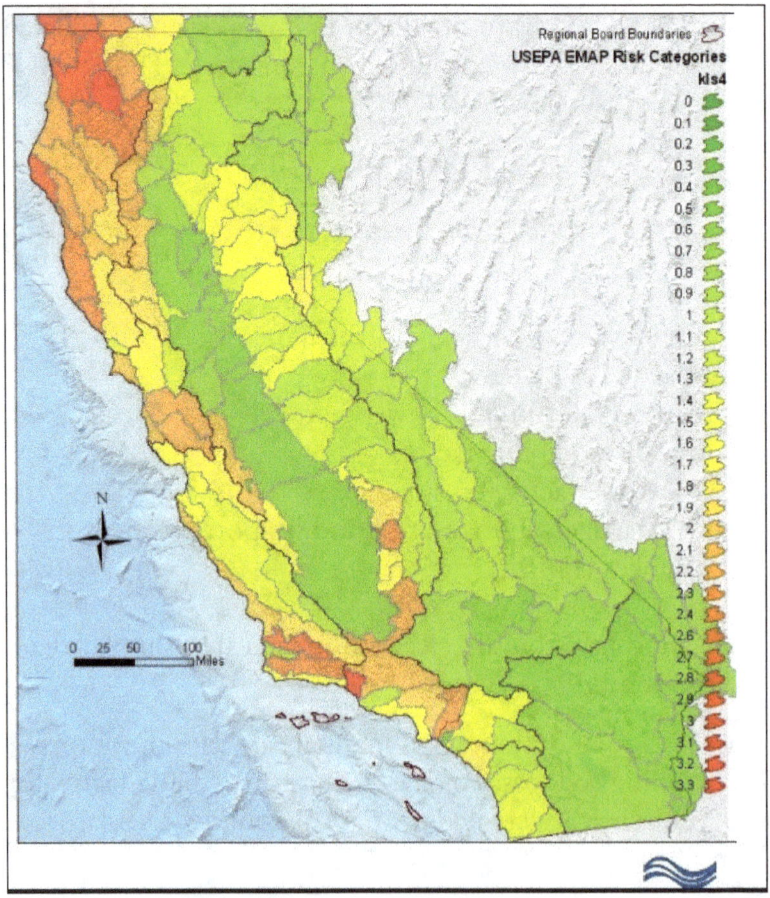

Figure 2-1 Statewide Map of K * LS Source: Water Resources Control Board, January 15, 2008

PROJECT RECEIVING WATER RISK

Receiving water risk is based on whether a project drains to a sediment-sensitive waterbody. A sediment-sensitive waterbody is either on the most recent list for waterbodies impaired for sediment; has a USEPA-approved Total Maximum Daily Load implementation plan for sediment;

or has the beneficial uses of COLD, SPAWN, and MIGRATORY.

A map of California sediment-sensitive waterbodies has been presented in Figure 2-2. The Risk Levels have been altered by eliminating the possibility of a Risk Level 4, and expanding the constraints for Risk Levels 1, 2, and 3. Therefore, projects with high receiving water risk and high-sediment risk will be considered a Risk Level 3 to water quality, see Figure 2-3.

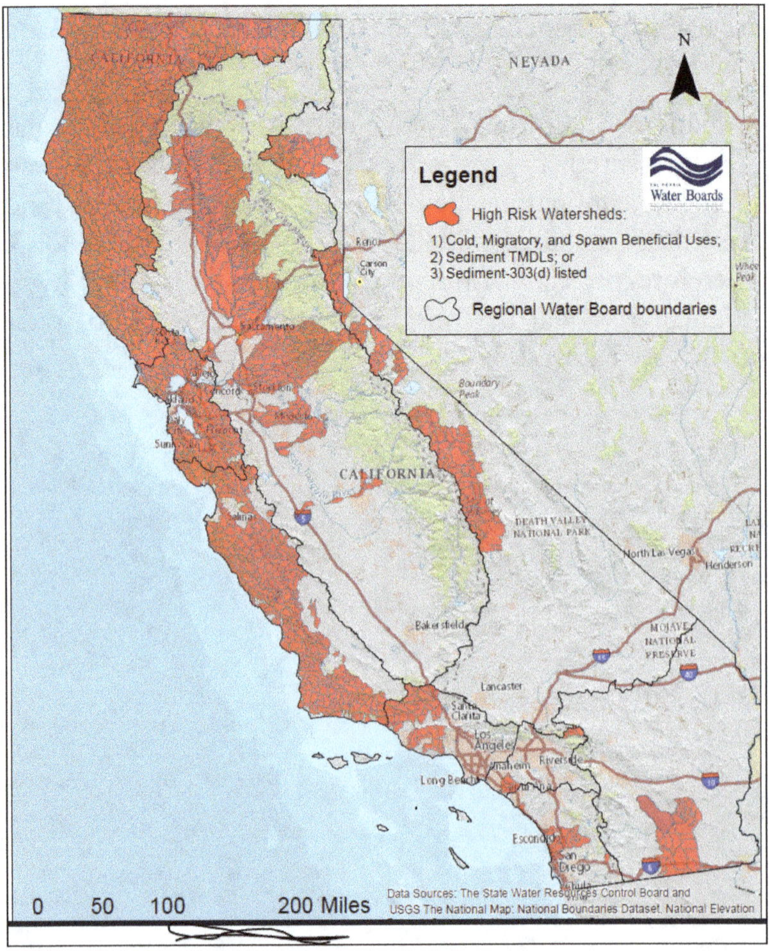

Figure 2-2 Source: California State Water Resources Control Board

In response to public comments, the Risk Level requirements have also been changed such that Risk Level 1 projects will be subject to minimum BMP and visual monitoring requirements, Risk Level 2 projects will be subject to Numeric Action Levels (NALs) and some additional monitoring requirements, and Risk Level 3 projects will be subject to NALs, and more rigorous monitoring requirements such as receiving water monitoring and, in some cases bioassessment.

Combined Risk Level Matrix				
		Sediment Risk		
		Low	Medium	High
Receiving Water Risk	Low	Level 1	Level 2	
	High	Level 2		Level 3

Figure 2-3, Identification of Risk Levels

All dischargers are subject to the narrative effluent limitations specified in the General Permit. The narrative effluent limitations require storm water discharges associated with construction activity to meet all applicable provisions of Sections 301 and 402 of CWA. These provisions require controlling of pollutant discharges that utilize Best Available Technology (BAT) and the Best Conventional Pollution Control Technology (BCT) to reduce pollutants and any more stringent controls necessary to meet water quality standards.

Risk Level 2 dischargers that pose a medium risk to water quality are subject to technology based NALs for pH and turbidity. Risk Level 3 dischargers that pose a high risk to water quality are also subject to technology based NALs for pH and turbidity.

DEFINITIONS

Hydrologic cycle: The continuous cycling of water between planetary reservoirs such as the ground, water bodies, and the atmosphere.

Stormwater: The Rainwater that does not infiltrate into the ground and runs off into waterbodies or enter the stormwater drainage system, is called as stormwater.

Therefore, State and Federal Law requires the development of a Stormwater Pollution Prevention Plan (SWPPP) before commencement of construction activities by a Certified Qualified Stormwater Developer (QSD) and implemented by a Qualified Stormwater practitioner (QSP).

Many California Board of Professional Engineers, Land Surveyors and Geologists (CBPELSG) have self-certified. In addition, the California Stormwater Quality Association (CASQA) has a lookup tool to find licensed QSDs and QSPs: CASQA QSP/QSD Lookup Tool.

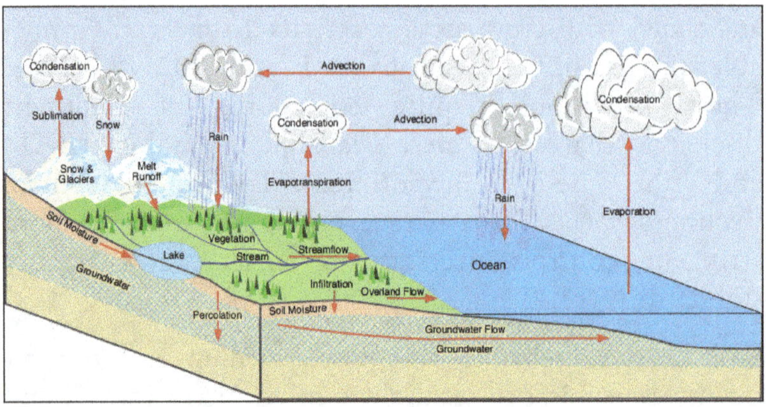

Figure 2-4 Hydrologic cycle

The following construction activities trigger the development of SWPPP:

- Clearing
- Grading
- Surface Disturbances
- Stockpiling

- Excavation and Trenching.

As discussed earlier, dischargers whose projects disturb one (1) or more acres of soil or whose projects disturb less than one acre but are part of a larger common plan of development that in total disturbs one or more acres, are still required to obtain coverage under the General Permit for Discharges of Storm Water Associated with Construction Activity.

A SWPPP may be called many things. Some states may use terms like:

- Construction Best Practices Plan
- Sediment and Stormwater Plan
- Erosion, Sediment and Pollution Prevention Plan
- Construction Site Best Management Practices Plan
- Erosion Control Plan and Best Management Practices
- Best Management Practices Plan
- Erosion and Sediment Control Plan

If you are responsible for erosion and sediment control and stormwater management at a permitted construction site, then you should prepare SWPPP.

What's SWPPP? It is a document for your use and regulatory agencies use who is responsible for overseeing the stormwater controls. A guide how to prepare an effective SWPPP can be obtained from U.S EPA, a document entitled "Development Your Stormwater Pollution Prevention Plan".

WHAT IS SOIL EROSION?

Soil erosion is a gradual process that occurs when the impact of water or wind detaches and removes soil particles, causing the soil to deteriorate. Soil deterioration and low water quality due to erosion and surface water run-off have become severe problems worldwide. The rate at which soil erosion causes the removal of topsoil under different land use practices varies significantly. Figure 2-5 shows different soil loss amounts for different land use types.

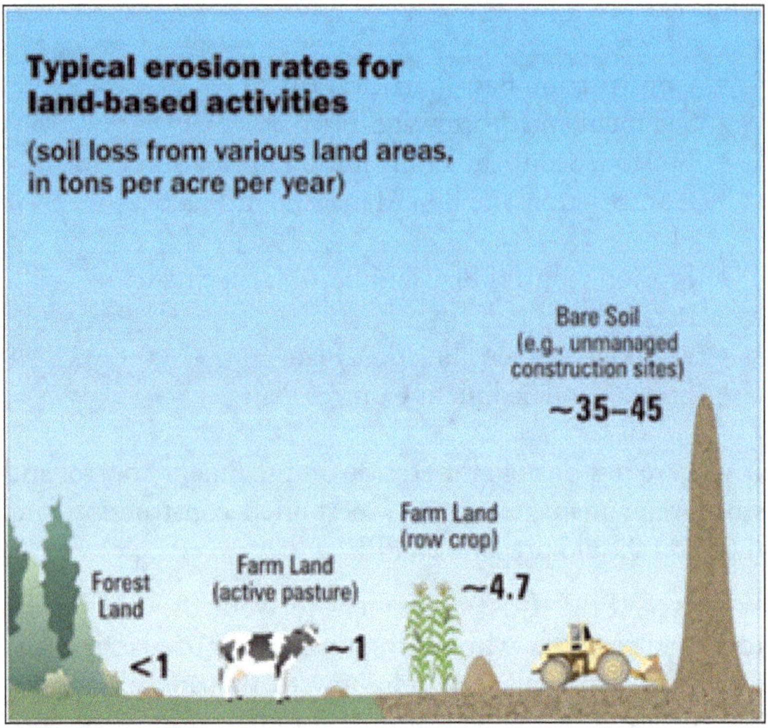

Figure 2-5 Soil Erosion rate per acre for different vegetation/land use type

As presented in Figure 2-5, during the construction activity the rate of soil loss/deterioration is maximum. As such, during construction activities the protection of soils and preventing its displacement become a major undertaking that the construction crew should have a plan to minimize the soil loss and its deterioration during construction activities. Stormwater runoff is rain or snowmelt that flows over land and does not percolate into the soil. Stormwater runoff occurs naturally, in small amounts from almost any type of land surface during large storm events.

Impervious surfaces such as buildings, homes, roads, sidewalks, and parking lots, can significantly alter the natural hydrology of land by increasing the volume, velocity, and temperature of runoff and by decreasing its infiltration capacity. Increasing the volume and velocity of stormwater runoff can cause sever stream bank erosion, flooding and degrade the biological habitat of these streams.

Reducing infiltration can lower groundwater recharge and affect drinking water supplies. As stormwater runoff moves across surfaces it picks up trash and pollutants such as sediment, oil and grease, pesticides, and other toxins. Soils exposed by construction activities are especially vulnerable to erosion. Runoff from un-stabilized construction sites can result in the loss of approximately 35-45 tons of sediment/acre/year.

Excess sediment can cloud the water reducing the amount of sunlight reaching aquatic plants, clog fish gills, smother aquatic habitat and spawning areas, and impede navigation in our waterways. The way to stop soil erosion and sediment transport is to keep the soil in place through vegetation, erosion control blankets, or other methods that prevent the soil from becoming dislodged during rain events.

FACTORS INFLUENCE EROSION

Erosion process is typically influenced by:

1. Climate
2. Topography
3. Soils
4. Vegetative cover

Understanding how these factors influence erosion will help you select and design appropriate controls to minimize erosion from the construction sites.

Climate: The frequency, intensity and duration of rainfall are the principal factors influencing erosion from a construction site. Know the weather patterns in your area and, if possible, plan your soil disturbances activities for periods of historically lower rainfall.

Topography: The longer and steeper a slope, the greater the potential for erosion from that slope. Use practices such as diversions or fiber rolls to break up long slopes. Consider minimizing soil disturbance activities on steeper slopes.

Soil: Soil Type can also impact erosion. Soil texture, structure, organic matter content, compaction and permeability can all influence erosion rates.

Vegetative Cover: Vegetative cover provides several critical benefits in preventing erosion- it absorbs the energy of raindrops, slow velocity of runoff, increase infiltration, and helps bind the soil. Soil erosion can be greatly reduced by maximizing vegetative cover at a construction site.

HOW TO PREVENT STORMWATER POLLUTION

An effective SWPPP is the key. If sediment and erosion control and good housekeeping practices are not followed, construction activity can result in the discharge of significant amounts of sediment and other pollutants.

The Best Management Practices or BMPs is often used to describe the controls and activities used to prevent stormwater pollution.

In SWPPP document the word "BMPs" is used more often to describe both structural and non-structural controls and practices.

SWPPP should include the following:

1. Cover/Title Page
2. Project and SWPPP contact information,
3. Site and Activity Description, Including a Site Map,
4. Identification of Potential Pollutant Sources,
5. Description of Controls to Reduce Pollutants,
6. Maintenance/Inspection Procedures,
7. Records of Inspections and Follow up maintenance of BMPs,
8. SWPPP Amendments, and
9. SWPPP Certification.

SWPPP should describe:

1. Who is the stormwater pollution prevention team?
2. Who is going to install structural stormwater controls?

3. Who will supervise and implement good housekeeping programs, such as site cleanup and disposal of trash and debris, hazardous material management and disposal, vehicle and equipment maintenance and so on?
4. Who will conduct routine inspections of the site to ensure all BMPs are being implemented and maintained?
5. Who will maintain the BMPs?
6. Who is responsible for documenting changes to the SWPPP?
7. Who is responsible for communicating changes in the SWPPP to people working on the site?

When you apply for your stormwater permit, the application may ask for a SWPPP contact. This could be construction site operator, but in many cases it's a staff person who is responsible for conducting inspections, ensuring BMPs are installed and maintained, and updating the SWPPP when necessary.

The following are the topics that need to be discussed in preparation of SWPPP:

1. How long will it take to stabilize your site?
2. How are the channels and slopes to be protected?
3. How will you reduce impermeable surfaces and promote infiltration?
4. How will you control the perimeter of your site?
5. How will you protect receiving water adjacent to your site?
6. How will you protect and keep the hazardous material at the site?
7. How will you minimize the area and duration of exposed soils and material?

To better answer the above questions, you need to conduct a complete site assessment or review the results of Phase I and Phase II Environmental Site Assessment that have been done, if any. Without site assessment, preparation for a successful SWPPP will not be possible.

The first step in developing your SWPPP is to evaluate your proposed construction site. Your SWPPP should describe the undeveloped site and identify features of the land that can be incorporated into the final plan and natural resources that should be protected. This initially can be done by a thorough walk-through. Remember each construction site and associated SWPPP is unique avoid copying SWPPP from one site to another. The walk-through will help you to understand the unique features of your site.

Assess the existing conditions at the construction site, including, topography, drainage, and soil type. This assessment is referred to as fingerprinting is the foundation for building your SWPPP. Use the site info to evaluate:

1. How stormwater is currently draining from the site,
2. Identify slope and slope length,
3. Identify soil type(s) and any highly erodible soils,
4. Identify past soil contamination, spills, etc.,
5. Identify natural features, such as trees, streams, wetlands, etc.

In most cases, the site designer can compile all this information on a digitized drawing that can then be adapted to show the planned construction activities, the phases of construction, and the final site map.

Topographic maps are readily available on the Internet or (www.mapquest.com). Or by conducting the U.S.

Geological Survey Store (http://store.usgs.gov). If you need to determine your soil type, contact local Natural Resource Conservation Service (NRCS) (http://soils.usda.gov).

SWPPP should clearly identify the receiving waters and stormwater systems through which stormwater from your site could flow. If your site's stormwater flows into a municipal storm drain system, you should determine the ultimate destination of that system's discharge. This might be obvious and easy to document. However, in some systems, you may have to consult with the local agency responsible for the storm drain system to determine the water body to which you are discharging. Remember, stormwater can travel long distances over roads, parking lots, downslopes, across fields and through storm sewers and discharge ditches.

The SWPPP should contain a brief description of the construction activities, including:

- Project type or function (for example, low density residential, shopping mall, highway)
- Project location includes latitude and longitude,
- Estimated project start and end dates,
- Sequence and timing of activities that will disturb soils at the site,
- Estimated total area expected to be disturbed by grading, excavation, or other construction activities,
- Percentage of impervious areas before and after construction.

Identify the pollutants and sources that are likely to be found on the site. The principal pollutant of concern is sediment. There are, however, other pollutants that may be found, usually in substantially smaller amounts, in stormwater runoff from construction sites.

These can include nutrients, heavy metals, organic compounds, pesticides, oil and grease, bacteria, viruses, trash and debris and other chemicals.

After identifying the pollutants and sources, be as specific as possible in your SWPPP about the BMPs you will use to address them.

NON-STORMWATER DISCHARGES

Most permits will require you to identify any non-stormwater discharges in your SWPPP. You should identify these sources in your SWPPP and identify pollution prevention measures to ensure that the pollutants are not introduced to these discharges and carried to nearby waterbodies.

EPA's Construction General Permit (CGP) identifies these allowable non-stormwater discharges which will include the following:

- Discharges from fire-fighting activities,
- Fire hydrant flushing,
- Water used to wash vehicles, buildings and pavements where detergents are not used,
- Water used to control dust,
- Potable water,
- Uncontaminated air conditioning condensate,
- Uncontaminated groundwater or spring water among others.

The main goal of any SWPPP is to protect rivers, lakes, wetlands, and coastal waters that could be affected by the construction project.

Keep in mind that there are many BMP options available to you. However, selecting a combination of BMPs would help you effectively to control runoff and soil erosion. For instance, the combination of silt fence along with slope stabilization gives you a more effective alternative to backup BMPs.

Erosion controls apply to certain techniques which keep the soil particles in place and prevent their movement.

SELECTING EROSION AND SEDIMENT CONTROL BMPS

Ten keys to effective Erosion and Sediment Control (ESC) principles are as follows:

Erosion Control (keeping the dirt in place):

1. Minimize disturbed areas and protect natural features,
2. Phase construction activity,
3. Control stormwater flowing onto through the project,
4. Stabilize soil promptly,
5. Protect slopes and continually evaluate your BMPs, are they performing well? Using BMPs in series also gives you some protection in case one BMP fails.

Sediment Controls (the second line of defense):

6. Protect Storm drain inlets,
7. Establish perimeter controls,
8. Retain sediment on-site and control dewatering practices,
9. Establish Stabilized construction exit,

10. Inspect and maintain controls to ensure the sediment does not leave your site.

MINIMIZE DISTURBED AREA

Carefully delineating and controlling the area that will be disturbed by grading or construction activities, you can greatly reduce the potential for soil erosion and stormwater pollution problems. Limit the disturbance area to only those necessary for the construction of your project. Natural vegetation is your best and cheapest erosion control BMP. Preserving and protecting topsoil is also a good BMP. Removing topsoil exposes underlying layers that are often more prone to erosion and have less infiltration capacity.

PHASED CONSTRUCTION ACTIVITY

Another technique for minimizing the duration of exposed soil is phasing. By scheduling or sequencing your construction work and concentrating on certain areas, you can minimize the amount of soil that is exposed to erosion elements at any given time.

Limiting the area of disturbance to places where construction activities are underway and stabilizing them as quickly as possible can be one of your most effective BMPs.

CONTROL STORMWATER FLOWING ONTO AND THROUGH PROJECT SITE

Plan for any potential stormwater flows coming onto the project area from upstream locations and divert (and slow) flows to prevent erosion. Likewise, the volume and velocity of on-site stormwater runoff should be controlled to minimize soil erosion.

Diversion ditches or berms direct runoff away from unprotected slopes and may also direct sediment-laden runoff to sediment-trapping structure, see Figure 2-6.

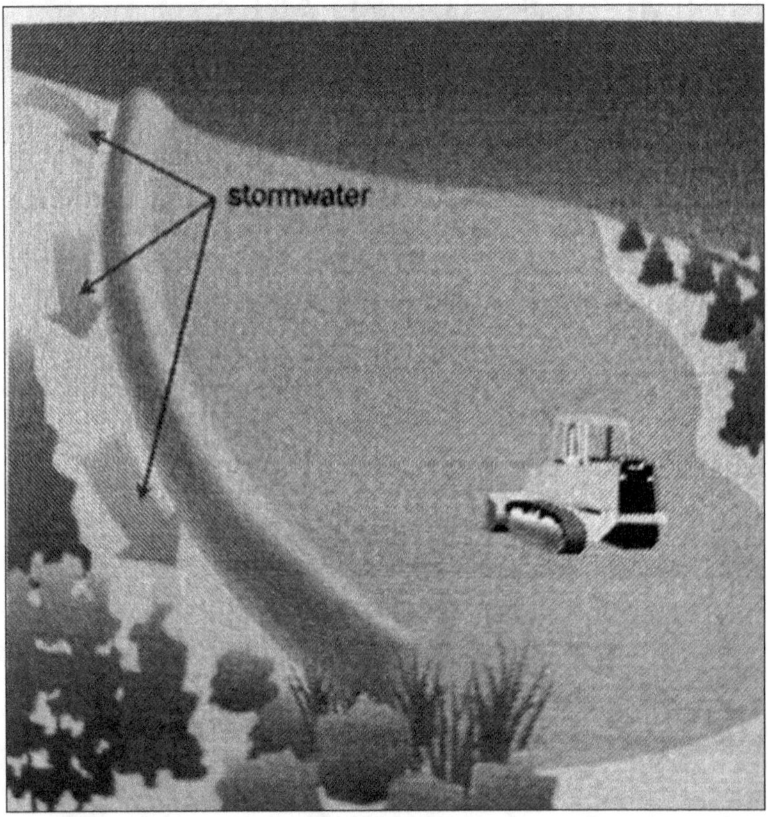

Figure 2-6

STABILIZE SOILS PROMPTLY

Where construction activities have temporarily or permanently ceased, you should stabilize exposed soils to minimize erosion. Many permits require stabilization within a specified time frame. Temporary measures are necessary when an area of a site is disturbed but where activities in that area are not completed or until permanent BMPs are completed. Topsoil stockpiles should also be protected to minimize erosion. Temporary-cover BMPs include:

- Temporary seeding,
- Mulches,
- Matrices,
- Blankets and mats, and
- Soil binders.

Permanent-cover BMPs include permanent seeding, and planting, sodding, channel stabilization, and vegetative buffer strips. Silt fences and other sediment control measures are not stabilization measures.

Temporary seeding: Temp seeding an area to establish vegetative cover is one of the most effective, and least expensive methods of reducing erosion. This approach as a single BMP might not be appropriate on steep slopes when vegetation cannot be established quickly enough to control erosion during a storm event.

PROTECT SLOPES

Protect all slopes with appropriate erosion controls. Steeper slopes, slopes with highly erodible soils, or long slopes require a more complex combination of controls. Erosion control blankets, bonded fiber matrices, and turf reinforcement mats are very effective options. Silt fence or fiber rolls may also be used to help control erosion on moderate slopes and should be installed on level contours spaced at 10 – 20-foot intervals. You can also use diversion channels and berms to keep stormwater off slopes.

Example BMP: Rolled erosion control products include mats, geotextiles, and erosion control blankets, and products that provide temporary stabilization and help to establish vegetation on disturbed soils.

PROTECT STORM DRAIN INLETS

Protect all inlets that could receive stormwater from the project until final stabilization of the site has been achieved. Install inlet protection before soil-disturbing activities begin. Maintenance throughout the construction process is important. Upon completion of the project, storm drain inlet protection is one of the temporary BMPs that should be removed. Storm drains inlet protection should be used not only for storm drains within the active construction project, but also for storm drains outside the project area that might receive storm discharges from the project. If there are storm drains on private property that could receive stormwater runoff from your project, coordinate with the owner of that property to ensure proper inlet protection.

Storm drains inlet protection prevents sediments from entering the inlet with filtering material. Several types of filters are commonly used for inlet protection: silt fence, rock-filled bags, or block and gravel.

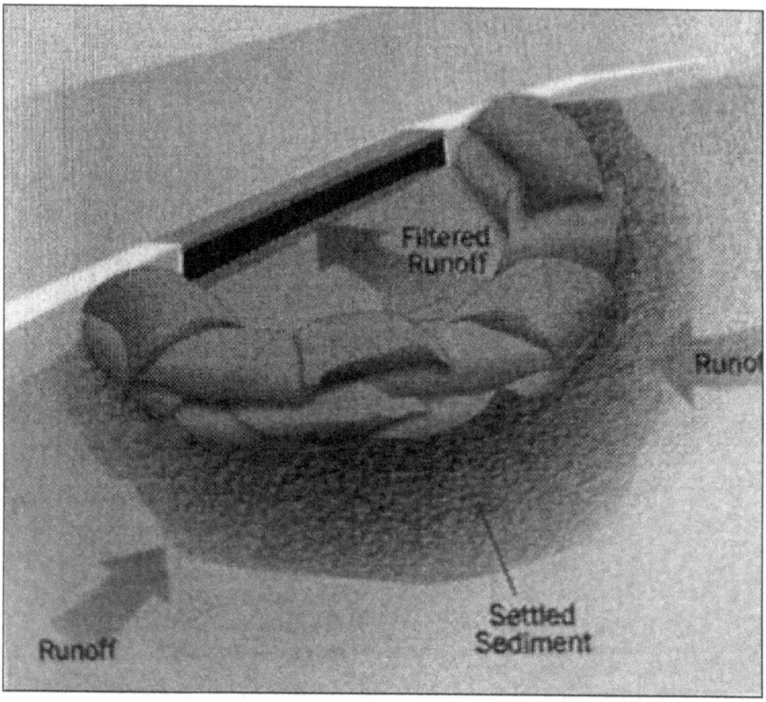

Figure 2-7

ESTABLISH PERIMETER CONTROLS

Maintain natural areas and supplement them with silt fence and fiber rolls around the perimeter of your site to help prevent soil erosion and stop sediment from leaving the site.

Install controls on the downslope perimeter of your project (it is often unnecessary to surround the entire site with silt fence). Sediment barriers can be used to protect stream buffers, riparian areas, wetlands, or other waterways.

A silt fence is a temporary sediment barrier consisting of a geotextile attached to the supporting posts and trenched into the ground. Silt fencing is intended to retain sediments that have been dislodged by stormwater, see Figure 2-8.

Figure 2-8

Silt fences are designed only for runoff from small areas, and it is not intended to handle flows from large slopes or in areas of concentrated flow. Fiber rolls serve the same purpose and consist of an open mesh tubular sleeve filled with fibrous material which traps sediment.

INSTALLATION TIPS OF PERIMETER CONTROLS

Do:

1. Use silt fence or fiber rolls as perimeter controls, particularly at the lower or downslope edge of disturbed area,
2. Leave space for maintenance between the toe of slope and silt fence or roll,
3. Trench in the silt fence on the uphill side (6 inches deep by 6 inches wide),

4. Curve the end of silt fence or fiber roll up-gradient to help contain runoff.

Don't:

1. Do not install a silt fence or fiber rolls in ditches, channels, or areas of concentrated flow,
2. Install it running up and down a slope or hill,
3. Use silt fencing or fiber rolls alone in areas that drain more than a quarter-acre per 100 feet of fence.

RETAIN SEDIMENT ON-SITE AND CONTROL DEWATERING PRACTICES

Sediment barriers described in ESC Principle 7 can trap sediment from small areas, but when sediment retention from a larger area is required, consider using a temporary sediment trap or sediment basin. These practices detain sediment-laden runoff for some time, allowing sediment to settle before the runoff is discharged. Proper design and maintenance are essential to ensure that these practices are effective. You should use a sediment basin for common drainage locations that serve an area with 10 or more acres disturbed at any one time.

Your permit should include the exact basin-sizing requirement. As a rule of thumb, it should be able to store 3,600 cubic feet of runoff per acre drained. The basin should be in low-lying areas of the site and on down-gradient side of the bare soil areas where flows converge.

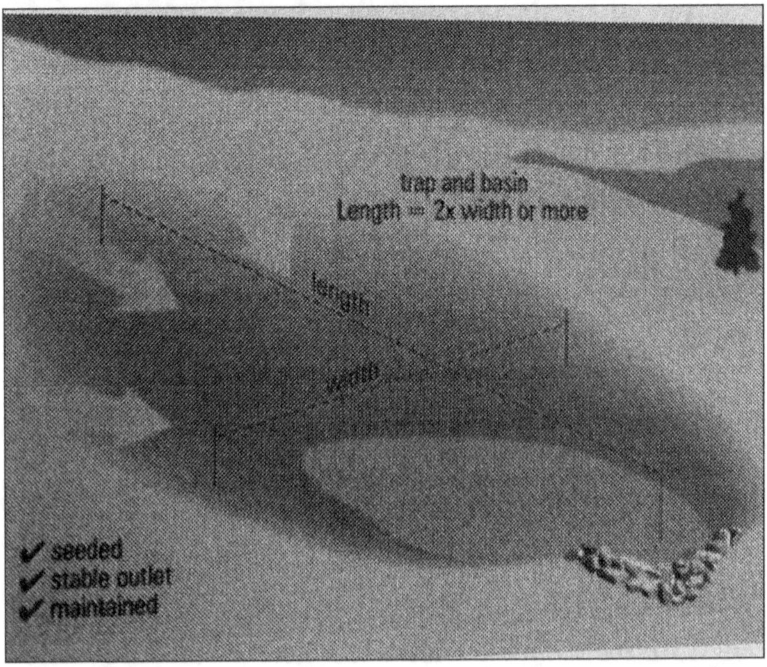

Figure 2-9

Do not put sediment traps or basins in or immediately adjacent to flowing streams or other waterways.

Dewatering practices are used to remove groundwater or accumulated rainwater from excavated areas. Pump Muddy water from these areas to a temporary or permanent sedimentation basin or to an area completely enclosed by silt fence in a flat vegetated area where discharges can infiltrate into ground. Never discharge muddy water into storm drains, streams, lakes, or wetlands unless the sediment has been removed before discharge.

ESTABLISH STABILIZED CONSTRUCTION EXIT

Vehicles entering and leaving the site have the potential to track significant amounts of sediments onto streets. Identify and clearly mark one or two locations where vehicles will enter and exit the site and focus stabilizing measures at those locations. Construction entrances are commonly made from large, crushed rock. They can be further stabilized using stone pads or concrete. Also, steel wash racks and a hose-down system will remove even more mud and debris from vehicle tires. Divert runoff from wash areas to a sediment trap or basin. No system is complete, sweeping the street regularly completes this BMP.

A rock construction exit can reduce the amount of mud transported onto paved roads by vehicles. The construction exit does this by removing mud from vehicle tires before the vehicle enters a public road. It is also a good idea to clean vehicles before leaving the site by pressure washing to remove mud from their tires, see Figure 2-10.

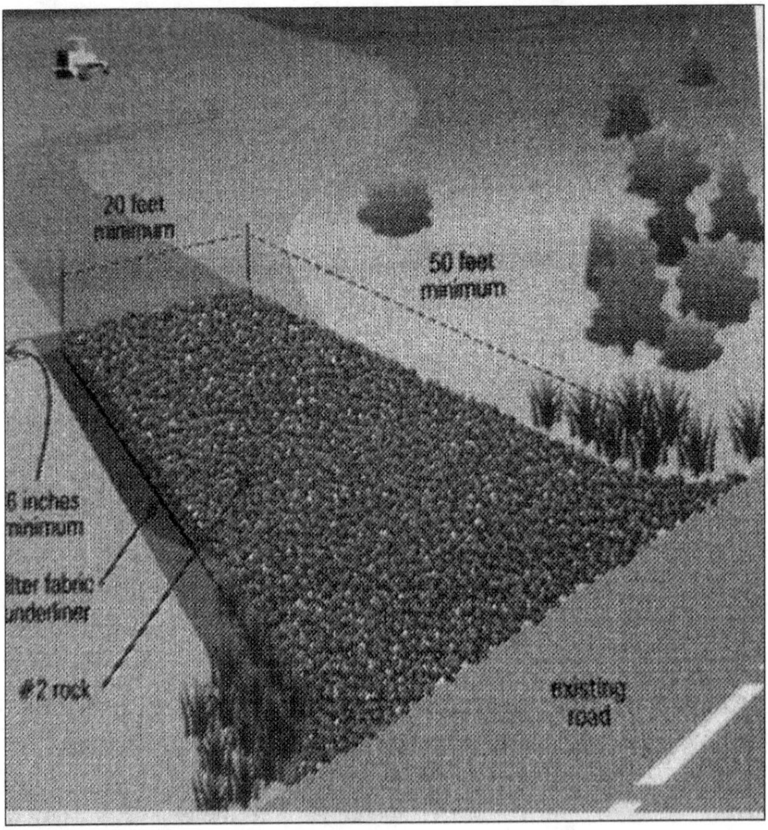

Figure 2.10

INSPECT AND MAINTAIN CONTROLS

Inspection and maintenance are just as important as proper planning, design, and installation of controls. Without adequate maintenance, erosion and sediment controls will quickly fail, sometimes after just one rainfall, and cause significant water quality problems and potential violations of the NPDES construction general permit. Your permit requires you to always maintain your BMPs. To do this effectively, you should establish an inspection and maintenance approach or strategy that includes both regular and spot inspections.

POLLUTION PREVENTION PRINCIPLES AND SELECTING GOOD HOUSEKEEPING BMPS

Six Key Pollution Prevention Principles for Good Housekeeping are:

1. Provide for waste management,
2. Establish proper building material staging areas,
3. Designate paint and concrete washout areas,
4. Establish proper equipment/vehicle fueling and maintenance practices,
5. Control equipment/vehicle washing and allowable non-stormwater discharges,
6. Develop a spill prevention and response plan.

PROVIDE FOR WATER MANAGEMENT

Design proper management procedures and practices to prevent or reduce the discharge of pollutants to stormwater from solid or liquid wastes that will be generated at your site. Practices such as trash disposal, recycling, proper material handling, and cleanup measures can reduce the potential for stormwater runoff to pick up construction site wastes and discharge them to surface waters. Provide convenient, well-maintained, and properly located toilet facilities. Provide for regular inspections, service, and disposal.

Locate toilet facilities away from storm drain inlets and wasteways to prevent accidental spills and contamination of stormwater. Treat or dispose of sanitary and septic waste in accordance with state or local regulations.

Proper material use, storage, waste disposal, and training of employees and subcontractors can prevent or reduce the discharge of hazardous and toxic wastes to stormwater. Implement a comprehensive set of waste-management practices for hazardous or toxic materials, such as paints, solvents, petroleum products, pesticides, wood preservatives, acids, roofing tar and other materials.

ESTABLISH PROPER BUILDING MATERIAL HANDLING AND STAGING AREAS

Your SWPPP should include comprehensive handling and management procedures for building materials, especially those that are hazardous or toxic. Paints, solvents, pesticides, fuels and oils, other hazardous materials or any building materials that have the potential to contaminate stormwater should be stored indoors or under cover whenever possible or in areas with secondary containment. Secondary containment techniques should also ensure the protection of groundwater.

DESIGNATE WASHOUT AREAS

Concrete contractors should be encouraged, where possible, to use the washout facilities at their own plants or dispatch facilities. If it is necessary to provide concrete washout areas on-site, designate specific washout areas and design facilities to handle anticipated washout water. Washout areas should also be provided for paint and stucco operations. Because washout areas can be a source of pollutants from leaks or spills, EPA recommends that you locate them at least 50 yards away from storm drains and watercourses whenever possible.

ESTABLISH PROPER EQUIPMENT/VEHICLE FUELING AND MAINTENANCE PRACTICES

Performing equipment/vehicle fueling and maintenance at an off-site facility is preferred over performing these activities on the site. For grading and excavating equipment this is usually not possible or desirable. Create an on-site fueling and maintenance area that is clean and dry. The on-site fueling area should have a spill kit, and staff should know how to use it. If possible, conduct vehicle fueling and maintenance activities in a covered area; outdoor vehicle fueling, and maintenance is a potentially significant source of stormwater pollution.

CONTROL EQUIPMENT/VEHICLE WASHING AND ALLOWABLE NON-STORMWATER DISCHARGES

Environmentally friendly washing practices can be practiced at every construction site to prevent contamination of surface and groundwater from washing. Procedure and practices include using off-site facilities; washing in designated, contained areas only; eliminating discharges to the storm drain by infiltrating the wash-water or routing to the sanitary sewer; and training employees and subcontractors in proper cleaning procedures.

DEVELOP A SPILL PREVENTION AND RESPONSE PLAN

Most state and EPA construction general permits require the preparation and response plans. Generally, these plans can be included or incorporated into your SWPPP. The plan should clearly identify ways to reduce the chance of spills, stop the source of spills, dispose of materials contaminated by spills, and train personnel responsible for spill prevention and response. The plan should also specify material handling procedures and storage requirements and ensure that clear and concise spill cleanup procedures are provided and posted for areas in which spills may potentially occur.

When developing a spill prevention plan, include, at a minimum, the following:

- Note the location of chemical storage areas, storm drains, tributary drainage areas, surface waterbodies on or near the site and measures to stop spills from leaving the site,
- Specify how to notify appropriate authorities, such as police, firefighters, hospitals, other authorities,
- Describe the procedures for immediate cleanup of spills and proper disposals,
- Identify personnel responsible for implementation of the plan.

The effectiveness of erosion and sediment control BMPs and good housekeeping and pollution prevention measures depend on consistent and continual inspection and maintenance.

This step focuses on developing and planning for BMP inspections and maintenance to ensure that a schedule and procedures are in place. Your responsibility does not stop after BMPs are installed. Your BMPs must always be maintained in good working order. Further, your permit requires that you conduct regular inspections and document the findings of those inspections in your SWPPP.

INSPECTION, MAINTENANCE AND RECORDKEEPING

In developing your inspections schedule consider the following:

- Consider using spot inspections. You may want to inspect certain parts of your site more frequently or even daily, such as storm drain inlets at nearby streets.
- Consider using informal inspections. Besides the formal inspections you can add several informal ones
- Consider adding inspections before or during rain events.
- Train staff and subs. Use your staff and subs to identify problems with BMPs, etc.,

Figure 2-12 Major violation of stormwater runoff in a neighborhood

PREPARING INSPECTION REPORTS

Complete an inspection report after each inspection. Keep the inspection report with your SWPPP. The following information is required to be included in your inspection report:

- Inspection date,
- Inspector information, including the names titles and qualifications of personnel conducting the inspection,
- Weather information for the period since the last inspection, including the best estimate of the beginning of each storm. Its duration, approximate amount of rainfall for storm in inches, and whether any discharges occurred,
- Current weather information and a description of any discharges occurring at the time of the inspection,
- Description of evidence of previous or ongoing discharges of sediment or other Location(s) of BMPs that need to be maintained,
- Location(s) of BMPs that failed to operate as designed or proved inadequate for a location,
- Location(s) where additional BMPs are needed but did not exist at the time of inspection,
- Corrective action required, including any necessary changes to the SWPPP and implementation dates,
- Reference to past corrective actions documenting follow up actions taken, and
- Consider taking digital photographs during inspections.

REFERENCES

Better Site Design: A handbook for Changing Development Rules in Your Community Available from http://www.cwp.org

EPA 833-R-06-004, May 2007. "Developing Your Stormwater Pollution Prevention Plan. A Guide for Construction Sites

EPA Stormwater Construction Website: http://www.epa.gov/npdes/stormwater/construction

EPA Nonpoint Source Low Impact Development Site http://www.epa.gov/owow/nps/lid/

National Pollutant Discharge Elimination System (NPDES) General permit for Storm Water Discharges Associated with Construction and Land Disturbance Activities. Order No. 2009-0009DWQ. NPDES No. CAS000002

Protecting Water Resources with Smart Growth

http://epa.gov/smartgrowth/pdf/waterresources_with_sg.pdf

THREE
SUSTAINABILITY

INTRODUCTION

In 1987, the United Nation defined sustainability as "meeting the needs of the present without compromising the ability of future generations to meet their own needs." Today, there are almost 140 developing countries in the world seeking ways of meeting their development needs, but with the increasing threat of climate change, concrete efforts must be made to ensure that development today does not negatively affect future generations.

The sustainable development goals form the framework for improving the lives of populations around the world and mitigating the hazardous man-made effects of climate change, meanwhile bringing economic development and prosperity around the globe.

In the broadest sense, sustainability refers to the ability to maintain or support a process continuously over time. In business and policy contexts, sustainability seeks to prevent the depletion of natural or physical resources, so that they will remain available for the long term.

KEY TAKEAWAYS

- Sustainability is the ability to maintain or support a process over time,
- Sustainability is often broken into three core concepts: economic, environmental, and social,
- Many businesses and governments have committed to sustainable goals, such as reducing their environmental footprints and conserving resources,
- Some investors are actively embracing sustainability investments, known as "green investments",
- Skeptics have accused some companies of "greenwashing", the practice of misleading the public to make a business seem more environmentally friendly than it is.

HOW SUSTAINABILITY WORKS

Accordingly, sustainable policies emphasize the future effect of any given policy or business practice on humans, ecosystems, and the wider economy. The concept often corresponds to the belief that without major changes to the way the planet is run, it will suffer irreparable damage.

As concerns about anthropogenic climate change, biodiversity loss, and pollution have become more widespread, the world has shifted to embrace sustainable practices and policies, primarily through the implementation of sustainable business practices and increased investments in green technology.

Sustainable construction is the response of construction industry to the rapid negative changes in earth's environment and its ecosystem. Three major changes are motivating construction industry to develop an ethical response:

- Accelerated destruction of planetary ecosystem including massive destruction of life and biodiversity and fear of GW,
- Enormous increase in population and demand for natural resources by both developed and developing countries such as BRIC,
- Green Building movement is coinciding with similar transformation in manufacturing, tourism, agriculture, medicine, which adopted various approaches toward greening their activities.

3 PILLARS OF SUSTAINABILITY

The idea of sustainability is often broken down into three pillars: economic, environmental, and social—also known informally as profits, planet, and people.

In that breakdown, the concept of "economic sustainability" focuses on conserving the natural resources that provide physical inputs for economic production, including both renewable and exhaustible inputs. The concept of "social sustainability" refers to the well-being of human health and enhancing air quality and elongating the human life span. While the issue of the "environmental sustainability" which includes preventing the generation of greenhouse gases such as carbon dioxide, methane, and nitrous oxide by cutting the consumption of non-renewable energy sources such as fossil fuels which ultimately will prevent occurrence of global warming.

ENVIRONMENTAL IMPACT ANALYSIS

Back in the 1972, three major thinkers on the global environment; Barry Commoner, Paul Ehrlich and John Holdren came up with this conceptual model to help us understand the major driving forces contributing to the degradation that they witnessed occurring in the US and beyond. It was called the IPAT equation (IPAT 1972). It simply says the impact that is occurring to our planet is the results of three parameters; population, consumption rate of resources per capita, and the technology to create that resource.

The IPAT Equation

The IPAT equation is a widely used simplification of the factors causing environmental degradation. The equation is I = P x A x T. This is short for environmental **Impact** = **Population** x **Affluence** (consumption per person) x **Technology** (impact per unit of consumption). It's crucial to remember that the three factors are intermediate causes, not root causes.

$$\text{Impact} = \text{Population} \times \frac{\text{consumption}}{\text{per person}} \times \frac{\text{impact}}{\text{per unit of consumption}}$$

I P A T

Population:

The world population growth rate is shown in Figure 3.1. As it shows since 1970s the world population is showing a downward trend, however, the total world population is showing a steady upward pattern, and it is expected to reach 12 billion by 2100.

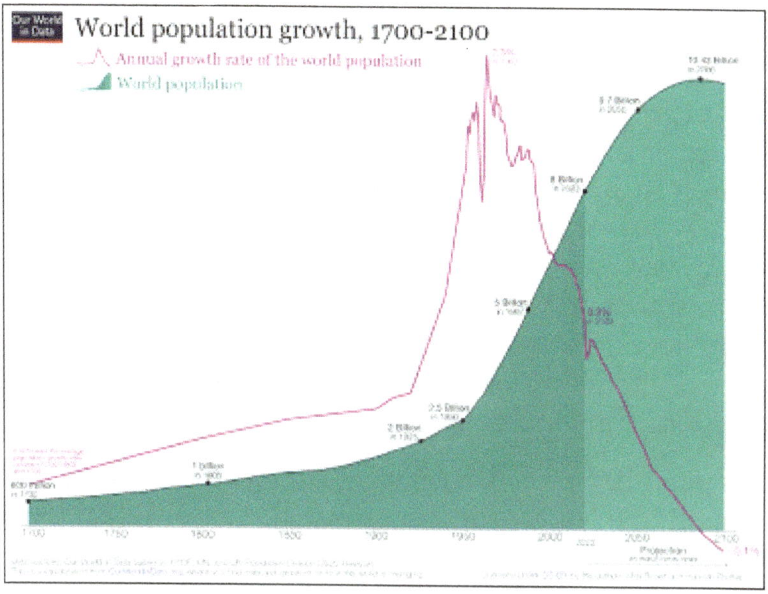

Figure 3.1 World population growth rate and trend
Source: UN and UN Population Division 2022

Consumption Rate

As time goes on, not only does our planet's population increase but the lifestyle of people keeps changing and there is an ever-increasing pattern for people to eat more, travel more, and consume luxurious goods. This change in lifestyle, especially among the people of developed countries as well as underdeveloped nations increases the demand for food, clean air, water, luxuries goods as well as energy.

Impact per unit consumption:

The impact of using each unit of resources such as water, fuel, and energy on our planet will be dependent upon the type of technology in creating those resources. Therefore, the type of technology employed in creating those resources plays a major role in calculating the type and the magnitude of the environmental impact that the IPAT equation will reveal. For instance, if consumable resource is energy, then magnitude and type of the impact in producing that energy will depend on the type of technology used to produce that energy. If the energy is derived from non-renewable energy sources such as fossil fuel or coal, the impact will be much bigger than if that energy is produced from renewable energy sources such as solar, biofuel or wind turbines. Throughout this book, we will discuss and recommend using renewable sources of energy in construction of our built environment for reducing the concentration of greenhouse gases in the atmosphere.

It should be noted that building operations and construction are responsible for approximately 40% of global CO_2 emissions. That's why the construction industry is concerned with environmental issues such as global warming that will be inevitable if our future built environment does not use a clean source of energy such as photovoltaic, solar energy panels, biofuel, or wind turbines for generation of energy.

Use of fossil fuels to generate the required energy or business as usual will bring adverse environmental conditions which will be hard or impossible to reverse. The examples of the adverse environmental impacts will include:

- Global warming and climate change,
- Local, regional, and national water shortages – water stress,
- Declining non-renewable resource stocks and unsustainable use of conditionally renewable stocks,
- Land conversion and habitat loss,
- Biodiversity reduction and endangered species,
- Environmental contamination.

The direct impact of adverse environmental conditions on construction industry will at least be as follows:

- Increased energy costs and decreased supply and reliability
- Fewer, more expensive resources for construction and higher interest rate for redevelopment
- Tighter regulations and more restrictions/requirements for design and site development.

As global warming kicks in, and our climate becomes more drastic, buildings and their occupants will have to deal with higher peak temperatures, both heat and cold. The major role of the construction industry is preventing the occurrence of global warming by adopting new technologies to reduce the construction industry's dependency on fossil fuels. In addition, the building industry is contemplating choosing the proper building material with high recycle content, high durability, less emissivity with less energy demand during extraction, transportation, manufacturing, installation, maintenance, and end of life disposal processes though using the Life Cycle Analysis (LCA).

In the next chapters of this book, we will address how the todays' technologies (the "T" factor in the IPAT equation) should strive to adopt new techniques to reduce overwhelming demand for production of the building material and improve their environmentally friendly characteristics in creating a better indoor air quality condition, reducing demand on drinking water resources, using passive design strategies by constructing a "High Performance Building".

HIGH-PERFORMANCE BUILDING

Definition of Green Building

Green building is the practice of creating structures and using processes that are environmentally responsible and resource-efficient throughout a building's life cycle from sitting to design, construction, operation, maintenance, renovation, and deconstruction. This practice expands and complements the classical building design concerns of economy, utility, durability, and comfort. Green buildings are also known as sustainable or high-performance buildings.

Figure 3-2
Environmental Group

Impacts of the built environment

Aspects of Built Environment:	Consumption:	Environmental Effects:	Ultimate Effects:
• Sitting • Design • Construction • Operation • Maintenance • Renovation • Deconstruction	• Energy • Water • Materials • Natural Resources	• Waste • Air pollution • Water pollution • Indoor pollution • Heat islands • Stormwater runoff • Noise	• Harm to Human Health • Environment Degradation • Loss of Resources

Green buildings are designed to reduce the overall impact of the built environment on human health and the natural environment by:

- Efficiently using energy, water, and other resources
- Protecting occupant health and improving employee productivity
- Reducing waste, pollution, and environmental degradation

For example, green buildings may incorporate sustainable materials in their construction (e.g., reused, recycled content, or made from renewable resources); create healthy indoor environments with minimal pollutants (e.g., reduced product emissions); and/or feature landscaping that reduces water usage (e.g., by using native plants that survive without extra watering is called xeriscaping).

There are number of reasons to build green, including potential environmental, economic, and social benefits. They involve many different notions including, but not limited to:

- Design for the full life cycle,
- Efficiency of resource use (including reduce, reuse, and recycle content),
- Lower (zero or better) carbon and energy footprints,
- Lower monthly utility bills,
- Neutral or positive environmental impacts.

It should adjust to changing political, economic, and ecological conditions. Sustainability is a moving and complex target. Note that sustainability also asks us to consider a fair distribution of human welfare today, which is already highly unequal, and strive at a minimum, not to make it worse and preferably to make it less extreme. When it comes to the construction industry, sustainability clearly encompasses many different notions and actions.

Buildings last for generations thus sustainability requires us to not only look at the present condition, but the full-cycle episode, including the end-of-life destiny when the building materials could either become a problem waste, or a beneficial resource depending on decisions made today.

High-performance buildings are becoming the default for renovation and construction worldwide. They have the capability to reduce:

- Operating costs
- Maintenance costs
- Environmental costs

High-performance buildings can generate better long-term economic values and higher human satisfaction. The challenge consists of taking the green building logic and methodology and making it the norm. It can apply for both new as well as existing buildings. It can be safely argued that the building professional today has the responsibility to deal with local, regional, and global problems through the construction and renovation of buildings to the higher performance levels in comparison with traditional buildings.

The construction of high-performance building should not be looked at as a burden, rather as a business opportunity – a win-win situation because, when the life cycle of a building is considered not only it does provide a better living environment, but it reduces monthly utility costs. Although the number of buildings that can be labeled green using one of the recognized building rating assessment tools are growing, this is not yet the norm across the industry – one of the challenges for our industry is to get to the point where many buildings could be labeled green.

Cost-wise the construction of a green building may require more resources/capital in comparison with the traditional buildings, however, the payback period after it was occupied may be a few years. This is primarily due to rental income and the reduced monthly utility bills.

SUSTAINABILITY AND ETHICS

In the context of sustainable development and sustainable construction Kibert (2022) indicates that the idea of ethics must be broadened to address a wide range of concerns that are not usually considered. Ethics addresses relationships between people by providing rules of conduct that are generally agreed to govern the good behavior of contemporaries. Sustainable development requires a more extensive set of ethical principles to guide behavior because it addresses the relationships between generations, calling for what is sometimes referred to as intergenerational justice. Such fact already imbedded the in definition of sustainability by saying "meeting the needs of the present without compromising the ability of future generations to meet their needs". Therefore, the rights of any unborn individuals in every decision that we make today should be considered in our sustainability processes. The idea of sustainability and ethics not only covers the right of every human being, but it goes beyond and considers the right of every living creature/organism.

In another word, humans are certain to create materials and develop procedures or processes that have not evolved in a natural sense, which have no precedent in nature. The question then becomes: what constraints should society place on the development of new materials, products, and processes? The ongoing debates about genetically modified organisms (GMOs) and cloning are indicative of the uncertainty about the outcomes of human tinkering with the blueprints of life, not to mention the creation of materials that have uncertain long-term impacts. Other major developments such as biotechnology, genetic engineering, nontechnology, robotics, and nuclear energy, to name but a few, present fundamental challenges to human society. Decisions about implementing technologies with no precedent in nature and with potentially unprecedented negative and irreversible impacts must be considered carefully, especially since, once a technology is deployed, it is extremely difficult to reverse course if negative consequences are discovered. Decision about how to move forward must be based on:

1. An ethical framework that represents society's general moral attitudes toward life and future generations,
2. An understanding of and willingness to accept risk, and
3. The economic costs of implementation and resulting impacts.

The current generations can affect the health and quality of life of remote future generation. This choice of today's generation will directly affect the quality and quantity of resources remaining for future inhabitants of Earth. This concept of obligation that crosses temporal boundaries is referred to as intergenerational justice. It implies a chain of obligation between generations.

For example, depletion of Ozone layer by current generation may cause catastrophic mutation in the future generation due to excessive radiation of ultra-violet light and depriving them having the same quality of life as the current generation, which is in violation of sustainability definition.

Ethics is really, to a large degree, about taking responsibility. With respect to sustainability, this reflects the belief that we have an obligation to consider the welfare of future generations, to look outside our immediate generation and, indeed, our immediate family or community and consider the effects on others across space and time. We have a massive inequity in our economic system at all scales – within a given city, a nation, and across the globe. And this inequity not only reflects access to resources, goods, and services, but also to the impacts of the negative effects of this unequal access. A classic example, or course, would be the small island nations in danger of total annihilation from sea level rise yet who have done little or nothing to contribute to the causes of this sea level rise; the introduction of Greenhouse gases (GHGs) into the global atmosphere.

INTERGENERATIONAL JUSTICE AND CHAIN OF OBLIGATION

Kibert (2016) states that there is an asymmetry of power between present and future generations because, while today's people can make choices that will severely affect people 100 to 200 years into the future, for example, ignoring the long-term impacts of climate change, the same cannot be said of future generations. This is simply no mechanism for future, remote generations to influence the past. Therefore, the current generation can affect the health and quality of life of these remote generations. It becomes clear that the choices of today's generations will directly affect the quality and quantity of resources remaining for future inhabitants of Earth and its environmental quality. This concept of obligation that crosses temporal boundaries is referred to as intergenerational justice. Furthermore, the concept of intergenerational justice implies a chain of obligations between generations that extends from today into the distant future.

The actions and decisions of the current generation affect not only the welfare but also the composition of future generations. When we create conditions that change resource availability or that alter the environment, future populations will be compositionally different than if the resource base and environmental conditions had been passed on, from one generation to future generations, unchanged. For instance, mutation caused by excessive ultraviolet radiation through an ozone layer depleted by human activities, or by synthetic toxic chemicals used without adequate safeguards, certainly will result in different people and conditions. Consequently, the chain of obligations that underpins the key sustainability concept of intergenerational justice includes parents' responsibility for enabling their offspring to meet their moral obligations to their children and beyond. Clearly, doing this would include educating the offspring about these obligations and the basis for them.

DISTRIBUTIONAL EQUITY

There is an obligation to ensure the fair distribution of resources among people currently alive so that the life prospects of all people are addressed. This obligation can be referred to as distributional equity or distributive justice and refers to the right of all people to an equal share of resources, including goods and services, such as materials, land, energy, water, and high environmental quality (Kibert 2022).

Distributional equity is based on principles of justice and reasonable assumption that all individuals in each generation are equal and that a uniform distribution of resources must be a consequence of intergenerational equity. One of the challenges of sustainability is increasing prosperity for the billions who are barely able to survive on a day-by-day basis. The principle of distributional equity requires that Earth's resources be more fairly allocated so that everyone has at least a decent quality of life.

Figure 3-3

Distributional equity also underpins the chain of obligation concept. Distributional equity is a complex concept, and several principles underpin and are related to it:

1. Difference principles,
2. Resource-based principles,
3. Welfare-based principles,
4. Desert-based principle,
5. Libertarian principles, and
6. Feminist principles.

PRECAUTIONARY PRINCIPLES

One ethical principle that is helpful in addressing this responsibility "to do no harm" is the so-called precautionary principle. It applies when we lack the knowledge to fully understand the cause-and-effect relationships of our potential actions. While the danger of the precautionary principle is an excess of conservatism but looking at the world and some of the problems we have witnessed, a little more precaution would have been a good thing to do. Kibert (2022) indicates that the precautionary principle requires the exercise of caution when making decisions that may have adverse effect on nature, natural ecosystems, and global biochemical cycles. When an activity raises threats of harm to human health or the environment, precautionary measures should be taken even if some cause and affect relationship are not fully established scientifically.

List of four tenets of the precautionary principle includes the following:

1. People have a duty to take anticipatory actions to prevent harm,
2. Burden of proof of harmlessness of a new technology, process, activity, or chemical lies with proponents not the public.
3. Before using the new technology, process or chemical or starting a new activity, people have an obligation to examine a full range of alternatives including the alternative of not doing it.
4. Decisions applying the precautionary principle must be open, informed, and democratic and must include the affected parties.

Kibert (2022) believes that despite the wisdom of exercising caution when addressing complex issues that may have unknown, far-reaching effects, the precautionary principle is controversial and sometimes is perceived as a threat to progress, since it fails to consider the positive consequences of its application. For example, refusing to use new drugs because society has not fully established their effects on human and nature may foreclose options for advancing human health. Nonetheless, the consequences of not applying the precautionary principle are becoming apparent in several areas. Most notably, the widespread use of estrogen-mimicking chemicals is believed to damage the reproductive systems of animal species and notably of humans. With these concerns in mind, the debate surrounding the application of the precautionary principle has focused great attention on the environment impacts of technology and has pressured technologies to acknowledge the potential consequences of their efforts on human and nature.

REVERSIBILITY PRINCIPLE

Reversibility Principal is: Making decisions that can be undone by future generations is the foundation of the reversibility principle. In 1965 Arthur C. Clark suggested a rule that well describes this principle: "Do not Commit the Irreversible". This principle calls for a wider range of options to be considered in decision making.

Addressing an issue of energy choices is an excellent example because a rapidly growing global economy is faced with looming energy shortages, exacerbated by the depletion of finite oil supplies. Per Kibert (2016), in the United States, a shift is underway to consider nuclear plants as a major source of energy because they probably can generate electricity at an acceptable cost and be a source of thermal energy for producing hydrogen from water for use in fuel cells.

The reversibility principal would force today's society to confront the issue of whether the choice of nuclear energy as an option is reversible by a future society. Two questions would immediately emerge from this consideration. First, is technology safe enough for widespread use? The nuclear industry suggests that over the past two decades of national hiatus from building new plants, the technology was advanced to the point that Chernobyl, Three Mile Island or Fukushima Daiichi incident can be eliminated. The second question is: How would a future society cope with the nuclear waste from these plants? Converting the waste into harmless materials via new technology is highly unlikely, and the power plants built today would force future generations to store and be put at risk by the radio nuclides in the spent fuel rods. A subset of questions on the same subject would result because of assuming that, if storage of the radioactive waste for periods of time in the 10,000-year range is feasible, what are the storage options? In addressing this question, Gene I. Rochlin (1978) suggested that there are two options. One is to deposit the waste deep in a stable rock formation where it could be recovered if, for example, leaks in the storage containers were detected by future generations. A second option is to deposit it in inaccessible locations – for example, by placing the waste deep in the ocean, where sliding continental plates would gradually cover it. The former solution allows future generations access the waste to take corrective action, while the latter does not allow that just not allowed that option.

The reversibility principle is related to the precautionary principle because it lays out criteria that must be observed prior to the adoption of a new technology. It is less stringent than the precautionary principle in some respects because it suggests reversibility as a primary criterion for deciding to employ the technology; the precautionary principle, in contrast, requires that technology not be implemented if its effects are not fully understood and if the risks are unacceptable.

THE POLLUTER PAYS PRINCIPLE AND PRODUCER RESPONSIBILITY

Both the precautionary and reversibility principles are forward looking, and they deal with future developments. The polluter pays principle and producer responsibility principle, also discussed by Kibert (2016), deals with existing developments and current activities. Many technologies today – like power plants, vehicles, building heating systems – use energy and emit pollutants.

The polluter pays principle addresses existing technologies that have not been subject to other principles and places the onus for mitigating damage and consequences on the individuals causing the impacts. Historically, the polluter pays principle has focused on retrospective liability for pollution; for example, an industry causing pollution would pay for cleanup costs arising from it.

The polluter pays principle has recently shifted to the avoidance of polluting and addressing wider environmental impacts through producer responsibility. It is an extended version of the polluter pays principle, as it applies to waste and resource management, placing responsibility for the environmental impact associated with a product on the producers of that product.

The producer responsibility is intended to address the whole life-cycle environmental problems of the production process from initial minimization of resource use, through extended product life span, to recovery and recycling of products once they have been disposed of as waste. Producer responsibility is used increasingly throughout the world to address the environmental impacts of certain products. The European Union the has applied producer responsibility through directives on packaging and packaging waste, waste electronic and electrical equipment, end of life value (ELV).

PROTECTING THE VULNERABLE

What does Vulnerability mean? Vulnerability is the inability to resist a hazard or to respond when a disaster has occurred. For instance, people who live on plains are more vulnerable to floods than people who live higher up. Being a vulnerable person or a living creature such as animal or plant, vulnerability means these people or living organism are in a position where other people can hurt them.

Per Kibert (2022), there are populations, including those of the animal world, that are vulnerable to the actions of portions of human species, due to the destruction of ecosystems under the guise of development, introduction of technology (including toxic substances, endocrine disruptors, and GMOs), general pattern of conduct (war, deforestation, soil erosion, eutrophication, desertification, and acid rain, to name a few). People who are powerless due to governing economic structures are vulnerable to the decision of those who are powerful because of their wealth or influence. This imbalanced power structure is governed by moral obligation. Those in power have a special obligation to protect the vulnerable, those dependent on them. In a family children dependency on parents gives them the rights against their parents. Future generations are also vulnerable because they are subject to the effects of decision we make today. In a technological society, many portions of the human population and certainly the animal world can be exposed to harm by the actions of individuals or companies performing medical research or because the government that is charged with protecting them fails in its responsibility when it comes to pollution, the use of toxic substances, and a wide variety of other poorly controlled actions.

Breaches of ethics are common when it comes to vulnerable populations, such as prisoners, people with mental disabilities, women and people in developing countries. As noted earlier, today's actions have consequences for future generations that have considered only recently. Future people are certainly vulnerable to our actions, and both their existence, and their quality of life are potentially compromised by short-term thinking and decisions based solely on the comfort and wealth of past populations. The ethical principle of protecting vulnerable places an enormous responsibility on Earth's current population, one even made more difficult due to rampant global poverty.

PROTECTING NON-HUMAN

Living species such as animals, plants and to some extent, biota including bacteria, viruses, mold, algae, and other living organisms, are being affected by human decisions and actions. The principle of protecting this world is an extension of the principle of protecting vulnerable, particularly animals but also plants that are in danger of extinction.

Animal rights fall under this principle. It should be emphasized that the non-living portion of Earth is essential in supporting life, and a set of sustainability principles should address the requirements for this key element of the life support system. Some would argue that ethics should require the character of beautiful places, such as Grand

Canyon and Yellowstone Park. This principle is an important one because humans have become disconnected from both the living and the non-living world when, in fact, are utterly dependent on them for our survival. Indeed, the biophilia hypothesis, will be described later in this chapter, states that humans crave a connection with nature and that our health, at least in part, is dependent on being able to connect with nature on a routine basis. Human ingenuity in the form of technology is having quite the opposite effect.

Kibert (2016) states that putting nature on an equal footing with humans is a difficult leap for many people, but vigorously protecting nature is in the best interest of humanity. Indeed, simply protecting nature does not quite meet the imperative of the principle of protecting the rights of a non-human world. Rather humans should consider restoring nature in all activities, righting the wrongs of the past, and in the process restoring the badly damaged link between humans and nature.

Figure 3-4 (Photo by mr. Yin: https://tinyurl.com/4fzkf73x)

Protecting the life of non-human world is important not only because each species or organism is a node in a web of life but also because each species is a teleological center of life, meaning each organism has a purpose and a reason for being and is therefore, inherently good, or valuable See Figure 3-4.

RESPECT FOR NATURE AND LAND ETHICS

A land ethic is a philosophy or theoretical framework about how, ethically, humans should regard the land. The term was the first time used by Aldo Leopold, in his book entitled "A Sand County (1949), a classic text of the environmental movement. There he argues that there is a critical need for a "new ethic", an ethic dealing with humans' relation to land and to the animals and plants which grow on it.

Leopold offers an ecologically based land ethic that rejects strictly human-centered views of the environment and focuses on the preservation of healthy, self-renewing ecosystems. A Sand County Almanac was the first systematic presentation of a holistic or eco-centric approach to the environment. Although Leopold is credited with coining the term "land ethics", there are many philosophical theories that speak to how humans should treat the land. Some of the most prominent land ethics include those rooted in economics, utilitarianism, libertarianism, egalitarianism, and ecology.

Economics-based land ethics has been criticized by Leopold. Leopold sees two flaws in this type of ethic. First, he argues that most members of an ecosystem have no economic worth. For this reason, such an ethic can ignore or even eliminate these members when they are necessary for the health of the biotic community of the land. And second, it tends to relegate conservation necessary for healthy ecosystems to the government and these tasks are too large and dispersed to be adequately addressed by such an institution. This ties directly into the context within which Leopold wrote A Sand County Almanac.

Finally, some critics have questioned whether Leopold's land ethic might require unacceptable interferences with nature to protect current, but transient, ecological balances. If the fundamental environmental imperative is to preserve the integrity and stability of natural ecosystems, wouldn't this require frequent and costly human interventions to prevent naturally occurring changes to natural environments? In nature, the "stability and integrity" of ecosystems are disrupted or destroyed all the time by drought, fire, storms, pests, newly invasive predators, etc. Must humans act to prevent such ecological changes, and if so, at what cost? Why should we place such high value on current ecological balances? Why think it is our role to be nature's steward or policeman? According to these critics, Leopold's stress on preserving existing ecological balances is overly human-centered and fails to treat nature with the respect it deserves.

ECONOMICS-BASED LAND ETHICS

This theory has been criticized by Leopold. Leopold sees two flaws in this type of ethic. First, he argues that most members of an ecosystem have no economic worth. For this reason, such an ethic can ignore or even eliminate these members when they are necessary for the health of the biotic community of the land. And second, it tends to relegate conservation necessary for healthy ecosystems to the government and these tasks are too large and dispersed to be adequately addressed by such an institution. This ties directly into the context within which Leopold wrote A Sand County Almanac.

ECOLOGICALLY BASED LAND ETHICS

Land ethics may also be based upon the principle that the land (and the organisms that live off the land) has intrinsic value. These ethics are, roughly, based on an ecological or systems view. This position was first put forth by Ayers Brinser in Our Use of the Land, published in 1939. Brinser argued that white settlers brought with them "the seeds of a civilization which has grown by consuming the land, that is, a civilization which has used up the land in much the same way that a furnace burns coal. Later, Aldo Leopold's posthumously published A Sand County Almanac (1949) popularized this idea.

Finally, some critics have questioned whether Leopold's land ethic might require unacceptable interferences with nature in order to protect current, but transient, ecological balances.[16] If the fundamental environmental imperative is to preserve the integrity and stability of natural ecosystems, wouldn't this require frequent and costly human interventions to prevent naturally occurring changes to natural environments? In nature, the "stability and integrity" of ecosystems are disrupted or destroyed all the time by drought, fire, storms, pests, newly invasive predators, etc. Must humans act to prevent such ecological changes, and if so, at what cost? Why should we place such high value on current ecological balances? Why think it is our role to be nature's steward or policeman? According to these critics, Leopold's stress on preserving existing ecological balances is overly human-centered and fails to treat nature with the respect it deserves.

BASIC CONCEPTS AND VOCABULARY

We are back to our definitions and to this notion of "sustainability". The most famous definition of this concept, sustainable development, coined in the Brundtland Commission, formerly the World Commission on Environment and Development, was sub-organization of the United Nations (UN) that aimed to unite countries in pursuit of sustainable development. The Brundtland Commission which was published in 1987 says: "The sustainable development is the one that meets the needs of the present generation without compromising the ability of future generations to meet their needs". Business-as-usual, the way we've done things in the past does not reflect the sustainable development definition – not only we have failed to protect the interests of future generations, but we've also failed to equitably meet the needs of millions, nay billions of people on the planet today. We act as if materials are infinite – that they'll last forever and support the systems we have developed to exploit them. We fail to recognize that many of the resources we currently depend upon are non-renewable and that others are only conditionally renewable, subject to collapse should we overexploit them past their sustainable yields. At the philosophical core of the green building movement is this notion that not only is there financial capital, but that there is natural capital and if we spend it on current consumption, rather than relying on the annual interest it generates, soon we'll be bankrupt!

The following is the definition of new terminology which describes sustainable development. Without understanding these definitions and these terminologies our understanding of sustainable development will not be possible. The most notable and important ideas include:

- The concept of sustainable development,
- Industrial ecology
- Construction ecology
- Biomimicry
- Design for environment
- Ecological economics
- Carrying capacity
- Ecological footprint
- Ecological rucksack
- Embodied energy
- The biophilia hypothesis
- Eco-efficiency
- The Natural Step
- Life cycle assessment
- Life cycle costing
- Factor 4
- Factor 10

These concepts are described briefly in the following sections.

SUSTAINABLE DEVELOPMENT

Sustainable development is the one that "meets the needs of the present without compromising the ability of future generations to meet their own needs".

This implies intergenerational justice; highlights the responsibility of the current population for the welfare of millions of unborn. It also implies that we are borrowing the planet, its resources, and its environmental functions and quality from future generations.

How far in the future should we consider the new generations; 100 years? 200 years? How long our buildings will last, will they be able to dispose of them?

Although no clear answer to this important question is readily apparent, the Native American philosophy of thinking seven generations, or 200 years, into the future is instructive. If in two centuries few contemporary buildings will be standing, we must ask whether our current stock of materials will provide recyclable resources for future generations or saddle them with enormous and difficult waste disposal problems (Kibert 2022).

INDUSTRIAL ECOLOGY

Industrial ecology is the study of industrial systems that operate more like natural ecosystems. A natural ecosystem tends to evolve in such a way that any available source of useful material or energy will be used by some organism in the system. Animals and plants live on each other's waste matter. Industrial ecology is a newer subfield of ecology that examines the interaction between industries and the environment. It looks for ways to help reduce the environmental impact of different industries.

An example of industrial ecology is Denmark's Kalundborg Eco-Industrial Park. At Kalundborg, waste products are cycled between different industries so that the waste produced by one company can be used as the raw materials for another industry. Industrial ecologists study the life cycle of resource usage by different industries and develop recommendations for helping industries become more sustainable. Some of the tools used by industrial ecologists include carbon footprint analysis and life cycle analysis.

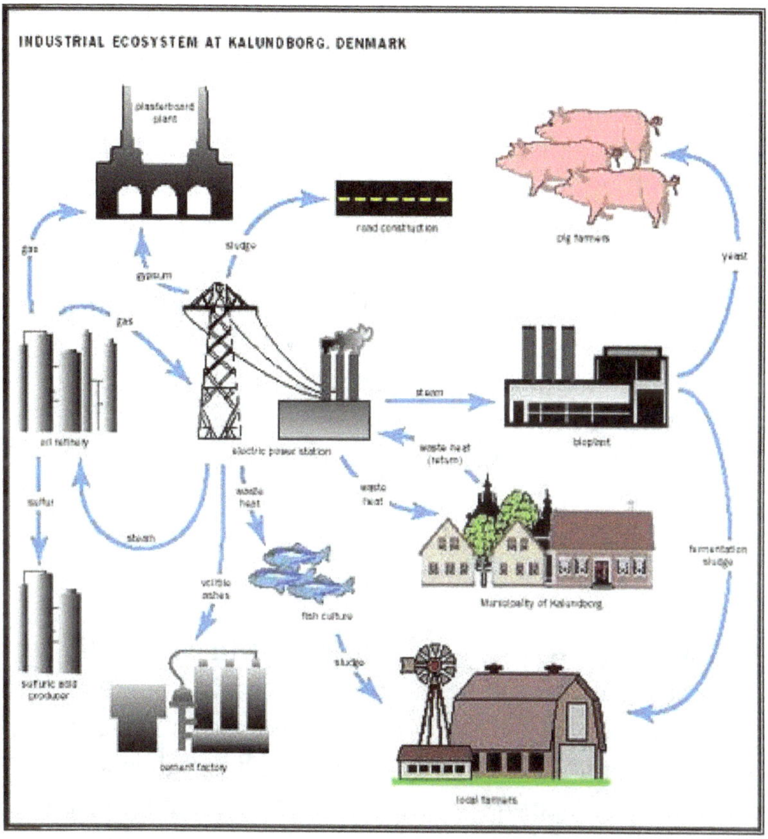

Figure 3-5 The industrial complex in Kalundborg, Denmark, exchanges energy, water, and materials among its member companies and organizations demonstrating industrial symbiosis, one of the basic concepts of industrial ecology. (Source: International Journal of Digital Innovation in the Built Environment - Jan 202)

This is a classic example of industrial ecology, where excess heat energy, waste, and water are shared among five major partners:

1. The Asnaes Power Station, Denmark's largest coal-fired power plant, with a 1,500-megawatt capacity,

2. The Statoil refinery, Denmark's largest, with a current capacity of 4.8million tons (4.4 million mt) per year,
3. Gyproc, a plaster factory producing 150 million ft^2 (14 million m^2) of gypsum wallboard annually (roughly enough to build all the houses in the six towns the size of Kalundborg)
4. Novo Nordisk, an international biotechnological company, with annual sales of over $2 billion, which manufactures industrial enzymes and pharmaceuticals, including 40 percent of world's supply of insulin,
5. The city of Kalundborg District has a heating system, which supplies heating to 20,000 residents and water to its homes and industries.

The Kalundborg complex diagrammed in Figure 3-5 was the first eco-industrial park; since its inception, similar waste exchange complexes have been created around the world. Since the early 1990s, the concept of industrial ecology has expanded to encompass issues of design for environment, product design, closing material loops, recycling, and other environmentally conscience practices that we will be covering later in this chapter. Industrial ecology is a comprehensive approach to implementing sustainable industrial behavior (Kibert 2022).

CONSTRUCTION ECOLOGY

Construction ecology involves using as few resources while meeting design requirements. Sometimes referred to as "green building", this construction style is efficient in terms of the energy generated by its occupants, and energy used by the structure itself.

Construction ecology is a subset or subcategory of industrial ecology that applies specifically to the built environment. The applications of construction ecology principles should result in buildings with:

- Components that are decoupled from the building for easy replacement,
- Composed of recyclable products,
- Built with recyclable and bulk materials,
- Slow metabolisms (due to their durability and adaptability),
- Consideration of occupant's health,
- Easy deconstruction at the end of their useful lives.

BIOMIMICRY

Biomimicry simply says imitation of living. It aims to take inspiration from natural selection solution adopted by nature and translate the principles to human engineering. The biomimicry approach aims to favor "choices" tested by nature which had millions of years to understand what works best and what does not.

This concept demonstrates direct application of ecology/nature to the production of industrial objects. Benyus in her book entitled "Biomimicry: Innovation inspired by nature (1997) and has since received widespread attention as a concept that demonstrate the direct application of ecological concept to the production of industrial objects. According to Benyus, biomimicry is the "conscience imitation of nature's genius", and it suggests that most of what we need to know about energy and materials has been developed by natural systems over almost 4 billion years of trial and error. Benyus also laid out 10 lessons for corporations that are based on the emulation of nature as the model for human-designed systems:

1. Use waste as a resource.
2. Diversify and cooperate for fully use the habitat.
3. Gather and use energy efficiently.
4. Optimize rather than maximize.
5. Use materials sparingly.
6. Don't foul the nest.
7. Don't drawdown resources.
8. Remain in balance with the biosphere.
9. Run on information.
10. Shop locally.

Benyus also suggested "four steps for a biometric future":

1. Quieting. Immerse ourselves in nature.
2. Listening. Interview with flora and fauna of our own planet.
3. Echoing. Encourage biologists and engineers to collaborate, using nature as a model and measures.
4. Stewarding. Preserve life's diversity and genius.

Biomimicry suggests the creation of strong, tough, and intelligent materials with no waste and using the sunlight, natural materials under ambient temperature during the manufacturing process, for instance, creation of seashell vs. ceramic clay tiles.

DESIGN FOR THE ENVIRONMENT

Design for the environment (DfE) is a design approach to reduce the environment impact of products, processes, or services. It considers the whole lifecycle, going beyond just use of recycled materials or proper packaging or disposal.

DfE is the process of investigating the possible environmental impacts of a product and refining the product design as necessary to reduce or negate the potential impacts. The environmental impact of a product covers a wide range of potential topics. It considers the entire life cycle of the product.

DfE concept invests a greater effort during design phase to ensure the recovery, reuse and/ or recycling of the product's components at the end of its useful life.

Example: A window assembly designed using DfE strategies, would be easy to disassemble into its basic metal, glass, and plastic components. As implied to built environment, DfE implies that entire buildings should be designed to be taken apart, deconstructed to recover components for further disassembly, reuse, and recycling.

ECOLOGICAL ECONOMICS:

Ecological economics is a fundamental requirement of sustainable development that specifically addresses the relationship between human economies and natural ecosystems. The ecological economics philosophy counters the human's tendency to ignorantly or deliberately degrading ecosystems by extracting useful, high-quality matter and energy which ultimately are transformed into useless, low-quality waste and heat.

For example, the fracking that could increase emission of greenhouse gases during its implementation is still a worthwhile effort to produce natural gas rather than using coal for production of natural gas for heating our homes.

Fracking, which is the process of pumping water and sand between layers of rocks to release natural gas and oil, has played a key role in the reduction of greenhouse gas emissions in the United States. The nation's fracking boom drove down the price of natural gas, which is used to heat homes and create electricity. The cheap natural gas began to replace the coal that had been used to produce electricity for generations. This transition has benefited the environment because natural gas emits half as much greenhouse gas as coal.

According to report from the International Energy Agency, replacing coal with natural gas led the U.S. to be the world leader in cutting carbon dioxide emissions in 2019. The agency specifically credited "record-low natural gas prices" for the dip in emissions.

Natural gas has also proven to be a strong back-up for renewable energy. Unlike coal, natural gas obtained through fracking allows energy producers the flexibility to match demand. If a cloudy day hampers the amount of energy created in solar panels, natural gas can be used to supplement a power grid with ease. Renewable energy production has doubled over the past decade in large part because natural gas can fill the efficiency gaps that are present in most renewable energy sources.

Fracking has been a benefit to the American economy, as well. In the past, oil prices in the U.S. were reliant on what was taking place in the Middle East or Venezuela, two of the least stable regions of the globe. Fracking has created an energy buffer between the U.S. and the rest of the globe that has led gas and energy prices to remain stable over the past few years despite continued chaos in oil-rich countries.

Additionally, fracking has created jobs in many states throughout the country. The fracking industry alone has created 2.1 million jobs, but that number could grow to 3.9 million jobs by 2025. Fracking has also contributed to a growth in jobs in associated industries. A report from

Reuters estimated that natural gas added $2.08 trillion to the U.S. manufacturing sector in 2013 alone.

Some fringe critics have continued to call for fracking to be banned despite the clear benefits. New York Rep. Alexandria Ocasio-Cortez, for example, has suggested that all fuels that emit greenhouse gases should be banned. But the fact remains that the technology behind most renewable energy sources is not stable enough to meet electricity demand in the U.S. However, despite the economic incentives of using fracking, it has detrimental environmental consequences beyond imagination. Upon implementation, it pollutes groundwater resources, create soil contamination as well as degrading air quality in the surrounding areas and communities.

CARRYING CAPACITY:

The number of people, other living organisms, or crops that a region can be supported without environmental degradation. The species population size is limited by environmental factors like adequate food, shelter, water, and mates. If these needs are not met, the population will decrease until the resource rebounds.

The carrying capacity of an area refers to the maximum number of people who can be realistically sustained by the geography of that area. This number can be affected by access to food, water shelter, and other significant factors.

According to the carrying capacity network (www.carryingcapacity.org):

The Carry capacity for any given area is not fixed. It can be altered by improved technology, but mostly it is changed for the worse by pressures which accompany a population increase. As the environment is degraded, carry capacity shrinks, leaving the environment no longer able to support even the number of people who could formerly have lived in the area on a sustainable basis. No population can leave beyond the environment's carry capacity for very long.

In general, living at full carrying capacity of at any given region is not environmentally a desirable thing.

ECOLOGICAL FOOTPRINT:

The impact of human activities measured in terms of the area of biologically productive land and water required to produce the goods consumed and to assimilate the waste generated. The ecological footprint is the inverse of the carrying capacity. It represents the land area needed to support a given population. For example, London requires a land area 125 times greater than its physical footprint.

The entire planet cannot live with the North American lifestyle. Americans are human species apart; their lifestyle need more land and resources than any other nation in the world. However, the ultimate problem must be resolved through sustainable development.

THE ECOLOGICAL RUCKSACK

An ecological rucksack is the total quantity (in Kg) of materials moved from nature to create a product or service, minus the actual weight of product. It considers the entire production process, from the cradle to the point when the product is ready for use. It is calculated by adding up all the resources and energy consumed, and waste generated throughout the life cycle.

The environmental stress caused by an activity is proportional to the quantity of materials moved. The greater the masses moved, the higher is the environmental impact. The concept of ecological rucksack focuses on this large displacement of earth and rock rather than on minute quantities of materials. It has been the large land transformations occasioned by increasing demand, coupled with depleted deposits of rich materials, that have been historically neglected by environmentalists and policy makers (Kibert 2016). Table 3-1 shows the ecological rucksack of some materials. Figure 3-6 explains the concept of ecological rucksack.

Table 3.1 Ecological Rucksack of some Well-known Materials

Material	Ecological Rucksack
Rubber	5
Aluminum	85
Recycled Aluminum	4
Steel	21
Recycles Steel	5
Platinum	350,000
Gold	540,000
Diamond	53,000,000

Figure 3-6 Schematic Diagram of the Concept Behind Ecological Rucksack

THE BIOPHILIA HYPOTHESIS

The term biophilia was used by German-born American psychoanalyst Erich Fromm in The Anatomy of Human Destructiveness (1973), which described biophilia as "the passionate love of life and of all that is alive." The term was later used by American biologist Edward O. Wilson in his work Biophilia (1984), which proposed that the tendency of humans to focus on and to affiliate with nature and other life-forms has, in part, a genetic basis.

Biophilia is a human relationship with nature. Anecdotal and qualitative evidence suggests that humans are innately attracted to nature. For example, the appearance of the natural world, with its rich diversity of shapes, colors, and life, is universally appreciated. This appreciation is often invoked as evidence of biophilia. The symbolic use of nature in human language, in idioms such as "blind as a bat" and "eager beaver," and the pervasiveness of spiritual reverence for animals and nature in human cultures worldwide are other sources of evidence for biophilia. Such spiritual experience and widespread affiliations with natural metaphors appear to be rooted in the evolutionary history of the human species, originating in eras when people lived in much closer contact with nature than most do today. Human divergence from the natural world appears to have occurred in parallel with technological developments, with advances in the 19th and 20th centuries having the most significant impact, fundamentally changing human interactions with nature. In its most literal sense, this separation was made possible by the construction of enclosed and relatively sterile spaces, from homes to workplaces to cars, in which modern humans were sheltered from the elements of nature and in which many, particularly people living in more-developed countries, now spend most of their time.

Some of the most powerful evidence for an innate connection between humans and nature comes from studies of biphobia (the fear of nature), in which measurable physiological responses are produced upon exposure to an object that is the source of fear, such as a snake or a spider. These responses are the result of evolution in a world in which humans were constantly vulnerable to predators, poisonous plants and animals, and natural phenomena such as thunder and lightning. Fear was a fundamental connection with nature that enabled survival, and, as a result, humans needed to maintain a close relationship with their environment, using sights and sounds as vital cues, particularly for fight-or-flight responses.

BIOPHILIA AND CONSERVATION

Genes that influence biophilia have not been identified, and it is suspected that the increased dependence of the human species on technology has led to an attenuation in the human drive to connect with nature. Wilson and others have argued that such declines in biophilic behavior could remove meaning from nature, translating into a loss of human respect for the natural world. In fact, the loss of desire to interact with the natural world, resulting in a decreased appreciation for the diversity of life-forms that support human survival, has been cited as a potential factor contributing to environmental destruction and the rapid rate of species extinction. Thus, reestablishing the human connection with nature has become an important theme in conservation.

In Biophilia, Wilson introduced a conservation ethic based on multiple dimensions of the innate relationship humans share with nature. His notion of environmental stewardship drew on various concepts, including the practical dependence of humans on nature, which centers on the ecological services (e.g., clean water and soil) nature provides; the satisfaction derived from direct interaction with nature, such as through exploration and development of outdoor skills; the physical appeal of nature, evident in its role as a source of inspiration and peace; and the human attachment to nature in the form of emotional connections to landscapes and animals.

BIOPHILIA AND TECHNOLOGY

Biophilia has been explored by researchers in a wide range of fields, and, as a result, its meaning and significance have been variously interpreted. Juxtaposed to the notion that biophilia competes with the human technological drive is the notion that technology is in itself an extension of human evolution and biophilia. Both perspectives were offered in The Biophilia Hypothesis (1993), a work coedited by Wilson and American social ecologist Stephen R. Kellert. Among the collection of views the work presented were those of American biologists Lynn Margulis and Dorion Sagan and Indian ecologist Madhav Gadgil, who considered the possibility that the human attraction to other life-forms is reflected in the diversity of technological developments that exist in the world today. Some of these technologies, including those employed in molecular biology and genetic engineering, have enabled scientists to develop entirely new forms of life, with which humans are wholly fascinated. The idea that technology feeds the human biophilic drive also finds support in the search for life on other planets (see extraterrestrial life).

Regardless of the extent to which individuals feel or perceive biophilia, research has indicated that simply spending time in nature is beneficial for human health. In one study, persons who spent even just two hours per week interacting with nature reported greater satisfaction and better health than persons who spent less time in natural environments. Among adults, spending time in natural environments is associated with improved mental and spiritual health; for children, being out in nature has been found to encourage physical activity and play.

ECO-EFFICIENCY

Eco-efficiency refers to the delivery of goods and services to meet human needs and improve quality of life while progressively reducing their environmental impacts of goods and resource intensity during their life cycle.

The eco-efficiency is concerned with the efficiency of the production process, that is, the efficiency of the technologies involved in it. It focuses on finding solutions that improve the efficiency of resource use and of waste/pollution generation, mainly through investment in research and development.

In 1992, the World Business Council for Sustainable Development (WBCSD) developed the concept of eco-efficiency which includes environmental impacts and cost as a factor in calculating business efficiency. The WBCSD considers the term eco-efficiency to describe the delivery of competitively priced goods and services that satisfy human needs and enhance the quality of life while progressively reducing ecological impacts and resource intensity throughout the products' life cycles to a level commensurate with earth estimated carrying capacity. The following is seven elements of eco-efficiency as defined by the WBCSD:

1. Reducing the material requirements of goods and services,
2. Reducing the energy efficiency intensity of goods and services,
3. Reducing toxic dispersion,
4. Enhancing material recyclability,
5. Maximizing sustainable use of renewable resources,
6. Extending product durability,
7. Increasing the service intensity of goods and services.

As a concept, eco-efficiency describes most of the fundamental principles underpinning the concept of sustainable development. Its promotion by the WBCSD, essentially an association of major corporations, is a positive sign that the business community is beginning to take sustainability seriously.

The Natural Step

Natural step provides a framework for considering the effects of materials selection on human health. In another word, the concept of natural step is about finding ways to satisfy our basic needs in a way that does not undermine the balance of social and ecological systems.

The Natural Step articulates four system conditions, or basic principles, that should be followed to eliminate the effects of materials practices on our health. The four system conditions are as follows:

1. For a society to be sustainable, nature's functions and diversity are not systematically subject to increasing concentrations of substances extracted from Earth's crust,
2. For a society to be sustainable, nature's functions and diversity are not systematically subject to increasing concentrations of substances produced by society,
3. For a society to be sustainable, nature's function and diversity are not systematically impoverished by overharvesting or other forms of ecosystem manipulation,
4. In a sustainable society, resources are used fairly and efficiently to meet basic human needs globally.

Life Cycle Assessment

Life Cycle Assessment (LCA) is a process of evaluating the effects that a product has on the environment over the entire period of its life thereby increasing resource-use efficiency and decreasing liabilities. It can be used to study the environmental impact of either a product or the function the product is designed to perform. LCA is commonly referred to as a "cradle-to-grave" analysis. LCA's key elements are (1) identify and quantify the environmental loads involved, e.g. the energy and raw materials consumed, the emissions and waste generated; (2) evaluate the potential environmental impacts of these loads; (3) assess the options available for reducing these environmental impacts.

LCA is the most important tool in choosing building material that will have a minimal impact on the environment. Throughout this book we will discuss and recommend using this technique for evaluation of environmental impact of building materials.

Life Cycle Costing

Life cycle costing (LCC) is the total discounted dollar cost of owning, operating, maintaining, and disposing of a building or a building system over a period. It is an ability to model a building's financial performance over its life cycle. It may require greater initial capital investment but yield significantly lower capital operational costs overtime.

For example, the financial return for installation of a photovoltaic (PV) system would be determined by amortizing the system's costs over its probable life; the worth of the energy generated each year would then be calculated to determine the net annual benefit. Application of LCC may determine whether the payback for this system meets the owner's economic criteria. LCC analysis can also be combined with LCA results to weigh the combined financial and environmental impact of a particular system (Kibert 2022).

Embodied Energy

Embodied energy is the energy consumed by all the processes associated with the production of a building. From the mining and processing of natural resources to manufacturing, transporting and product delivery.

Embodied energy does not include the operation, maintenance or disposal of the building material.

Products with greater embodied energy usually have greater environmental impact due to the emissions and greenhouse gases associated with energy consumption. However, another calculation, which divides the embodied energy by the product's time in use, yields a true indicator of the environmental impact. More durable products will have lower embodied energy per time in use. For example, a product with high embodied energy, such as aluminum, could have a very low embodied energy per time in use because of the product's extremely high durability. In addition, certain products have relatively low embodied energy when recycled. Recycled aluminum has just 10 percent of the embodied energy of aluminum made from bauxite ore.

Similarly, recycled steel has about 20 percent of the embodied energy of steel made from ores. A list of typical embodied energies for common construction materials is presented in Table 3-2.

Table 3-2 Embodied Energy of Common Construction Materials

Material	Embodied Energy BTU/lb	Embodied Energy MJ/kg	Embodied Energy MJ/m³
Aggregate	43	0.1	150
Concrete	559	1.3	3,180
Lumber	1,075	2.5	1,380
Brick	1,075	2.5	5,170
Cellulose Insulation	1,419	3.3	112
Mineral Wool Insulation	6,277	14.6	139
Fiberglass insulation	13,026	30.3	970
Polystyrene Insulation	50,298	117.0	3,770
Gypsum Wallboard	2,622	6.1	5,890
Particleboard	3,439	8.0	4,400
Plywood	4,471	10.4	5,720
Aluminum	97,587	227.0	515,700
Aluminum (recycles)	3,482	8.1	21,870
Steel	13,757	32.0	251,200
Zinc	21,925	51.0	371,280
Copper	30, 351	70.6	631,164
Polyvinyl Chloride (PVC)	30,093	70.0	93,620
Linoleum	49,868	116.0	150,930
Carpet (synthetic)	63,625	148.0	84,900
Paint	40,110	93.3	117,500
Asphalt Shingles	3,869	9.0	4,930

Source: Canadian Architect website www.dnarchitect.com
Note: MJ/kg = megajoules per kilogram of material
MJ/m³ = megajoules per cubic meter of material

Factor 4, Factor 5, and Factor 10

These concepts are based on the principles of reducing material needs in the construction industry. For instance, Factor 4 suggests that for humanity to live sustainably today, we must rapidly reduce resource consumption to one-quarter of its current levels.

The Factor 4 or Factor 10 concept visualizes a quadruple or 10-fold increase in resource efficiency using existing methodologies while avoiding negative impacts on the overall quality of life. The concept aims for society to last twice as long or enjoy twice as much whilst using half or 10 percent of the resources. Factor 4 moves away from labor productivity and towards resource productivity and sustainability. By using the best available technology, advanced engineering and improved production methods, fewer resources are required to produce more products and services. As a result, the life span resources is stretched and future generations are provided for.in other words, four times as much wealth can be extracted from the resources we currently use.

Factor 4 or Factor 10 are used in decision making, production; and product-oriented environmental protection. Fundamentally, Factor 4 is an economic idea. Reducing resource use by a factor of 4is not a fixed target. Rather technologies and processes should aim to increase resource efficiency.

Per Kibert 2022, an example of applying Factor 4 principle to the built environment is provided by Lee Eng Lock, Chinese engineer in Singapore. He has challenged many of the fundamental assumptions made by mechanical engineers in their systems design and layout. Rather than oversizing chillers, air handlers, pumps, and other equipment, he ensures that they are precisely the correct size for the job. This commonsense approach achieves the same cooling and comfort while using only 10 percent of energy of conventional designs, thus accomplishing a factor 10 reduction in energy. The Factor 10 concept has had a significant effect internationally and now is being implemented by the European Union. The factor 4 concept has revisited by Enst von Weizsacker and several Australin colleagues in 2009, and they concluded there was a potential Factor 5 of available efficiency improvements for the entire sector of the economy, without losing the quality of service or wellbeing.

Figure 3-7 The Factor 4 concept originated in the book Factor Four by Ernst von Weizsacker, Amory Lovins and Hunter Lovins

MAJOR ENVIRONMENTAL AND RESOURCE CONCERNS

Due to the following genuine concerns about our planet the construction industry is obligated to do its share toward saving our planet. The major environmental concerns that threaten our ecosystem and we have witnessed almost all these crises which are the manifestation of global warming are as follows:

- Climate Change
- Ozone Depletion
- Soil Erosion
- Desertification
- Deforestation
- Eutrophication
- Acidification
- Loss of Biodiversity

- Land, Water and Air Pollution
- Dispersion of Toxic Substances
- Depletion of Fisheries

The above issues will be discussed in the following sections.

Climate Change

The global warming that's causing our climate crisis is already having dire consequences. In just the past few decades the rising temperature in atmosphere has worsened extreme weather events. Chunks of ice in the Antarctic have broken apart, wildfire seasons are months longer, coral reefs have been bleached of their colors and mosquitoes are expanding their territory, to spread disease.

Who has caused such unpleasant events? It's mainly us. Humans are the main cause of climate change. We burn fossil fuels and chop down forests, causing average temperatures to rise worldwide. The global warming trend is increasingly disrupting our climate — the average weather over many years.

Earth has already warmed by about 1 degree Celsius, or 1.8 degrees Fahrenheit, since the 19th century, before industry started to boom. While we experience the effects, we're on our way toward 1.5 degrees C (2.7 F) by as early as 2030.

A warmer world — even by half-degree Celsius — has more evaporation, leading to more water in the atmosphere. Such changing conditions put our agriculture, health, water supply and more at risk.

Picture a North Carolina cotton farm that's been around since 1960, with global average temperatures rising by roughly half a degree since it grew its first crop. The increased evaporation and additional moisture to the atmosphere has led to 30% more intense rain during heavy downpours in that part of the U.S. Then a hurricane like 2018's Florence — already strengthened by warmer oceans and higher seas — dumps this excess rainfall on the farm. The crops get more flooded and damaged than they did half a century ago. It's how you go from half-degree of warming to economic hardship. Imagine the rise in temperature will promote Antarctic and North Pole ice to melt, this will cause the water levels to rise about 5-10 feet in oceans and cause some islands to go completely under the water.

In addition to climate change, certain chemicals used in building construction and facility operations have been thinning the ozone layer, the protective sheath of atmosphere consisting of three-molecule oxygen (O_3), which is located 10 to 25 files above Earth and serves to attenuate harmful ultraviolet radiation. In 1995, scientists discovered a vast hole the size of continental United States in the ozone layer over Antarctica. By 1999, the size of the hole had doubled. Ozone depletion caused by the interaction of halogens such as chlorine and bromine-containing gases such as chlorofluorocarbon (CFC) used in refrigeration and halons used for fire suppression., see Table 3-3.

Table 3.3 Gases Used for Typical Building Functions

Halogen Gas	Lifetime (years)	Global Emissions (1,000 of metric tons/yr)	Ozone Depletion Potential (ODP)*
Chlorine			
CFC-12	100	130-160	1
CFC-113	85	10-25	1
CFC-11	45	70-110	1
HCFC	1-26	340-370	0.02-0.12
Bromine			
Halon 1301	65	3	12
Halon 1211	16	10	6

Source: Excerpted from "twenty Questions and Answers about the Ozone layer" (2002) at www.esrl.noaa.gov/csd/
* ODP = Ozone Depletion Potential

Deforestation, Desertification, and Soil Erosion

Throughout geological time, the development of deserts has occurred naturally. However, in recent times human activity, such as improper land management, deforestation and climate change have accelerated the process of desertification in many areas.

When land is losing its capacity for biological productivity there is a loss of biodiversity, and a loss of potential food production capacity. Deforestation is not only threatening the local animal species, but there is also ultimately a risk to human life too. The threat to human life may begin as a disruption to the food supply, and the damaging effects this has on the local economy. Ultimately this can lead to the total collapse of any human settlements in that area and can contribute to poverty in the surrounding areas.

Causes of Desertification: Part of the problem is that when biological productivity is consumed in an unsustainable way, not only are the local plant species greatly reduced or lost, but also the animals which have been helping to maintain that local ecology are also lost. Animals help to spread the seeds of plants via their normal activities of feeding, foraging, burrowing and so on.

In order alleviate this situation and to offset Desertification, it is up to us humans to intervene and take on the role of "seed spreading" and "seed planting" of the indigenous animals once did.

A Rewilding Stick lets you plant about 500 seeds per hour directly into rough ground, while walking upright at a normal pace. There is no need to bend over while planting with a Rewilding Stick, so less chance of a sore back at the end of it.

In many situations this saves all the time and effort needed to set up and maintain a plant nursery, and all the attendant activities. Instead, the seeds are simply planted into the ground where they are intended to grow. If an individual decided they wanted to plant, say, 10 thousand trees, it would normally be an expensive, time-consuming (and back-breaking) proposition to do so with seedlings. They would either grow or buy the seedlings. Then, because they spent so much time and / or money on the seedlings they would feel that they must protect them from predation by using fences or Spiral Tree Protection Guards. This would take even more time and add even more to the costs. Yet, using a Rewilding Stick the same amount of planting can be done in just 20 hours of "walking". If a percentage of those plants are consumed by the indigenous animals, it does not matter (and is arguably a good thing) because when the plants that survive mature the animals will help spread their seeds.

Acid rain that we will discuss in this book is another reason for deforestation. Acis rain which is formed due to air pollution is a number one enemy of rain forests through which the foliage of trees gets burned by acid rain and causes the entire tree to die. Consumption of fossil fuels, which is the main reason for creation of air pollution, is main cause of air pollution and forming acid rain.

Soil erosion, which may be caused by water or wind, is the major reason for soil loss in agricultural areas as well as construction sites during rainy periods. Chapter two of this book addresses how soil erosion and sediment transport can be controlled during construction activities.

Eutrophication and Acidification

Excessive richness of nutrients in a lake or other body of water, frequently due to runoff from the land, which causes a dense growth of algae and death of fish and other animals from the lack of oxygen is referred to as eutrophication. Uncontrolled stormwater runs off through the farmlands, and landscaped areas pick up nutrients such as nitrogen and phosphorus and upon discharge into the lakes and estuaries promoting algal growth. Algal growth consumes dissolved oxygen in water and deprives fish population growth, see Figure 3-8.

Figure 3-8

Acidification is the process formed by dissolution of chemicals in air with the rainwater. The air pollution is mainly caused by burning fossil fuels. When the rainwater dissolves carbon dioxide (CO^2) or sulphur dioxide (SO^2), it generates carbonic acid or sulphuric acid which upon reaching to ground damages leaves and foliage of trees and vegetation. This is main reason for deforestation and desertification. Acidification and eutrophication are environmental conditions that frequently threaten water supplies as well.

Loss of Biodiversity

Biodiversity or biological diversity is the variety and variability of life on Earth. Biodiversity is a measure of variation at the genetic, species and ecosystem levels. Biodiversity is not distributed evenly on Earth. It is usually greater in the tropics because of the warm climate and high primary productivity in the region near the equator. Tropical forest ecosystem covers less than 10 percent of Earth's terrestrial surface and contain 50 percent of the world's species. There are latitudinal gradients in species diversity for both marine and terrestrial taxa.

Species biodiversity is the level of biodiversity most discussed. An estimated 1.7 million species have been described out of total estimated 5 – 7 million species. However, deforestation and climate change are causing such a rapid extinction of many species that some biologists are predicting loss of 20 percent of existing species over the next 20 years.

Deforestation is particularly devastating, especially in rainforests, which comprise just 6 percent of the world's land but contain more than 500,000 of its species.

Figure 3-9 An Ecosystem Symbolic Diagram

Impact of Chemicals on Our Life

While it is not always possible to know the cause of an individual disease, studies have found that some decease especially birth defects can be linked to exposure to prescription or over the counter medications, chemical exposure in the home or workplace, or toxic substances that contaminate the air or groundwater in communities.

For instance, when a child has birth defects that may have been caused by an exposure to a certain chemical, parents may seek answers to the cause that may lead them to the main cause of such devastating incident. Study has shown that there are many chemicals in our living environment that may have a devastating site effect that has not been fully discovered. These chemicals that may have been used on over-counter drug shelves of popular supermarkets are readily accessible to the public. The following is an example of certain drugs that are being prescribed by physician or are readily available to the public.

Antidepressant Drugs

These selective serotonin reuptake inhibitors such as Prozac, Paxil, Zoloft, Celexa, Lexapro are used for depression for a long time and may have been associated with:

- Heart defects
- Cleft lip and palate
- Omphalocele

Pesticides

Pesticides contain many toxic chemicals that can cause many decease and birth defects in children whose parents were exposed through agricultural work, groundwater contamination, or airborne drift. Parents who work in the fields for the farming industry — or simply live near an agricultural area — are at high risk of exposure to pesticides used on the fields. Pesticides may damage the parents' chromosomes and lead to birth defects among children who are conceived during or just after exposure. Birth defects and complications linked to pesticides include:

- Brain cancers
- Endocrine disruption
- Neurological disorders
- Lower birth weight and size
- Premature death
- Limb defects

Industrial Chemicals Used in High Tech and Electronics Manufacturing

Toxic chemicals used in many industries, including high tech and electronics manufacturing can cause birth defects and injuries to children whose parents are exposed to these dangerous chemicals. Such chemicals are often used in clean-room environments to make semiconductor parts. These chemicals include, but are not limited to, etching materials, solvents, cleaning fluids, and ethylene glycol ethers (EGE).

These, along with other toxic substances, can cause reproductive harm to both men and women, disrupting the parents' abilities to reproduce or causing them to pass damaged and defective DNA onto their children. Birth defects linked to exposures in high tech industries include:

- Skeletal and limb abnormalities
- Cognitive impairment
- Blindness
- Developmental Delays

- Missing or deformed organs
- Shortened or missing limbs
- Heart defects
- Cranial defects
- Missing or shortened limbs
- Organ malformation
- Facial abnormalities

Hydraulic Fracturing ("Fracking") and Birth Defects:

Recent studies have found that glycol ethers, toluene and several other well-established chemicals known to cause birth defects and miscarriages are prominent in the gas extraction process known as fracking:

Solvents or Chlorinated Hydrocarbons

Epidemiologic studies, animal studies and case reports strongly support the causal relationship between these birth defects and exposures to chemicals contained in industrial solvents.

- Cleft lip and/or palate
- Neural tube defects
- Anencephaly
- Spina bifida
- Hydrocephalus
- Cerebral Palsy
- Cardiovascular defects
- Digestive anomalies.

The following industries involve potential exposures to solvents:

- Furniture making

- Manufacturing of industrial machinery and equipment
- Medical and scientific laboratories
- Commercial printing
- Cosmetology and hairdresser products
- Plastics industry
- Textile industry
- Dry-cleaning facilities

Lead and Its Constituents:

Toxic heavy metals such as lead can increase the risk of birth defects. Lead is a hazardous material used in the production of hundreds of products including batteries, metals, paint, ceramic glazes, cable covering, ammunition, and more. The Environmental Protection Agency banned the use of lead in paint, gasoline, and other products in the 1970s. However, the substance is still present in older products and in the environment including drinking water from contaminated lead pipes, scented candle wicks, plastic handles on certain tools, and leaded crystals and ceramics. High or even low levels of lead exposure can cause a variety of illnesses and have deleterious reproductive effects in men and women including lower sperm count and spontaneous abortion. Furthermore, maternal exposure to lead, which is easily transferred from mother to child through the placenta as early as the first trimester, poses significant risks of lead birth defects. While there is no specific pattern of birth defects linked to lead exposure during pregnancy, some common lead birth defects may include:

- Neurological damage,
- Developmental delays,
- Low birth weight,
- Skin markings including skin tags,

- Slowed postnatal neurobehavioral development,
- Premature delivery/shortened gestation,
- Undescended male testicles,
- Miscarriage/spontaneous abortion,

Anti-Seizure Drugs and Birth Defects:

Research has consistently shown that anticonvulsant medications significantly increase the risk of seizure drug-related birth defects and other developmental disorders. Women who take anticonvulsants during pregnancy, particularly in the first trimester, are twice as likely as other women to have a baby with a birth defect including:

- Heart defects
- Defects of the skeletal system
- Cleft lip and cleft palate
- Urogenital and neural tube defects (including spina bifida)

Seizure drugs are also known to cause neurological, intellectual, and behavioral delays, although these may not become evident for several years after birth.

The notorious example occurring in the human population is was the use of diethylstilbestrol (DES) a synthetic estrogen prescribed until 1971 to prevent miscarriage in pregnant women. DES has since been linked to numerous health problems in offspring exposed to the drug in the womb, including reproductive complications and in fertility in DES daughters.

For more information, please refer to website (www.birthdefects.org) and information about endocrine disruptors can be found on the website of the National Defense Council (NRDC), www.nrdc.org/health/effects/qendoc.asp.

GREEN BUILDING MOVEMENT

The green building movement in the U.S. originated from the need and desire for more energy efficient and environmentally friendly construction practices. There are several motives for building green, including environmental, economic, and social benefits. The green building movement was the response of construction industry to the environmental and resource impacts of the built environment. The term green building refers to the quality and characteristics of the actual structure created using the principles and methodologies of sustainable construction.

Figure 3-10

Bullitt Center in Seattle Washington

In 1980, the concept of "sustainable development" initiated but did not go far enough. Except a few developed countries had begun widely implementing energy-saving building systems. In 1990, the United Kingdom introduced the world's first green building standard, followed by the formation of the U.S. Green Building Council in 1993.

The Bullitt Center in Seattle, Washington, Figure 3-10 was a concept conceived by Bullitt Foundation president Dennis Hays. The building was designed to have an ideal life span of 250 years. This sustainable building was also formulated to be carbon and energy neutral. In another words, no excess carbon dioxide was released to the environment and no imported electricity needed from outside sources during the operation of this building.

HISTORY OF GREEN BUILDING IN U.S.

The following has been adopted and summarized from Kibert 2016 in connection with history of green building emergence in the United States.

The green building movement has a long history in the United States, with its philosophical roots traceable to the late 19th century. Notable dates include:

1970 – The year first Earth Day was celebrated, and U.S. Environmental Protection Agency (USEPA) was founded.

Concern over resource availability, particularly reliance on fossil fuels, was magnified by the oil shocks of the early 1970s, which resulted from the Arab Israeli conflict. This further focused public interest to search other forms of energy sources such as solar, retrofitting homes and commercial buildings with insulation and energy recovery systems.

1980 – Although many energy-efficient buildings emerged after the oil crises of the 1970s, the first US buildings that considered a wider range of environmental and resource issues did not emerge until 1980.

1985 – William McDonough was hired by EDF to design its New York offices. The design featured natural materials, daylighting, and excellent indoor air quality.

1990 – A renewed interest in energy and resource conservation as humans began to seriously consider more complex global environmental issues, such as ozone depletion, global climate change and destruction of major fisheries.

1992 – Renovation of Audubon House, also in New York City, was a significant early effort in the contemporary green building movement. During this year, the Building Research Establishment Environmental Assessment Method (BREEAM), the British green building rating system, was also developed.

1993 – The first highly publicized green building project in the United States, the "Greening of the White House", was initiated in 1993 and included renovation of the Old Executive Office Building, 600,000-square-foot structure across from the White House. This drew national attention and resulted in dramatic energy cost savings of about $300,000 per year. The establishment of the US Green Building Council (USGBC) also occurred in 1993. It took four years to present the first version of LEED V-1.0.

1998 – Test version known as LEED 1.0 was developed. It was enormously successful, and the Federal Energy Management Program sponsored a pilot effort to test its assumptions.

2000 – A greatly improved LEED 2.0 was launched and provided for a maximum of 69 credits and four levels of building certification: Platinum, Gold, Silver, and Bronze.

2003 – A further refined version of LEED was established in 2003. and labeled LEED for New Construction (NC) version 2.1. The name of the lowest levels of certification, "Bronze" was changed to "certified".

2005 LEED version 2.2 (LEED NC-2.2), major improvements such as moving the rating system online.

2009 – Major revision and launching LEED v-3.0

2013 – LEED – v-4.0 was developed including reweighting credits and restructuring the rating system.

2018 – LEED v-4.1

LEED v4.1 helps the construction industry to focus on efficiency and leadership to deliver the triple bottom line returns of people, planet, and profit. LEED v4.1, raises the bar on building standards to address energy efficiency, water conservation, site selection, material selection, day lighting, waste reduction and improving indoor air quality.

LEED is the world's leading green building project and performance management system, delivering a comprehensive framework for green building design, construction, operations, and performance:

- The rigorous focus on material selection, human comfort, air quality and human health features of a building rightly prioritizes the most important asset of the building: human beings.
- The specific focus on social equity ensures that buildings are not considered in isolation of their communities but prioritize access and inclusiveness for all.
- To ensure a building is resilient to natural and unnatural disturbances a comprehensive set of design and construction strategies has been established in LEED v4.1.

LEED v4.1:

- Ensures that all building stakeholders – developer, property manager, occupant and the community benefit from sustainable design, construction, operations, and performance,
- Supports projects to implement sustainable and healthy building practices to realize environmental, economic, social and community benefits for decades to come,
- Emphasizes integrative design to ensure better design, translation of design into high quality construction, optimize operations and high performance of a building,
- Helps buildings deliver higher quality beyond market practices by incorporating innovative design, technologies, construction, and material selection strategies,
- Focuses on both performance oriented sustainable strategies and outcomes.
- Helps buildings consume fewer resources, reduce operating costs, increase value, and create safer and healthier environments for its occupants,
- Helps buildings reduce their GHG/carbon emissions,
- Helps buildings use toxin free materials to deliver cleaner indoor air to improve productivity, focus and reduce respiratory illnesses of its occupants,
- Prioritizes sustainable materials, helping manufacturers to design, produce and deliver building materials that reduce a building's environmental impact. LEED v4.1 also helps manufacturers reduce energy, water, waste during manufacturing, carbon footprint during

distribution and transportation and overall carbon emissions through the entire production lifecycle.

We will discuss further in Chapter 6 how to rate a building using LEED v4.1.

REFERENCES

Benyus, Janine 1997. Biomimicry: Innovation Inspired by Nature. New York: William Morrow

Clark, W. C. Sustainability science: A room of its own. Proceedings of the National Academy of Sciences of the United States of America **104**, 1737–1738 (2007b).

Clark, W. C. 12 questions to...William C. Clark. Gaia **15**, 164–169 (2006).

Clark, W. C. Sustainability science: The emerging research program. Proceedings of the National Academy of Sciences of the United States of America **100**, 8059–8061 (2003).

IPAT, A Dictionary of Environment and Conservation, Current Version 2017 (3rd ed.)

Developed by Paul Ehrlich, and John Holdren in 1972

Kibert, J., Charles. 2022 "Sustainable Construction, Green Building Design and Delivery", Fifth Edition, John Wiley and Sons, Inc. Hoboken, New Jersey.

National Research Council. Our Common Journey: A Transition Toward Sustainability. Washington, DC: National Academy Press, 1999.

Thompson, P. B. Agricultural sustainability: What it is and what it is not? International Journal Agricultural Sustainability **5**, 5–16 (2007).

von Weizsacker, Ernst, Amory Lovins , and Hunter Lovins, 1997 Factors Four: Doubling Wealth, Halving Resource Use. London: Earthscan

von Weizsacker, Ernst, K. Hargroves, M. Smith, C. Deshna, and P. Stasinopoulos. 2009. Factor 5: Transforming the Global Economy through 80% increase in Resource Productivity. London: Earthscan

World Commission on Environment and Development. Our Common Future. New York, NY: Oxford University Press, 1987. [This document is commonly referred to as The Brundtland Report: IUCN, UNEP, WWF. Caring for the Earth: A Strategy for Sustainable Living. Gland, Switzerland, 1991].

FOUR
ECOLOGICAL DESIGN

BACKGROUND

Ecological design provides a coherent framework for redesigning our landscapes, buildings, cities, and systems of energy, water, food, manufacturing, and dealing with waste disposal strategies. Ecological design is simply the effective adaptation to and integration with nature's processes.

Ecological design can be defined as any form of design that minimizes the environmentally destructive impacts by integrating itself with living processes, nature's own flows, cycles, and patterns. There are examples of eco-design in all kinds of consumer goods, such as biodegradable furniture, recyclable tableware, edible coffee cups, and bamboo sunglasses or toothbrushes, as well as clothes and shoes made with plastic recovered from the ocean and ecological gold jewelry. Ecological or eco-friendly gold jewelry is made with gold that has been sourced in a way that minimizes its environmental impact.

Eco-design involves designing or redesigning products, services, processes, or systems to avoid or repair damage to environment, society and the economy. Eco-design is present all around us, in sustainable flooring, green energy heating systems, eco-friendly packaging and even recyclable products. Eco-design is the same as sustainable design, whilst sustainable product design focuses on both social and environmental sustainability – Eco-design focuses only on reducing environmental impacts at every step of your product's life cycle.

Ecological design is an emerging design concept for high-performance green buildings – however, no clear template or set of fixed options has been developed. Ecological design is limited as much by the knowledge of designers as by the state of the construction industry. A net zero energy (NZE) building, for example, could be achieved by many different combinations of materials, design and technologies; multiple design elements are needed and to some degree they are site-specific. There are trade-offs as there always have been – that's not a new concept – e.g. bigger windows mean more daylight, less artificial light, but potentially more heat gain in summer and more heat loss in winter. It's all about trying to optimize the gains from each design aspect while minimizing the tradeoffs that cause a robbing Peter to pay Paul kind of situation. There're no one-size fits all.

Eco-design concept was popularized by Sim van der Ryn and Stewart Cowan, who wrote the book "Ecological Design" in 1996. They argued for integrating human activities with natural processes to minimize environmental impact. Prior to this, in 1971 Vicor Papanek was credited for outlining strategies for reducing environmental impact in product development, which he called "eco-design strategies" such as "design for disassembly, design for longevity, and design for dematerialization". Eco-design can be best achieved by collaboration of designers, ecologists and contractors during multiple meeting and discussions. The outcome of such collaborative approach will end up in a final design which are the reflective of following considerations:

- Reducing the construction footprint
- Revisiting the construction operation and
- Landscaping modification

During this collaborative framework, they must:

- Select approaches that maximize passive energy (heating, cooling, lighting, and ventilating)
- Consider the best choice of materials and products.

BENEFITS OF SUSTAINABLE DESIGN

Sustainable design for high-performance buildings is based on the following categories:

- Location and Transportation (siting),
- Water Efficiency,
- Energy Efficiency,
- Materials & Resources,
- Indoor Environmental Quality,
- Commissioning and Proper Operation & Maintenance.

SITTING OR SELECTING THE FUTURE PROJECT SITE:

Before the commencement of the construction activities at any given site, conducting an environmental due diligence activity including performing a Phase I Environment Site Assessment is very crucial. In Chapter 1 of this book, we discussed why the Phase I ESA is needed to protect the owner's financial interests as well as the health and well-being of the future building occupants. However, the other side of the coin is to maintain the sustainability principles in choosing the right location for the building in question and calling it a green building. This section discusses how choosing the right location to the building will or will not honor the sustainability principles. As we discussed, sustainability is based on three major principles. These principles must be economically attractive, socially desirable, and environmentally beneficial.

Let's discuss how choosing the right location for the building will help to meet the economic, social, and environmental requirements of sustainability issue. We have the option of choosing the building location in a downtown area or on the outskirts of town where the virgin land within the green field zone is available. Choosing the building location within the downtown area will help us economically since there are public transportation including bus and subway transit systems as well as public parking structures and connecting roads and streets have already been developed. However, choosing the green fields does not provide public transportation, roads, or parking structure. As such, it will add economically to the investment amount if the new roads and parking facilities need to be built to provide these facilities to the future building tenants. In addition, the building in the heart of downtown will serve the future tenants as well and meet the social aspect of sustainability. This is because it will be more convenient for the future tenants to do their daily shopping and be close to medical facilities and take a walk and do their daily exercise. Being the building within the downtown area will also meet the environmental aspect of sustainability since the future tenants or employees will enjoy the public transportation and avoid driving and reduce carbon emission. Choosing the building location in the middle of the downtown area will also help to receive the following environmental benefits:

- Land preservation and less air pollution,
- Reduce resource use and energy use,
- Protection of ecological resources,
- Soil and water conservation.

It appears that selecting the right location for the building is very important in achieving our sustainability goals.

Water Efficiency:

Reducing the consumption of drinking water, indoors and outdoors, has the following economic benefits:

- Consuming less water indoors and outdoors will reduce the first cost and save money.
- It will generate less wastewater and save additional waste disposal fees,
- fewer wastewater treatment systems will be built, which will have additional economic benefits.

The social benefits:

- Reducing drinking water consumption will preserve the water resources for future generations and will meet the social aspect of sustainability.

Environmentally cutting the consumption of drinking water indoors and outdoors will:

- Lower potable water use, will reduce discharge to waterways,

- Less strain on aquatic ecosystems
- Preservation of water resources for wildlife and agriculture.

Energy efficiency:

One of the major reasons for sustainable construction is reducing energy usage inside buildings and reducing carbon footprint which eventually will prevent global warming. Reducing energy consumption and carbon footprint has the following economic, social, and environmental benefits:

Economically, it will reduce fuel and electricity costs. This will also lower the peak power demand as well as demand for constructing new power plants and energy infrastructure.

Socially, it will improve comfort conditions for occupants, it will require fewer power plants and transmission lines.

Environmentally, it will have the following benefits:

- Lower electricity and fossil fuel use
- Less air pollution and fewer carbon dioxide emissions
- Lower impacts from fossil fuel production and distribution

Materials and Resources:

Using the right material following the 3 "R" Rules of sustainability will have the following economic, social, and environmental benefit of sustainable construction. The 3 R rules simply implies to the following:

- Reuse
- Reduce

- Recycle

Let's review the economic, social, and environmental benefits of 3 "R" rule in construction:

Economically, it will reduce the first cost by reducing, reusing, and recycling of the building material. This will also lower the waste disposal cost of material in landfills. In addition, by reusing and selecting durable materials it also saves the replacement cost of material in the long run.

Socially, there will be a fewer demand for new landfills, and it will create a greater market for environmentally preferable products. In addition, it will decrease traffic due to decreased use of transportation for material hauling.

Environmentally, it will reduce strain on landfills and use of virgin resources. As a result, it will create a better management system for sustainable forestry operation for production of certified wood for building industry by using reduced energy and lesser air pollution.

Indoor Environmental Quality:

Maintaining a healthy indoor air quality condition inside the built environment has many benefits and meets the underlying sustainability principles.

Economically, healthy indoor air quality condition results in a higher productivity among the building inhabitants, and lower medical incidences. This lowers the staff turnover, litigation costs and insurance premiums.

Socially, it will reduce adverse health impacts of absenteeism and will create an improved occupant's comfort and satisfaction.

Environmentally, it will cause a reduced emission rates of

volatile organic compounds (VOCs), carbon dioxide and carbon monoxide.

Commissioning, Proper Operations and Maintenance:

Commissioning and proper operation and maintenance of the built environment has many advantages which satisfies the three basic principles of sustainability issue as follows:

Economically, it lowers energy costs, reduced occupant/owner complaints and longer building and its equipment lifetimes.

Socially, it increases occupant's productivity, increases satisfaction as well as their health and safety.

Environmentally, it lowers energy and water consumption as well as the carbon footprint. In addition, it reduces air pollution and other emissions.

EARLY THOUGHTS ON DESIGN CONSIDERATIONS FOR SUSTAINABLE BUILDING

Historically, when different municipalities started to design green buildings, and follow the sustainability principles, there were no given guidelines to follow. As such, the major cities and local municipalities started to come up with initial guidelines. The following are some of the very first thoughts on the direction of green building guiding lights were adopted from Local Government Sustainable Buildings Guidebook, 1993:

- Resources should be used only at the speed at which they naturally regenerate and should be discarded only at the speed at which local ecosystems can absorb them.

- Material and energy resources must be understood as a part of a balanced human/natural cycle. Waste occurs only to the extent that it is incorporated back into that cycle and used for the generation of more resources.
- Site planning should incorporate resources naturally available on site, such as solar and wind energy natural shading, and drainage.
- Resource-efficient materials should be used in the construction of the building and in furnishings to lessen local global impact,
- Energy and materials waste should be minimized throughout the building's life cycle from design through reuse or demolition.
- The building shell should be designed for energy efficiency.
- Material and design strategies should strive to produce excellent indoor environmental quality, of which indoor air quality is a major component.
- The design should maximize occupant health and productivity.
- Operation and maintenance should support waste reduction and recycling.
- Location and systems should optimize employee commuting and customer transportations and minimize the use of single-occupancy vehicles. These include using alternative work modes such as telecommuting and teleconferencing.
- Water should be managed as a limited source.

Today, after the full development of the green building idea throughout the world, still the above-mentioned items are the cornerstone of the sustainability construction and green buildings doctrine.

INITIAL SCHOLARS WHO CONTRIBUTED TO ECOLOGICAL DESIGN

Among the philosophers and writers who contributed to the creation of ecological design consideration, two outstanding individuals deserve a big recognition. Ken Yeang, a Malaysian architect in 1995 published a book entitled "Designing with Nature". This book was the first publication to attempt to tackle the tremendous challenge of how to apply ecology directly to architecture. Yeang used the term green architect and sustainable architecture interchangeably. Defining them as "designing with nature and designing with nature is an environmentally responsible way." Kibert (2022), indicated that Yeang (1995) approached this problem by making several important assumptions:

- The environment must be kept biologically viable for people.
- Environmental degradation by people is unacceptable.
- Destruction of ecosystems by human must be minimized.
- Natural resources are limited.
- People are part of a larger closed system.
- Natural system processes must be considered in planning and design.
- Human and natural systems are interrelated and essentially one system.
- Changing anything in the system affects everything else.

Yeang continues his efforts to develop his concept of ecological design, and he is particularly well known for his work on tall green buildings. He has written several other books about ecological design and in 2008 published an updated work on the general subject of ecological design called Eco-Design: A manual for Ecological design. He has also written extensively on the greening of skyscrapers in The Green Skyscraper: The basis for Designing Sustainable Intensive Buildings (1999), Eco Skyscrapers (2007), and Eco Skyscrapers, volume 2 (2011).

Figure 4-1 The National Library of Singapore Designed by Ken Yeang Photograph courtesy of Ken Yeang

The latest book by Ken Yeang, was published in October 2019 entitled "Saving the Planet by Design". A summary of this latest book is as follows:

Can we 'save the Planet'? For a resilient, durable, and sustainable future for human society, we need to repurpose, reinvent, redesign, remake and recover our human-made world so that our built environment is benignly and seamlessly bio-integrated with Nature to function synergistically with it. These are the multiple tasks that humanity must carry out imminently if there is to be a future for human society and all lifeforms and their environments on the Planet.

Addressing this is the most compelling question for those whose daily work impacts on Nature, such as architects, engineers, landscape architects, town planners, environmental policy makers, builders, and others, but it is a question that all of humanity needs to urgently address.

Per Ken Yeang, designing with ecology is contended here as the authentic approach to green design from which the next generation of green design will emerge, going beyond current use of accreditation systems. For those who subscribe to this principle, this is articulated here, showing how it can be implemented by design. Adopting these principles is fundamental in our endeavor to save our Planet Earth, and changes profoundly the way we design, make, manage and operate our built environment.

The second distinguished authors in ecological design were Sim Van der Ryn and Stuart Cowan who delved into the subject of ecological design. They published their book named Ecological Design to provide a context for green design. Van Der Ryn published several books including Emerging Ecologies, Design for life, Design for Emphatic Life, and so on.

During the 1960s, as Western notions of endless progress and growth gave way to concerns over industrial pollution, resource depletion and ecological limits, attitudes toward the environment became social, political, and ideological. Published to accompany the first expansive survey of the history of environmental thinking in architecture, Emerging Ecologies: Architecture and the Rise of Environmentalism looks at the role architects have played in defining our understanding of "nature" and the "environment," specifically during the rise of environmental discourse.

THE GOLDEN RULES FOR ECO-DESIGN

To assist engineers, architects and planners in the production of an environmentally benign built environment, Stephan Bringezu (2002) of the Wuppertal Institute suggested five "golden rules of ecological design":

1. Potential impacts to the environment should be considered on a life-cycle basis (from cradle to grave).
2. The intensity of use of processes, products, and services should be maximized.
3. The intensity of resources use (material, energy, and land) should be minimized.
4. Hazardous material should be eliminated.
5. Resource input should be shifted toward renewable.

CRADLE-TO-CRADLE

Cradle-to-cradle (C2C) can be defined as the design and production of product of all types in such a way that at the end of their life, they can be truly recycled or used for another purpose rather than disposing them as a waste, imitating nature's cycle with everything either recycled or returned to the earth, directly or indirectly through food, as a completely safe and nontoxic material.

One example of C2C is a space station that uses a 3D printer to make tools to be used in space. When the astronauts are finished with the tools, they can break down and make them into new tools. 3D printers use reuseable materials. As we can see through this approach, raw materials are not thrown away (from cradle to grave) but are reused indefinitely or serve as food for new products. Cradle to cradle describes the safe and potentially infinite circulation of materials and nutrients in cycles. All constituents are chemically harmless and recyclable.

The definition of C2C design is a philosophy encouraging biomimicry in all life facets, revolutionized by Michael Braungart and William McDonough in their book "Cradle-to-Cradle: Remaking the Way We Make Things." In laying the framework for cradle-to-cradle concept, they suggested that people and industry should set out to create:

- Buildings that, like trees, produce more energy than they consume and purify their own wastewater,
- Factories that produce effluents that can be used as drinking water,
- Products that, when their useful life is over, do not become useless waste but can be tossed on the ground to decompose and become food for plants, animals, or alternatively, that can turn to industrial cycles to supply high quality raw materials for new products,
- Billions and trillions of dollars worth of materials accrued for human and natural purposes each year,
- A world of abundance, without pollution and waste.

Michael Braungart and William McDonough (2002) suggested that the solution is to follow nature's model of eco-effectiveness.

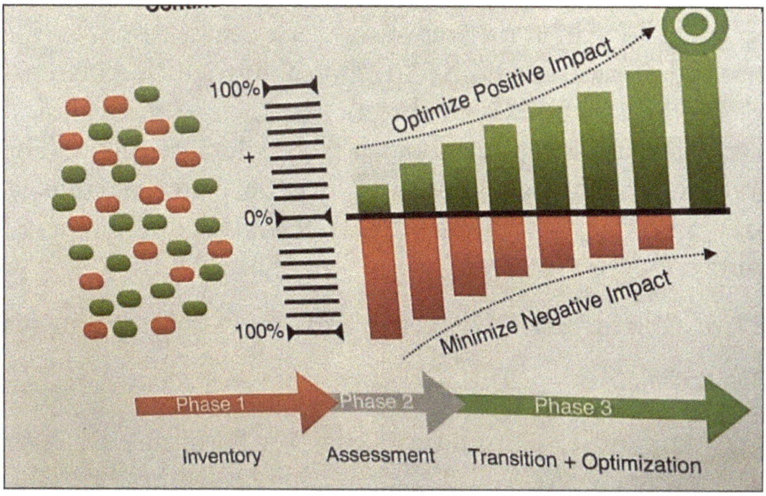

Figure 4-2 Cradle to Cradle Strategy is to remove negative impacts of products by focusing on removing known problematic substances and replacing them with benign materials (Photograph modified from McDonough Braungart, Chemistry LLC)

Doing this entails separating the materials we use in human activity into biological substances (which can be returned to natural ecosystem, where they can benefit other creatures as nutrients) and technical substances (which can with proper design, be 100 percent recollected and recycled or even upcycled, producing in second use, products of greater value than their original use, with zero waste), Kibert (2022). Carpet and shoes, for example, could be made of two layers, a biological outer layer that abrades over time, whose fiber could serve as nutrients in the soil for compost, and a much more durable technical inner layer that would be 100 percent recyclable, after its long life, into another identical product. A biological nutrient is a material or product that is designed to return to the biological cycle.

According to Michael Braungart and William McDonough, packaging for example can be designed as a biological nutrient so that, at the end of its use, it can be tossed on the ground or compost heap. A technical nutrient is a material or product that is designed to be returned to the technical cycle, the industrial metabolism from which it comes. The authors also defined a class of materials they refer to as unmarketable, which are neither technical nor biological nutrients.

Dave Pollard described this process more elegantly in his blog:

1. Free ourselves from the need to use harmful substances (e.g., PVC, lead, cadmium, chromium, and mercury);
2. Begin making informed design choices (materials and processes that are ecologically intelligent, respectful of all stakeholders, and which provide pleasure or delight);
3. Introduce substance triage: (a) phase out known and suspected toxins, (b) search for alternatives for problematic substances, and (c) substitute for them "known positive" substances.
4. Begin comprehensive redesigns to use only "known positives" separate materials into biological and technical and ensure zero waste in all processes and products,
5. Reinvent entire process and industries to produce "net positives" to improve the environment.

NATURAL CAPITALISM

Natural capitalism is a system of four interlinking principles, where business and environmental interests overlap, and in which business can better satisfy their customers' needs, increase profits and help solve environmental problems all at the same time. Natural Capitalism is based on the following four principles:

1. Resource productivity, using resources more efficiently, dramatically increase the productivity of natural resources,
2. Biomimicry, redesigning industrial systems along biological lines and shift to biologically inspired models,
3. Service and flow economy, aligning incentives between business and consumers and moving to solution-based business models,
4. Restoring or re-investing in stocks of natural capital and reinvesting in natural capitalism.

The goal of Natural Capitalism is to reduce resource extraction, which can be accomplished in three ways:

1. Dematerialization of the products.
2. Increasing the recycling rate of products at the end of their life cycle and,
3. Increasing the durability of products.

Natural Capitalism seeks not merely to reduce waste but to eliminate the very concept of waste. In closed-loop production systems, modeled on nature' designs, every output either is returned harmlessly to the ecosystem as a nutrient, like compost, or becomes an input for manufacturing another product.

The free market is the purest form of capitalism. Capitalism is an economic ideology in which the means of production is controlled by private business. This means that individual citizens run the economy without the government interfering in production or pricing. Instead, pricing is set by the free market. This means that value is based on supply and demand and the relationship between producers and consumers.

Capitalism is very different from socialism and communism, in which the government maintains tight control of the economy. The United States is arguably the most well-known country with a capitalist economy, which many citizens see as an essential part of democracy and building the "American Dream." Capitalism also taps into the American spirit, being a "freer" market when compared to the more government-controlled alternatives.

Despite this distinction, the United States falls short on the top 10 in terms of the of the most capitalistic countries in a 2021 report from the Heritage Foundation and Global Finance Magazine. According to the Heritage Organization's Economic Freedom of the World Index, the top 10 countries with the most capitalist economies. (ranked from highest scores to the lowest), see Table 4.1.

Country	Econ. Freedom Index – Heritage 2023
Singapore	83.9
Switzerland	83.8
Ireland	82.0
Taiwan	80.7
New Zealand	78.9
Estonia	78.6
Luxembourg	78.4
Netherlands	78.0
Denmark	77.6
Sweden	77.5

Table 4.1 Heritage Organization of World Index

There are benefits and drawbacks to a capitalist economy. For example, capitalism cultivates competition, which often drives both innovation and affordability, with best products selling at the best possible prices. Economic growth is another benefit. On the other hand, one of the most significant disadvantages of capitalism is that it does not promote equality of opportunity. In short, some people may not get the opportunities that others receive. Moreover, the unrestrained "supply and demand" nature of the free market can result in prices too high for lower income individuals to afford. We should know that absolute capitalism in a practical sense does not exist. For example, no country in the world has ever achieved a 100 percent capitalist or free market economy. All capitalist economies are mixed to one degree or another. The following are some high-profile examples.

America is a capitalist country but not completely. As of 2023, Singapore, Hong Kong (China), Switzerland, New Zealand and the United States ranked as of the world's five most economically free countries in the Cato Institutes Human Freedom Index, making them the world's top 5 market economies in terms of economic freedom.

BIOLOGICAL MATERIALS, BIOMATERIALS

Biomaterials science encompasses elements of medicine, biology, chemistry, tissue engineering and materials science. A biomaterial is different from a biological material, such as bone, that is produced by a biological system. However, "biomaterial" and "biological material" are often used interchangeably.

The modern field of biomaterials combines medicine, biology, physics, and chemistry and has grown significantly in the past decade. Biomaterials can be engineered to exhibit properties that enhance their performance in medical implants, drug delivery systems, tissue engineering scaffolds, and diagnostic tools. Biological materials are part of the structure and function of living organisms, contributing to their overall physiology and functionality.

Biological materials: Includes natural products such as wood, hemp, bamboo, cotton, etc.

Biomaterials are produced through processes that employ or mimic biological phenomena, such as biopolymers, proteins, cellulose, and starch. Biomaterials consist of long chain molecules synthesized by living organisms. Since biological resources are renewable, there is a tendency to think of them as unlimited. But it is not true. People think if the land is cultivated carefully, crops can be planted in perpetuity. Does this sound right? No, If the land is pushed past its carrying capacity or otherwise abused, permanent damage can be done.

A widespread shift to biological materials for both energy and materials has other implications because large quantities of land may be required. This would increase food prices and impact on the poor and hungry population of the world.

Finally, there is little evidence that biologically based materials can replace the synthetic materials that have become common in construction, especially structural materials such as steel and concrete, not to mention copper and aluminum wiring, glass, and wide variety of polymers used in many applications.

Finally, this chapter needs to be viewed very carefully and outstanding items that were considered included the following:

- The development and deployment of new materials must consider the following factors: ethics, risks, and economics,
- Changing the decision system is badly needed to ensure that the risks to nature and humans are minimized,

- A LCA must be applied and the disassembling of materials from buildings is necessary to close loops,
- Ethics will have to govern the decision systems, in part by reshaping the economics/ laws of the marketplace.
- Using the biologically based material is in its infancy and more research and development is needed to consider such material in construction industry.

REFERENCES

Bringezu, Stephen. 2002. "Construction Ecology and Metabolism." In Construction Ecology: Nature as the Basis for Green Building., C. J. Kibert, J. Sendizimir, and G.B. Guy, eds., 196-219. London Spon Press

Dave Pollard, 2008. "Finding the Sweet Spot: The natural Entrepreneur's Guide to Responsible, Sustainable, Joyful Work.

Ken Yeang, a Malaysian Architect (1995), Designing with Nature: The Ecological Basis for Architectural Design. New York: McGraw Hill

Ken Yeang, 1996. The Skyscraper Bio-Climatically Considered: A Design Primer. London: Academy Editors.

Ken Yeang, and Arthur Spector 2011. Green Design: From Theory to Practice. London: Black Dog.

Ken Yeang, 2019, Saving the Planet by Design: Reinventing our World Through Ecomimesis, Routledge, New York, NY.

Kibert J. Charles. 2022 "Sustainable Construction, Green

Building Design and Delivery. Fifth Edition, John Wiley and Sons, Inc. Hoboken, NJ.

McDonough, William, and Michaael Braungart, 2002. Cradle o Cradle: Remaking the Way Er Make Things. New York: North Point Press.

Van der Ryan, Sim and Stwart Covan (1996). Ecological Design. Washington DC: Island Press.

Victor Papanak, (19710 “Design for Real World: Human Ecology and Social Change.

The Local Government Sustainable Buildings Guidebook, 1993. Washington DC.

FIVE
GREEN BUILDING ASSESSMENT SYSTEMS

BACKGROUND

In the 1960s, American architect Paul Soleri proposed a new concept of ecological architecture. In 1980, the concept of "sustainable development" took hold, and a few developed countries had begun wildly implementing energy-saving building systems. In 1990, the United Kingdom introduced the world's first green building guideline, followed by formation of the U.S. green Building Council (USGBC) in 1993.

Before the advent of building assessment systems, environmentally friendly buildings were conceptualized by teams of architects and engineers who relied on their collective interpretation of what constituted the green building (Kibert, 2022). The question is "Why bother to rate a building?" The following is the answer to this question. The rating of a building will:

- Promote high-performance building,
- Increase market demand for sustainable construction,
- Increase market value of the building,
- Lower the operating cost and increase rental values,
- Promote positive health and productivity of occupants,
- Reduce fossil fuel consumption.

From a historical perspective, BREEAM, or the Building Research Establishment's Environmental Assessment Method, is often credited as the first green building certification program. BREEAM, like its counterparts in other countries, rates the performance of a building based on a set of criteria. Building assessment systems are generally organized into categories, such as energy, water, and materials, and award points for criteria that are met by the project team in the design of the facility.

Green Building Rating Systems (GBRSs) are typically third-party, voluntary, and market driven standards that measure building's sustainability level by multi-criteria assessment, and encourage the adoption of environmentally, socially, and economically sustainable practices in design, construction, and operation of the building.

As mentioned earlier, BREEAM is the oldest green building rating system created in 1990. It has certified projects in over 50 countries, and as of June 2020, has over 560,000 certified projects, and over 2 million registered buildings. Today, green buildings have become an increasing area of focus for the built environment, aiming to address environmental and sustainability issues relating to energy, water, and waste. Green building certifications and standards have developed both showcase and evidence that a high standard of sustainable and eco-friendly practices have been used to create and operate the building. Two of the most well-known certifications include BREEAM and LEED. LEED stands for Leadership in Energy and Environmental Design. These two rating systems assess, and reward ratings based on the score received. However, there are a range of new sustainable building accreditations which have been developed, some of which also focus more on the health and wellbeing of the people who use the buildings.

Rating systems allow a project team to use a simple checklist to verify if green building practices and requirements are met. The rating system that are being used in US includes:

- LEED (US)
- Green Globes (US)
- Living Building Challenge (US)

LEED

LEED, or Leadership in Energy and Environmental Design, is a green building certification used in the United States. LEED is managed by the USGBC and is completed by a third-party verification method. Many building types can apply for this certification program, including new construction, existing buildings, homes, and communities. LEED has four certification levels including, certified, silver, gold, and platinum. LEED has nine areas of focus, including location and transportation, sustainable sites, water efficiency, energy and atmosphere, material and resources, indoor environmental quality, innovation, regional priority, and integrative processes. The LEED certification program aims to have buildings use their resources more efficiently and create a safe environment for all its occupants throughout the building's life cycle. As of 2019,

80,000 projects were registered, with 32,500 projects having completed the certification process. Top users of LEED certification include Intel Corp., Colgate-Palmolive, Mars Inc., and Method Southside Soapbox Factory.

In Chapter 6 will discuss in detail how to rate the building using LEED rating system.

GREEN GLOBES

Green Globes is a building rating system used in US and Canada. Green Globe is structured so that it can be done as a self-assessment in-house with the project manager and design team. It uses a questionnaire that is aimed to helping the user make changes to complete the certification, but there are no prerequisites to complete this certification. Like LEED, Green Globes has four levels of certification. Green Globes can be used in new construction, existing buildings, and commercial interiors. This certification program focuses on energy usage, water, waste management, emissions, indoor environment, and environmental management.

The Green Globes is a revolutionary green management tool that includes a rating system, an assessment protocol, and a guide for integrating environmentally friendly design into commercial buildings. It is building assessment system with roots in UK and Canada. In 2004 Green Building Institute (GBI) acquired the rights to promote Green Globe Building Assessment System (GGBAS) in US. In 2005 adopted by American National Standards Institute (ANSI).

Progressively presented as an alternative to LEED in the US. It provides a rating of one to four green globes for building projects depending on the percentage of the maximum points that the project achieves.

The Green Globe Rating Levels are as follows:

Once an assessment is verified by a third party, properties achieving a score of 35% or more receive a Green Globes rating based on the percentage of total points (up to 1000) achieved.

85-100%		Reserved for select building designs which serve as national or world leaders in energy and environmental performance. The project introduces design practices that can be adopted and implemented by others.
70-84%		Demonstrates leadership in energy and environmental design practices and a commitment to continuous improvement and industry leadership.
55-69%		Demonstrates excellent progress in achieving eco-efficiency results through current best practices in energy and environmental design.
35-54%		Demonstrates movement beyond awareness and commitment to sound energy and environmental design practices by demonstrating good progress in reducing environmental impacts.

THE LIVING BUILDING CHALLENGE

The most demanding of all the North American building assessment systems is the Living Building Challenge (LBC). It is an ambitious standard that challenges buildings to innovate to create positive environmental impacts, rather than mitigating negative ones. To be classed as a Living Building, projects must demonstrate they have achieved Net Positive Energy, Water, and Waste. Essentially these buildings must generate more energy than they consume to offset any negative effects of the building. This measured through seven performance categories.

It was used since 2005. The Living Building Challenge is based on:

- A few simple but very powerful concepts,
- Seven performance areas: Site, Water, Energy, Health, Materials, Equity, and Beauty,
- Must produce own energy, provide all required water, process its own sewage,
- Uses SBTool, a complex assessment system,
- The LBC is awarded only after 12 months of continuous operation that are meant to demonstrate that the building has achieved its performance goals,
- LBC rating system requires drastic reduction of energy and water.

LBC rating system requires drastic reduction of energy and water.

Figure 5-1 shows the building rating systems in different Countries.

Country	Rating System
Peru	EDGE
Philippines	BERDE/Philippine Green Building Council
Portugal	Under A/SBToolPT*
Qatar	Qatar Sustainability Assessment System (QSAS)
Republic of China (Taiwan)	Green Building Label
Saudi Arabia	Saudi Arabia Accredited Fronds (Sa'af)
Singapore	Green Mark
South Africa	Green Star SA/Edge
Spain	Verde
Switzerland	Menergie
United States	LEED/Living Building Challenge/Green Globes/Build it Green/NAHB NGBS/International Green Construction Code (IGCC)/Energy Star
United Kingdom	BREEAM
United Arab Emirates	Estidama
Turkey	CEDBIK
Thailand	TREES
Vietnam	LOTUS Rating Tools/EDGE
Czech Republic	SBToolCZ

Country	Rating System
Australia	Nabers/Green Star/BASIX (in NSW only)
Brazil	AQUA/LEED Brasil/EDGE
Canada	LEED Canada/Green Globes/Built Green Canada
China	GBAS
Colombia	EDGE
Egypt	Green Pyramid Rating System (GPRS)
Finland	PromisE
France	HQE
Germany	DGNB/CEPHEUS
Hong Kong	BEAM Plus
India	Indian Green Building Council (IGBC)/Green Building Construction India (GBCIndia)/ GRIHA/EDGE
Indonesia	Green Building Construction Indonesia (GBCI)/Greenship/EDGE
Italy	Protocollo Itaca/Green Building Council Italia
Japan	CASBEE
Republic of Korea	Green Building Certification Criteria/Korea Green Building Council
Malaysia	GBI Malaysia
Mexico	LEED Mexico
Netherlands	BREEAM Netherlands
New Zealand	Green Star NZ
Pakistan	Pakistan Green Building Council

Figure 5-1 Names of some of the green building assessment systems being used in over 70 countries After Kibert 2022

The following is the description of some of the building rating systems that are being used in other countries around the globe.

BREEAM

We introduced BREEM in the beginning of this chapter. This green building rating system which was developed and administered by Unted Kingdom in 1989. There are 10 categories in BREEAM:

1)Management, 2) Health and Well Being, 3) Energy, 4) Transport, 5) Water, 6) Materials, 7) Waste, 8) Land Use and Ecology, 9) Pollution, 10) Innovation. BREEAM Rating benchmarks are: outstanding, excellent, very good, good, pass, and unclassified.

BREEAM is aimed at making buildings more sustainable, as well as improving building performance and efficiency. Many other green building certification programs, including Green Globes were inspired by the ideas and innovations of BREEAM. Top users of BREEAM include Netherland, Norway, Sweden and Spain.

DGNB

DGNB, a green building certification program created by the German Sustainable Building Council, focuses on promoting sustainable building practices across Europe.

Germany has a long history of designing high-performance buildings, but only recently (2006) there has been an effort to develop a green building rating certification program. There are two green building assessment systems in Germany. DGNB is for non-residential and commercial buildings, while BNB is for assessing government buildings.

DGNB allocates the points for three major areas of consideration for sustainability: Ecology, Economy, and Socioculture.

These are the issues:

- What are the sustainability issues relevant to construction?
- What needs to be protected?
- How does this protection occur?

The three major areas of sustainability were used as the basis organizing the DGNB/BNB building assessment system, and three other issues were also considered: Technical Quality, Process Quality, and the project's location Profile. The three major areas of sustainability were determined to be equally important and each of these was allocated 22.5 percent of the available points, a total of 90 points. The final 10 points were allocated to Process Quality, issues such as Integrated Design, commissioning, and Quality Assurance.

DGNB uses a holistic approach with an emphasis on performance. This green building rating system has three levels of certification, platinum, gold, and silver. For this certification program, buildings are evaluated on ecological quality, socio-cultural and function quality, technical quality, and process quality. As of today (2024), over 8,500 buildings have earned a DGNB certification.

GREEN STAR

Green Star is an international sustainability reporting and rating system, that is popular particularly in Australia, and South Africa. It was developed in 2003. All green star categories include an innovation category that rewards projects for creating and utilizing new approaches to sustainability.

Like the other certification programs on this list, Green Star can be used in a variety of building types and is assessed in categories such as indoor environmental air quality, energy, transportation, water materials, land use and ecology, and emissions. The main goal of Green Star is to guide project teams to make conscious decisions regarding energy usage and material selection. This green building rating system has six levels of certification from the lowest to highest: 3 Star, 4 Star, 5 Star and 6 Star. As of today, less than 1,500 buildings have earned Green Star certification.

CASBEE

CASBEE, a Comprehensive Assessment system for Building Environmental Efficiency is the Japanese building assessment stem launched in 2001. CASBEE is used for both new construction and existing buildings throughout Japan. Starting in 2005, earning a CASBEE certification became mandatory in 24 Japanese municipalities. CASBEE moved internationally in 2014 when a building in Tianjin, China earned their CASBEE certification. This program focuses on energy and resource efficiency, and local and indoor environments.

CASBEE was designed to utilize Life Cycle Analysis (LCA) technique to reduce energy requirement and carbon footprint and, also maintain a good environmental quality within the build environment.

CASBEE defines BEE (Building Energy & Environmental) as maximizing the ratio of building quality to environmental loadings.

Building quality (Q) is described by CASBEE as the amenities provided for building users and consists of several quantities:

Q= Q1 + Q2 + Q3,

where

Q1 = Indoor environment,
Q2 = Quality of service and
Q3 = Outdoor environment on-site, and
Q= total Quality.

Similarly, there are several categories of environmental loadings (L) in CASBEE:

L = L1 + L2 + L3 where
L1 = Energy
L2 = Resources
L3 = Off-site environment
L = Total Loading

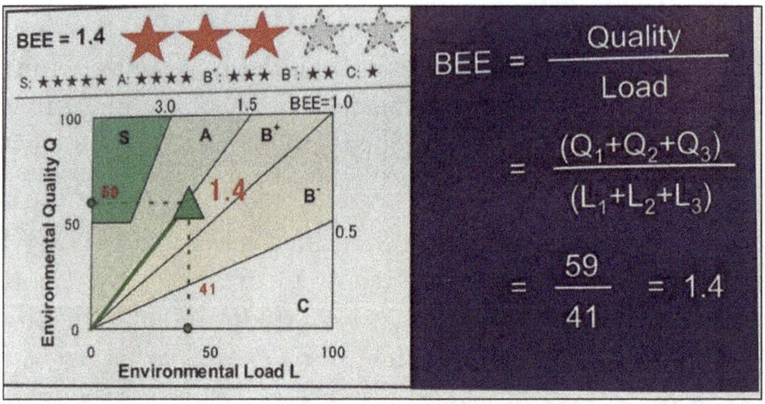

Figure 5-2 (Source: Japan Sustainable Building Consortium)

The BEE rating is determined by finding the intersection of Q (building Quality) and L (building loading). High ratings (S and A) are achieved by a building with high environmental quality and performance and low environmental loadings. Higher resource consumption and lower environmental quality produce below-standard ratings (B or C).

WELL

WELL, is a building certification program managed by the international Well Building Institute (IWBI). WELL, focuses mostly on building design attributes that impact occupant health and well-being. WELL, evaluates buildings on 11 concepts: air, water, nourishment, light, movement, thermal comfort, sound, materials, mind, community, and innovation. This green building system features a few preconditions or prerequisites to completing a WELL certification. Like LEED, WELL can be used for a wide variety of buildings and building spaces. As of 2019, 3,865 projects were registered, and 232 had earned their WELL certification. Top users of WELL include Wells Fargo, EY, Deloitte, Lenovo, and Fandango.

CASBEE Grade	Stars
S (Excellent)	★★★★★
A (Very Good)	★★★★
B^1 (Good)	★★★
B^2 (Rather Poor)	★★
C (Poor)	★

Figure 5-3 CASSBEE Building Rating Guide

In conclusion, the high-performance building movement worldwide is being propelled by the success of building assessment methods, in particular LEED and BREEAM. The GBI's Green Globes and the Living Building Challenge can help to bring the movement of the collective green building design concepts even further.

Currently there are over 40 green building councils that promote green building around the world.

REFERENCES

International Living Future Institute, 2010. Living Building Challenge Version 2.0.

Japan Sustainable Building Consortium, 2010. Comprehensive Assessment System for Built Environment Efficiency (CASBEE) Property Appraisal Manual. Tokyo: Japan Sustainable Building Consortium.

Kibert, J., Charles 2022. 'Sustainable Construction Green Building Design and Delivery' John Wiley and Sons, Inc. Hoboken, NJ.

World Green Building Council. 2015, 'Quality Assurance Guide for Green Building Tools'.

SIX
THE U.S. GREEN BUILDING COUNCIL LEED BUILDING RATING SYSTEM

BACKGROUND

The USGBC's LEED (Leadership in Energy and Environmental Design) is the most frequently used building rating system in the United States. The key elements of LEED rating system are as follows:

- Market-driven (is not regulated),
- Consensus Process and has balanced and transparent structure, driven by national building research organizations,
- USGBC is a non-governmental organization comprising wide range of collaborators from industry, academia, and government,
- Technical Advisory Group to ensure scientific consistency,
- Opportunities to Stakeholder comment and review,
- Member ballot of new rating system and periodical updates,
- Fair and open appeals process,

- LEED is not driven by product manufacturer's trying to sell the latest technologies, politicians, or the finance industry.

The US Green Building Council or USGBC is a non-governmental organization comprising a wide range of collaborators from industry, academia, and government.

- Membership is drawn from diverse public and private stakeholders. It facilitates the integrative/collaborative process,
- LEED comes from everyone that is involved.

The USGBC has developed LEED rating system in three years. The US Department of Energy (DOE) offered critical funding to support LEED's development. LEED certification has continued to grow since its beginning. As of 2023, there were over 105,000 LEED-certified buildings and over 205,000 LEED accredited professionals in 185 countries around the world.

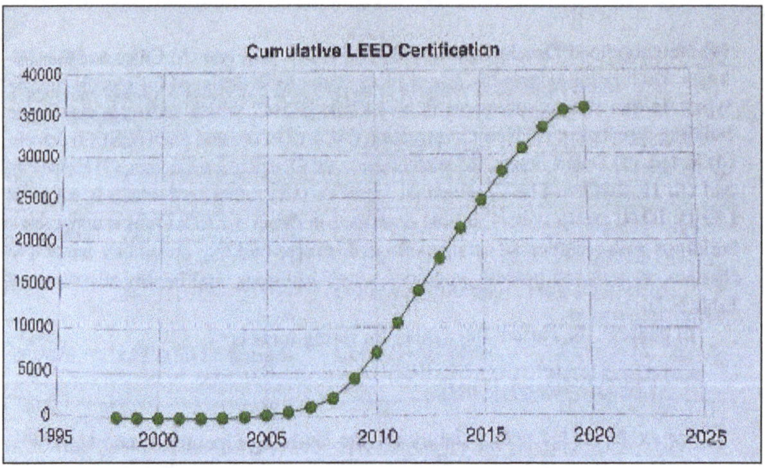

Figure 6-1 Cumulative square footage of LEED-certified projects through 2020 Source: Kibert 2022

LEED is not a single rating system but a suite of building rating systems. The current most popular is LEED-NC. NC stands for new construction which may include offices, hotels, governmental buildings, libraries, churches, etc., Only one LEED rating can be awarded to a building. It will depend on most of the building's type. Figure 6-2 shows different types of building pertaining to the LEED certification.

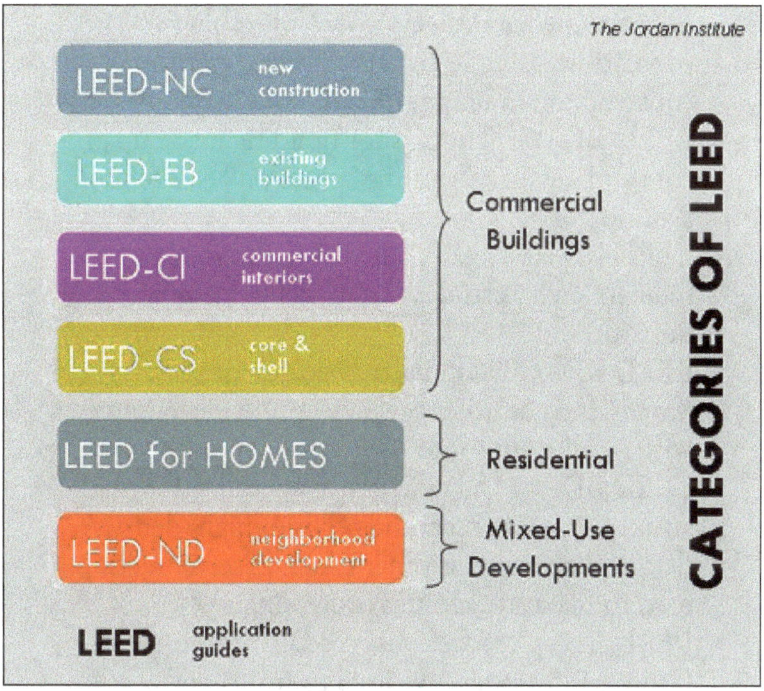

Figure 6-2 Different types of buildings pertaining to LEED certification

LEED is a flexible system- not a one-size-fits-all approach, and it not only applies to different building types (commercial vs. residential), but also beyond the building footprint. Because of this flexibility, there are different versions of LEED depending on the project type.

There are rating systems that address multiple project types such as:

- LEED BD+C: New construction, is for both new and major renovation of existing buildings (major HVAC improvements, significant building envelope, and interiors),
- LEED BD+C: Core & Shell is for projects where the developer controls the design and construction of entire mechanical, electrical, plumbing and fire protection system – but not the design and construction of tenant fit-out,
- LEED BD+C: School is for building made up of core and ancillary learning spaces on K-12 school grounds,
- LEED BD+C Retail addresses the unique needs of retailers such as banks, restaurants, electronics, big box stores,
- LEED BD+C Hospitality is dedicated to hotels, motels, inns or other businesses provide lodging with or without food.
- LEED BD+C: Data Centers is specially designed and equipped to meet the needs of high- density data computing equipment such as server racks, used for data storage and processing.
- LEED BD+C: Warehouses & Distribution Centers is for buildings used to store goods, manufactured products, merchandize, new material, self storage facilities
- LEED- BD+C Healthcare is hospitals that operates 24 hours a day, seven days a week and provide inpatient and long-term care.
- LEED BD+C Homes and Multifamily resident, it is focused on single-family houses, low-rise homes, affordable housing, of 1-3 stories.

- LEED BD+C Multifamily Midrise:

Is up to 4-8 occupiable stories above grade.

- Interior Design + Construction (ID+C)
    - LEED ID+C applies to projects that are a complete fit-out.
    - LEED ID+C Commercial Interior is for interior spaces dedicated to functions other than retail or hospitality.
    - LEED ID+C Retail guides retailers interior spaces used to conduct the retail sale of consumer products goods, such as showroom and storage areas for consumer services.
    - LEED ID+C Hospitality is designed for interior space dedicated to hotels, motels, inns, that provide transitional or short-term lodging with or without food.
- Building Operation + Maintenance (O+M)

LEED O+M applies to existing buildings that are undergoing improvement work or little or no construction.

- LEED O+M Existing Buildings is specifically projects that do not primarily serve K-12 educational, retail, data centers, warehouses and distribution centers, or hospitality uses.
- LEED O+M Retail guides existing retail spaces, both showrooms and storage areas.
- LEED O+M Schools is for existing buildings made up of core and ancillary learning spaces on K-12 school grounds.
- LEED O+M Hospitality is for existing hotels, motels, inns, and industries that provide short-term lodging with or without food services.

- LEED O+M Data Centers is for existing buildings specifically designed and equipped to meet the needs of high-density computing equipment such as server racks, used for data storage and processing.
- LEED O+M Warehouses and Distribution Centers is for existing buildings used to store goods, manufactured products, merchandise, raw materials or personal belongings such as self-storage.

Neighborhood Development (ND) applies to new land development projects or redevelopments projects containing residential uses, non-residential use or a mix. Projects can be at any stage of the development processes, from conceptual planning to construction.

LEED CREDENTIALS

The USGBC developed a system of credentials to identify individuals who are knowledgeable in green building practices. The Credential includes:

- LEED GA (Green Associate)
- LEED AP (Accredited Professional) with specialty
- LEED AP Fellow

LEED GA, a fundamental credential with basic knowledge of LEED. It can be pursued by anyone employed in a building or environmental field. For more info refer to:

http://www.usgbc.org/resources/leed-green-associate-candidate-handbook

LEED AP:

LEED rating systems provide one point for a project with a LEED AP involved. There are five specialities of the LEED AP credential as follows:

- LEED AP with a Building Design and Construction (BD+C), this is for projects using the New Construction and Major Renovations, Core and Shell, Healthcare, Retail, etc.
- LEED AP with a specialty of Interior Design and Construction (ID+C)
- LEED AP for Existing Buildings: Operations and Maintenance (EB:O&M),
- LEED AP for Homes and,
- LEED AP for Neighborhood Development (ND)

LEED AP is divided in two sections:

1. General knowledge of green building and the LEED rating systems.
2. In-depth knowledge of the building rating systems for the specialty being tested.

Credential Maintenance Program:

System used to expand the knowledge and experience of LEED professionals which includes:

- LEED professionals must maintain their status by paying $50 every 2 years.
- For LEED AP: 30 hours of coursework every two years
- For LEED GA: 15 hours of coursework every two years

LEED AP FELLOW:

LEED AP Fellow Program was developed to honor and recognize distinguished LEED APs who have made a significant contribution in green building and sustainability. For instance, a LEED AP can nominate LEED AP Fellow if that individual:

- After at least 10 years of relevant experience,
- The nominee must have held the LEED AP credential for at least 8 cumulative years.

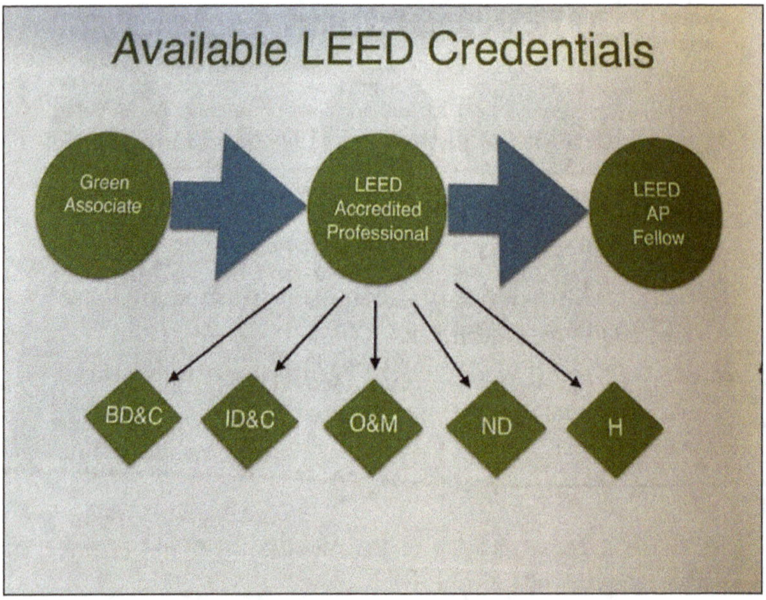

Figure 6-3 The available LEED credentials include the entry-level Green Associate, the five LEED AP specialty designations, and LEED AP Fellow. After Kibert 2016

GREEN BUILDING CERTIFICATION INSTITUTE RELATIONSHIP TO THE USGBC AND LEED:

Until 2008, the USGBC administered building certifications and professional designations in house. In 2008, a nonprofit organization, the Green Building Certification Institute (GBCI), was founded to provide a balanced third-party certification to be recognized by the American National Standards Institute (ANSI). GBCI is responsible for managing all aspects of LEED professional credentialing, including exam development, registration, delivery, and maintenance, to ensure ongoing excellence and that LEED professionals are proficient in the field. According to Kilber 2022, GBCI is responsible for managing the LEED project certification program by conducting technical reviews and analysis of submissions to verify and evaluate projects based on how well they have met the requirements of the various LEED rating systems. The project document is submitted through the LEED-online Internet portal.

USGBC retains the responsibility for creating and implementing new versions of LEED building rating system by integrating new green building technologies, systems, and strategies into the latest requirements. This responsibility includes establishing new reference guides and educational resources and outlining certification and accreditation requirements, among other things. The GBCI is arbiter of both building certification and LEED GA and AP accreditation, both of which are based on USGBC-generated rules. The relationship between the USGBC and the GBCI is visually depicted in Figure 6-4.

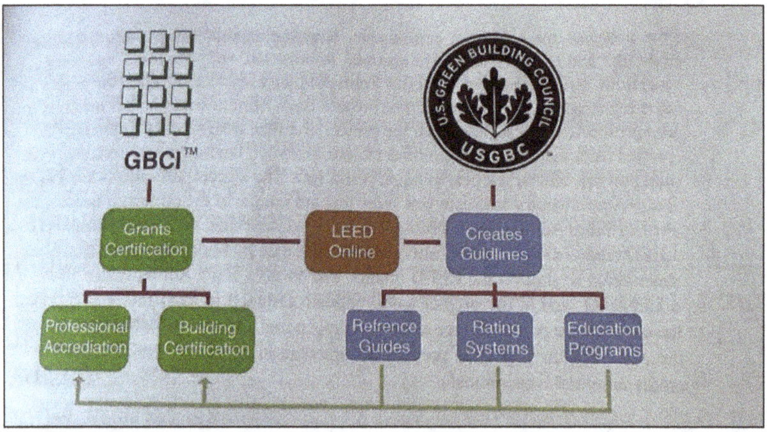

Figure 6-4 The relationship between USGBC and GBCI and their respective roles in the LEED Green Building system. The USGBC develops the requirement for LEED certification, and the GBCI ensures that the requirements have been met. The GBCI is also responsible for the testing and continuing education for LEED APs Source: Kibert 2022

LEED CERTIFICATION PROCESS

Prior to certification, the building is referred to as a LEED registered project. Achieving the LEED certification levels requires significant dedication from the professionals of the project team. This dedication must be maintained from design to the end of construction to successfully complete all steps in the certification process. These six steps include:

1. Ensuring that the building is eligible for certification,
2. Registering the project with GBCI,
3. Ensuring and documenting that the project meets the minimum project requirements (MPRs) and prerequisites and can attain at least the minimum number of points to achieve the LEED-certified level,

4. Submitting the required documentation via LEED-online,
5. If necessary, appealing points denied by the GBCI,
6. Receiving final notification from the GBCI that the project has achieved LEED certification.

LEED-Online

Over time, the LEED building rating system has shifted from requiring hard-copy documentation for certification to an Internet-based system known as LEED-Online. Therefore, project teams can submit all of their documentation online in an easy-to-use format. LEED-Online stores all LEED information, resources, and support in one centralized location. It enables team members to upload credit templates, track Credit Interpretation Rulings (CIRs), view documented responses to questions to questions posed by previous project team, manage key project details, contact the customer service department, and communicate with reviewers throughout the design and construction review process.

Registration:

The first step in LEED certification is project registration. Projects are registered by visiting the LEED Registration page of the USGBC website (www.usgbc.org), where information about the project is input and a registration fee is paid. Early registration is encouraged because starting the process as early as possible maximizes the potential for achieving certification. Registration establishes contact with the USGBC and provides access to potential information, software tools, and communications. Appointment of a Project Team Administrator occurs when the project is first registered at LEED-Online. The Project Team Administrator invites members of the project team to register with the project and then assigns roles to the individual project team members. Typical roles include architect, landscape architect, civil engineer, owner, and developer, to name just a few. The system also allows the Project Team Administrator to create new roles that are unique to the project, if needed. The Project Team Administrator develops a project description, assigns responsibility for LEED credits to the project team members, and then monitors the submission of documentation to support the LEED credits. The Project Team Administrator should be a LEED AP and is the project team member assigned to steer the project through the certification process. Once a project is registered and responsibilities are assigned, the project team begins to prepare documentation to satisfy the prerequisites and credit submittal requirements.

Credit Interpretation Ruling:

If a project team encounters difficulties applying a LEED prerequisite or credit to a specific project, the USGBC encourages the team to sort out the issue themselves first and contact the GBCI only as last resort. The CIR system ensures that rulings are consistent and available to other projects. If there is a gray area for which the project team requires clarification, the team can submit its query through LEED-Online and receive a ruling from the USGBC on the official interpretation of the situation. This latter response by the USGBC is the so-called CIR. All CIRs are contained in a database accessible from LEED-Online, and they can be used as precedents for addressing situations encountered by the project team.

Documentation and Certification

To earn LEED Certification, the project must meet all the MPRs, satisfies all the prerequisites, and earn the minimum number of points to at least attain the LEED-certified rating level. For LEED BD+C projects, the project team has the option to divide their review process into two separate reviews, design review and construction review, also known as a split review. The benefit of the split review is that it helps the project ream gauge whether the project is on track for achieving the anticipated LEED certification level. LEED-Online specifies which points are either design phase credits or construction phase credits, and project team must submit the complete documentation of design phase credits, and the project team must submit complete documentation of design phase credits for the design review. The GBCI will then respond with the Preliminary Design Review, including which credits are "Anticipated", "Pending" or "Denied" and if any documentation was "Approved" or "Not Approved". The project team then has the option to accept the results of the Preliminary Design Review as final or submit a response. If the project team determines that a response is necessary, they must submit a clear response to the Preliminary Design Review as well as appropriate documentation within 25 business day. The GBCI then will review the submitted documents along with the team's response and within 25 business days reply with the Final Design Review. The Final Design Review addresses each credit as "Anticipated", "Pending", or "Denied". During this period, no points are awarded to the project. Closer to Project completion, the project team can submit the remaining credits for the Preliminary Construction Review. The structure of this review is the same as the Preliminary Design Review. If project team members feel that they should be awarded specific credit, whether from the Final Design Review or Final Construction Review, they have the option to appeal.

Award of Certification

Upon notification of LEED certification, the project team has 30 days to accept or appeal the awarded certification level of platinum, gold, silver, or certified. Upon the project's acceptance, or the project team has not appealed the rating within 30 days, the LEED certification is final. GBCI will refer to the project as a LEED certified building, and the project team will receive an award letter and certificate specifying the LEED certification level. Although in the past, LEED certified projects were awarded plaques by the USGBC, currently a plaque can only be purchased online from the exclusive vender (www.greenplaque.com) , see Figure 6.5. the plaque exhibits the LEED certification level and year of achievement and typically is featured on either the exterior or interior of the high-performance building.

Figure 6-5 The Certification Plaque from the USGBC is made of recycled glass Source: Kibert 2016

Appeals

If the project team feels that sufficient grounds exist to appeal a credit denied in the Final LEED Review, it has the option to appeal. The appeal fee is $500 per credit appealed. A review of documentation for the appealed credits occurs within 25 business days, at which time an Appeal LEED Review will be provided to the applicant. All appeals are submitted via LEED-Online. If an appeal is pursued, a different review team will assess the appeal documentation.

LEED Registration and Documentation Fees

A registration fee must be paid for the LEED projects as part of the registration process. When the project team is prepared to initiate the review process, a certification fee must first be paid prior to action by the GBCI. As noted earlier, the project team has the option of submitting all documentation at the conclusion of the construction process or in two phases: (1) a design review, in which all credits that have been completely addressed by the designed team are put forward for review, and (2) a construction review, in which all the remaining credits are reviewed. The advantage of the two-phase review is that it speeds the certification process and allows the project team to decide on and act on appeals far earlier in the process.

LEED BUILDING DESIGN AND CONSTRUCTION RATING SYSTEM

As noted earlier, LEED refers to a wide variety of building assessment tools that are organized into five rating systems:

1. Building Design and Construction (BD +C),
2. Interior Design and Construction (ID+C),
3. Building Operation and Maintenance (O + M),
4. Neighborhood Development (ND),
5. Homes (H).

LEED (BD+C) is the most frequently used rating system; therefore, it is described in more detail. LEED (BD+C) rating system contains eight rating tools, namely, New

Construction, Core and Shell, Schools, Retail, Data Centers, Warehouses and Distribution Centers, Hospitality, and Healthcare. The prerequisite and credits listed in Table 6-1 through 6-3 apply to these rating tools within the LEED (B+C) rating system: (1) New Construction and Major Renovation, (2) Core and Shell, and (3) Schools. These three types of projects represent the most frequently used LEED rating tools. For other types of LEED (BD+C) projects, such as Healthcare, Data Centers, Warehouses and Distribution Centers, Hospitality, and Retail, plus projects in other LEED rating systems, such as LEED (O+M) or LEED (ID + C), the project team should consult the appropriate technical manuals for the specific type of project. The layout of LEED (BD+C) with respect to credits and points for New Construction, Core and Shell and School projects is shown in Table 6-1. For other BD+C project types, the LEED v4.1 technical manual should be consulted.

Table 6-1 **Points Assigned to LEED BD+C Projects**

Category for Credit name	New Construction (NC)	Core and Shell (CS)	Schools (S)
Integrated Project Planning & Design	1	1	1
Location and Transportation (LT)	16	20	15
Sustainable Sites (SS)	10	11	12
Water Efficiency (WE)	11	11	12
Energy and Atmosphere (EA)	33	33	31
Material and Resources (MR)	13	14	13
Indoor Environmental Quality (IEQ)	16	10	16
Innovation and Design (ID)	6	6	6
Regional Priority (RP)	4	4	4
Total Points	110	110	110
Certified	**40-49**	**40-49**	**40-49**
Silver	**50-59**	**50-59**	**50-59**
Gold	**60-79**	**60-79**	**60-79**
Platinum	**> 80**	**> 80**	**> 80**

Integrative Project Planning and Design

Points Assigned to LEED BD+C Integrated Project Planning and Design Category			
Category or Credit name	New Construction	Core & Shell	Schools
Integrated Project Planning and Design	1	1	1

Table 6-2 Point assigned to LEED (BD+C) Integrated Project Planning and Design Category

IP Credit: Integrative Process (1 point maximum)

One of the peculiarities of the newest version of LEED is that the first credit listed in LEED v4, Integrative Process (IP), is not identified as being in a major category and stands alone. This credit addresses the issue of collaboration among the project team members to enhance the sustainability qualities of the project.

The project team is required to collaborate in reducing energy and water consumption and to document efforts to improve the performance of the project compared to specific baseline cases.

LOCATION AND TRANSPORTATION

Location and Transport (LT) is a new category that first appeared in LEED v4. It focuses on the issue of building locations relative to ecologically sensitive land and access to transportation. Table 6-3 lists the six LT credits that apply to LEED (BD+C), New Construction Core and shell and School projects. To earn these points no prerequisite is applied.

Points can be earned if a project is in a development that has been previously certified under LEED (ND). The number of points earned varies depending on the rating (certified, silver, gold, or platinum) earned by the development. If the project team attempts this credit, the project is not eligible to earn points for other LT credits. In other words, this is an alternative path for the rest of LT category.

Table 6-3 Points Assigned to LEED BD+C Location and Transportation

Category for Credit name	New Construction (NC)	Core and Shell (CS)	Schools (S)
Location and Transportation (LT)	16	20	15
LEED for Neighborhood Development	16	20	15
Sensitive Land Protection	1	2	1
High Priority Site	2	3	2
Surrounding Density and Diverse Use	5	6	5
Access to Quality Transit	5	6	4
Bicycle Facilities	1	1	1
Reduced Parking Footprint	1	1	1
Green Vehicles	1	1	1

LT Credit: LEED for Neighborhood Development

This credit allows projects built on sites in a development that has been certified to LEED for Neighborhood Development (LEED-ND) to earn credits and is a replacement for all other credits under the LT category. That is, if a project has a LEED-ND certified location, it earns points solely from that designation and is not eligible to also earn points from the other individual LT credits. Table 6-4 indicates the number of points earned as a function of its LEED-ND rating.

LEED-ND Certification level	NEW Construction	Core and Shell	Schools
Certified	8	8	8
Silver	10	12	10
Gold	12	16	12
Gold	16	20	15

Table 6-4 Points Earned for LEED-ND Rating

LT Credit: Sensitive Land Protection: (2 points maximum)

Requires the projects that are not located on sensitive land earn points for avoiding impacts that otherwise would occur such as:

- Prime farmland
- Floodplains
- Habitat for endangered and threatened species in proximity of wetlands and water bodies

LT Credit: High Priority Sites (3 points maximum)

Option 1:

(Historic District). This option addresses infill locations in historic districts.

Option 2:

(Priority Designation) This option includes a range of additional possibilities, such as Federal Empowerment Zones, Low Income Areas, etc.

Option 3:

(Brownfield Remediation). This option earns points for locating the project on a brownfield site.

LT Credit: Surrounding Density and Diverse use (6 point maximum).

Locating a project in a dense urban environment can earn points.

Option 1:

(Surrounding Density), the points earned increase as the average density with ¼ mile of the project increases.

Option 2:

(Diverse Uses), a project located near publicly diverse uses, such as:

Schools,

Restaurants

Dry Cleaners and other services

Can earn points under this option.

LT Credit: Access to Quality Transit (6 point maximum)

Projects located within ¼ mile of bus, streetcar, or rideshare stops or within ½ mile of rail stations or bus rapid transit stops, can earn credit if they meet specific daily levels of service requirements. School projects have an option if specified percentage of students live within ¾ mile walking distance with pedestrian access.

LT Credit: Bicycle Facilities (1 point maximum).

If the project facilitates bicycle use, credit can be earned for this measure. Bicycle storage or a functional entry to the project must be within 200-yard walking or bicycling distance from the bicycle network with connections to diverse uses, schools or employment centers, or mass transit.

LT Credit: Reduced Parking Footprint (1 point maximum)

Minimizing the consumption of resources and land for parking is the target of this credit where projects cannot provide excess parking beyond local code requirements. If there is no local requirement, specific targets are provided for this purpose.

LT Credit: Green Vehicles (1 point maximum)

By designating a minimum 5 percent of parking spaces as a preferred parking for green vehicles, the project can earn points for encouraging the use of green vehicles. Green vehicles are defined as those earning a minimum score of 45 in the American Council for an Energy Efficient Economy automobiles.

SUSTAINABLE SITES

The sustainable site (SS) category rewards decisions about the environment surrounding the building with credits that emphasize the vital relationships among buildings, ecosystems, and ecosystem services. The SS credits are awarded for protecting natural habitat, keeping open spaces, dealing with rainwater, and reducing light pollution and heat island effects. In general, the purpose of SS is related to those decisions that make less impact to the environment.

The purpose of this category is to align land development and management with innovative sustainable design, and to protect, improve and regenerate natural systems. As we continue to develop the world for human purposes, it is crucial that we do not alter the natural habitat drastically violating the next generation rights. Table 6-5 presents the assigned credits for this category.

First SS Prerequisite: Construction Activity with Pollution Prevention Plan

This is a prerequisite to implement of an air born dust, soil erosion and sedimentation control (ESC) during the construction activity in California. For more details see Chapter 1. The prerequisite will prevent soil loss via water or wind and sedimentation of stormwater infrastructure. The best reference material is to use “Construction General

Permit Document" prepared by State Water Resources Control Board.

The project will require a Qualified Stormwater Designer (QSD) Specialist and Practitioner (GSP) to control sediment transport during rainfall events while the project is under construction. Since this is a prerequisite, no score can be obtained toward the rating.

Table 6-5 Points Assigned to LEED BD+C Sustainable Sites

Category for Credit name	New Construction (NC)	Core and Shell (CS)	Schools(S)
Sustainable Sites (SS)	10	11	12
Construction Activity Pollution Prevention	P	P	P
Environmental Site Assessment	NA	NA	P
Site Assessment	1	1	1
Site Development Protect or Restore habitat	2	2	2
Open Space	1	1	1
Rainwater Management	3	3	3
Heat Iland Reduction	2	2	2
Light Pollution Reduction	1	1	1
Site Master Plan	NA	NA	1
Tenant Design and Construction Guidelines	NA	1	NA
Joint Use of Facilities	NA	NA	1

Second SS Prerequisite: Prerequisite: Environmental Site Assessment (Phase I and Phase II)

Seeking LEED ratings. This prerequisite is a mandatory requirement for any school project in California seeking LEED rating. For more information, please refer to Chapter 2 of this book. If any recognized environmental condition (REC) is reported a Phase II ESA should be performed. For the contaminated sites a site closure letter (no further action (NFA)) should be received from the local over-sight agency before starting any construction activity on contaminated sites in California.

SS Credit: Environmental Site Assessment (1 point maximum)

ESA is mandatory for School projects (no credit). However, for other projects such as NC and CS it can earn 1 point by conducting Phase I and Phase II ESA using ASTM standard (E-1527-21) and (E1903-13) Guidelines, respectively.

SS Credit: Site Development – Protect or Restore Habitat

This addresses the protection of existing ecology or the restoration of the degraded natural systems. A detailed assessment of site condition is required and must include:

Topography, hydrology, climate vegetation, biological survey; and Evaluation of the above factors on human health and environment by construction of proposed built environment.

SS Credit: Open Space (1 points maximum)

The project team must document that outdoor space at least equal to 30 percent of the site area that have been set aside for environmental and human breathing room. In addition, at least 25 percent of this space must be vegetated. In densely populated areas the green roof can serve the purpose of open space requirements, provided that access to the eco-roof is provided by the project team.

SS Credit: Rainwater Management (3 points maximum)

Rainwater volume must be decreased, and its quality should be increased by preventing run-off and increasing groundwater recharge. In general, low-impact development and green infrastructures are to be used to accomplish these objectives.

SS Credit: Heat Island Reduction (2 points maximum)

SRI: Solar reflectance index (SRI): how well a material reflects solar energy and reduce the heat island affect. 0-100: (0) absorbs all thermal energy and (100) reflects all thermal energy.

For non-roof areas: use vegetation, shade from photovoltaic (PV) arrays and materials that can maintain SRI of at least 28 for three years. For roof areas: For low sloped roof SRI initial value of 82 with a 3-year value of at least 64.

For steep roof, SRI initial value of 39 and a minimum 3-year value of 32.

SS Credit: Light Pollution Reduction (1 point)

To improve night-time visibility and increase access to the night sky and prevent the disturbance of both wildlife and people by excess lighting. A methodology known as back-light, up light and glare (BUG) is used for this purpose. It is a function of where the project is located with the most stringent efforts required near a large natural or rural area. It applies to the illumination of exterior signage and light trespass.

SS Credit: Site Master Plan (1 point)

It applies only to school projects if the project is planned to expand its activities. By creating a master plan that addresses the expansion of school buildings, other facilities etc.

SS Credit: Tenant Design and Construction Guidelines (1point)

It encourages tenants regarding the potential implementation of sustainable design and construction features. Raising awareness of tenants may result in reduction in energy use and water consumption and improving indoor environmental quality.

SS Credit: Joint use of facilities (1point)

School Projects can earn one point for collaborating with community using the following options:

Option 1: Collaborate with the community to share three types of school facilities such as building, cafeteria, recreation fields for community events.

Option 2: Share at least two types of school spaces with the public or via contact with specific organizations such as the school library and parking lots.

Option 3: Sharing at least two spaces owned by organizations; for example, a YMCA-owned swimming pool for student physical training or athletic activities, an auditorium or a stadium.

WATER EFFICIENCY

As water becomes less accessible in many communities in the US and around the world, it imposes serious limitations in creating new developments and subdivisions. As such, cutting drinking water consumption within the residential as well as commercial buildings is becoming a priority. Considering the prolonged drought periods in the arid west, especially in California, reducing the indoor and outdoor water usage is quite important to follow the sustainability rules. One of the main characteristics of high-performance green buildings is their low water consumption in indoor and outdoor areas. As such, depending upon how efficient the water is being consumed in buildings the building will gain credits in LEED rating system. Table 6-6 shows how many credits can be obtained by reducing drinking water consumption within the built environment.

Table 6-6 Points Assigned to LEED BD+C Water Efficiency

Category for Credit name	New Construction (NC)	Core and Shell (CS)	Schools(S)
Water Efficiency (WE)	11	11	12
Outdoor Water Use Reduction	P	P	P
Indoor Water Use Reduction	P	P	P
Building-Level Water Metering	P	P	P
Outdoor Water Use Reduction	2	2	2
Indoor Water Use Reduction	6	6	6
Cooling Tower Water Use	2	2	2
Water Metering	1	1	1

Water efficiency (WE) category addresses water consumption reduction for both indoor and outdoor uses. It imposes 3 prerequisites for reducing water consumption in indoor and outdoor areas:

WE Prerequisite 1: Outdoor Water Use Reduction:

This prerequisite requires a landscape that does not require a permanent irrigation system beyond a two-year watering period. In another word, it requires an irrigation system that represents a 30 percent reduction in consumption below baseline calculated for a peak watering month.

To satisfy this prerequisite, EPA WaterSense Water Budget Tool should be used.

WE Prerequisite 2: Indoor Water Use Reduction:

Reduce indoor water consumption by 20 % below a baseline calculated based on building code fixture requirements. It applies to appliances such as dishwashers, clothes washers, and water used for processes such as water coolants or water towers and heat exchangers.

WE Prerequisite 3: Building-Level Water Metering:

All buildings are required to be supplied with water meters that determine monthly and annual water consumption. The meters can provide automated data or allow manual readings.

WE Credit: Outdoor Water Use Reduction (2 points)

The EPA WaterSence Water Budget Tool must be used to calculate the effectiveness of strategy of plant species selection and grouping and irrigation system design. This credit requires either a landscape that does not require a permanent irrigation system beyond a two- year watering or a landscape that represents a 50% reduction in consumption below a baseline calculated for the peak watering month.

WE Credit: Indoor Water Use Reduction (6 points)

Further reductions in drinking water consumption below baseline (25 to 50%). This can earn an increasing number of points. Additional performance requirements are specified for commercial washing machines and commercial kitchen equipment. EPA recommended faucets and showerheads is needed to receive this credit.

WE Credit: Cooling Tower Water Use (2 points)

Project with cooling towers can receive credit for conserving water while also controlling condenser water contaminants. This credit requires projects to perform a potable water analysis and then create a water treatment program for any cooling towers or evaporative condensers that will be installed.

The term “evaporative condensers” applies to closed-circuit water-cooling towers, closed circuit cooling towers that evaporate water to cool refrigerant, and rooftop HVAC units.

WE Credit: Water Metering (1 point)

Additional water metering for two or more water systems such as irrigation, domestic hot water, interior plumbing fixtures, boilers, reclaimed water, and process water, can earn points under this credit.

ENERGY AND ATMOSPHERE

About 40 percent of Us energy is consumed inside buildings. The energy derived from fossil fuels is the most notorious and harmful type of energy that produces carbon dioxide which is the main cause of global warming. Therefore, using the energy which has been generated from renewable sources is the prime objective of green building initiative that would not harm our planet. Therefore, this category carries the maximum number credits in comparison with other categories in LEED rating system. The LEED BD+C category of Energy and atmosphere (EA), addresses the issue of energy for the high-performance buildings; it also covers several issues that connect building systems to environmental impacts on air and the atmosphere, for example the elimination of hydrochlorofluorocarbons, which, due to their presence in chillers and other mechanical equipment, are implicated in climate change and ozone depletion, see Table 6-7. The following discusses LEED v4.1 credits and reporting requirements in the EA category. Note that the four prerequisites do not carry points, and all must be met for a building to be considered for certification.

Table 6-7 Points Assigned to LEED BD+C Energy & Atmosphere

Category for Credit name	New Construction (NC)	Core and Shell (CS)	Schools(S)
Energy & Atmosphere (EA)	33	33	33
Fundamental Commissioning and Verification	P	P	P
Minimum Energy Performance	P	P	P
Building-Level Energy Metering	P	P	P
Fundamental Refrigerant Management	P	P	P
Enhanced Commissioning	6	6	6
Optimize Energy Performance	18	18	16
Advanced Energy Metering	1	1	1
Grid Harmonization	2	2	2
Renewable Energy Production	5	5	3
Enhanced Refrigerant Management	1	1	1
Green Carbon and Carbon Offsets	2	2	2

P = prerequisite

EA Prerequisite1: Fundamental Commissioning and Verification.

The Mechanical Electrical and Plumbing (MEP) and the renewable energy systems must be assessed and tested by a qualified Commissioning Authority (CxA). The commissioning activity refers to testing and reviewing the performance of building equipment once the construction activities are concluded. To perform proper commissioning procedure, the American Society of Heating, Refrigeration and Air Conditioning Engineers (ASHRAE)'s guideline 0-2005 and ASHRAE Guideline 1.1-2007 relative to their energy, water indoor environmental quality, and durability performance.

For conducting the proper commissioning for exterior enclosures, the Owner's Project Requirement (OPR) and Basis of Design (BOD) follow National Institute of Building Sciences Guidelines 3-2012 for Exterior

Enclosures. Therefore, the CxA must review OPR and BOD, and compare them to project design for consistency.

The following should be performed by CxA:

1. Develop a Cx plan,
2. develop construction checklist,
3. develop a test procedure,
4. verify test execution,
5. maintain logs of activities, findings, and recommendations,
6. prepare a final Cx report.

The CxA is also responsible for preparing the facilities operations and maintenance plan for the facility operators to use in operating the building at optimum efficiency.

EA Prerequisite 2: Minimum Energy Performance

The project must comply with one of the pathways of ASHRAE 90.1-2016:

1. Prescriptive provisions as found in Section 5 through 10.
2. Section 11, energy cost budget method
3. Appendix G, Performance Rating Method, using the metrics of greenhouse gas (emission).

ASHRE 90.1-2016, Energy Standard for Buildings Except Low-Rise Residential Buildings, continues the movement to drastically reduce building energy consumption to the point that a substantial fraction of new construction will be net zero energy capable. Several states including New York, Viginia and Florida have adopted this standard as their commercial energy code. In addition to increasing thermal envelope performance, this version provides a cleaner compliance path for renewable energy treatment.

EA Prerequisite 3: Building Level Energy Metering:

Building-level energy meters or submeters that can be aggregated to determine total energy consumption are required. These meters include both meters that are the property of the building owners and meters owned by the utility company. The energy data must be tracked at least monthly and be provided to USGBC for a five-year period or until the ownership or lessee of the building changes.

EA Prerequisite 4: Fundamental Refrigerant Management.

Chlorofluorocarbons (CFCs) are ozone-depleting substances (air-conditioning equipment). This prerequisite has the intent of eliminating CFCs from buildings; therefore, it is protecting the ozone layer. It requires zero use of CFCs in new buildings HVAC&R systems. For existing equipment, a plan to phase out the CFCs must be submitted prior to project completion. However, there is some exemption to this rule such as standard refrigeration or water coolers containing 0.5 pounds of refrigerant.

EA Credit: Enhanced Commissioning (maximum 6 points)

This credit addresses increasing the level of commissioning to include the building energy systems (option 1) and envelop commissioning (option 2). Enhanced commissioning adds several additional requirements to the fundamental building commissioning category and provides two paths for compliance:

Option1. Enhanced system commissioning

Path 1. Enhanced system commissioning: The requirements include reviewing the energy systems design, reviewing contractor submittals, creating a system manual for building operators, verifying the training of operators, rechecking the building operation within 10 months of occupancy to verify performance, and developing an on-going commissioning plan.

Path 2. Monitoring: The project must complete path 1 and it can earn additional points by developing a monitoring procedure that includes the data points to be measured and evaluated for assessing the performance of energy and water systems.

Option 2: This option involves envelope commissioning and reviewing the energy systems design, contractor submittals, creating a system manual for building operators, rechecking the building operation within 10 months of occupancy to verify performance, and developing an on-going commissioning plan.

EA Credit 2: Optimize Energy Performance (18 points maximum)

Designing and building an energy-efficient building is important for sustainability reasons as well as for earning a LEED rating. Because Optimize Energy Performance has more than half of the EA credits assigned to it, it is by far the most important credit in the EA category—in fact, in the entire LEED rating system. ASHRAE 90.1–2016 is the basis for determining how well the high-performance building performs compared with the base case (i.e., the building that just meets the standard's minimum requirements). To be successful in obtaining a LEED rating for the building, the design team should ensure that the requirements of ASHRAE 90.1–2016 are exceeded by a substantial margin. The energy-related components of a typical building are its envelope (walls, roof, floor, windows, and doors); HVAC equipment; power distribution system; lighting system; and equipment such as pumps, appliances, refrigeration equipment, and elevators.

EA Credit 1 (EAc1) has three options for earning points toward LEED certification, as described below.

Option 1. Energy Performance Compliance (1–18 points for NC and CS projects and 1–16 points for S projects). ASHRAE 90.1–2016 introduced the concepts of Performance Cost Index (PCI) and Performance Cost Index Target (PCI_t). Points are earned based on the percentage improvement of the PCIt over the PCI as shown in Table 6.8. On-site renewable energy generation is included in the modeling to show a reduction in energy demand from external sources.

Option 2. Prescriptive Compliance Path: ASHRAE 50% Advanced Energy Design Guide (NC 1–6 points, CS 1–4 points). Projects must meet the scope requirements of the prescriptive ASHRAE 50 Percent Advanced Energy Design Guide for specific building types and follow the ASHRAE 90.1–2016 Prescriptive compliance path in the EA Prerequisite Energy Performance. Some of the available AEDGs include the 50 percent AEDG for Small to Medium Office Buildings and the 50 percent AEDG for K–12 School Buildings. The number of points awarded is based on the number of building components (envelope, glazing, interior lighting, exterior lighting, and plug loads) that follow the recommendations in these guides.

New Construction NC (%)	Major Renovation And CS (%)	NC Major Renovation And CS Points	School Points
5	2	1	1
10	5	2	2
15	10	3	3
20	20	4	4
25	25	5	5
30	30	6	6
35	35	7	7
40	18	8	8
45	40	9	9
50	45	EP*	EP*

Table 6.8 Points Versus Percent Improvement of PCI_t over the PCI

***EP = Exemplary Performance. A project can earn one exemplary performance poiny for an improvement of at least 5 percent above the highest point earning performance.**

EA Credit: Advanced Energy Metering

Metering of energy consumption to support energy measurement is an important strategy for reducing building energy use. This credit requites the installation of advanced energy metering for the whole building and submeters for use that consume 10 percent or more of total energy. The meters must be permanent and measure both energy use and peak powe, referred to as energy deman.there must be a data collection system, and it must be remtely accessible, able to store 36 month of data, and capable or reporting hourly, daily, monthly, and annual energy use. For CS projects,the requirements are slightly different in that the whole building must be meteredand provisions for tenat metering system must be included.

EA Credit: Grid Harmonization

Enable the facility to participate in demand response (DR) program that use load shedding or load shifting strategies earns points under this credit.

Option 1: Demand Response Program Available and Participation: Case 1: The utility has demand response program, the facility must be enrolled in the program for at least one year. Another option is to enroll with a DR provider with the intention of multi-year renewal. Case 2: Demand Response Capable Building: provide the infrastructure and plans for DR program and contact the local utility to discuss participation in their DR program. Case 3: Load Flexibility and Management Strategies: Use the USEPA Avoid Emissions and Generation Tool (AVERT) to determine local gris emissions and implement one of several strategies to address peak-load reduction. One of the strategies is on-site thermal or electric storage, for example ice storage systems that are used by many Florida schools for this purpose.

EA Credit: Renewable Energy Production

Green building rating systems should encourage the consumption of renewable rather than nonrenewable energy sources for building and provide points for on-site or site-recovered renewable energy systems. Eligible renewable energy systems include PV or solar thermal systems, active systems, bio-fuel-based electrical systems, geothermal heating/electrical systems, low-impact hydro, wave/tidal power systems, and wind-based electrical production systems. Table 6.9 indicates the relationship between renewable energy production and the number of points awarded.

Points	On-site Renewables (%)	Off-Site Renewables (%)	Existing Off-Site Renewables (%)	Green-e Certifies EACs &Carbon Offsets (%)	EACs & Carbon Offsets (%)
1	2	20	60	100	150
2	6	40	80	200	
3	15	60	100	300	
4	35	80			
5	60	100			
EP*	100				

Table 6.9 Points for Renewable Procurement

EP* = Exemplary Performance. A project can earn one exemplary performance point for an improvement greater than the level associated with the maximum number of points.

EA Credit: Enhanced Refrigerant Management

The EA prerequisite for Fundamental Refrigerant Management calls for zero use of CFC refrigerant in new building HVAC&R equipment. This credit provides for measures that further reduce the use of ozone-depletion refrigerants in buildings and addresses the climate change impacts of these substances. There are two options:

Option 1. Do not use refrigerants

Option 2. Select refrigerants that minimize contributions to the life-cycle ozone depletion potential and the life cycle direct global warming potential. To account for the potential damage of various refrigerants, a formula is used to quantify the combined impact in the building project.

EA Credit: Green Power and Carbon Offset

Any combination of on-site and off-site renewable energy or Energy Attribute Certificates (EACs) can earn points for renewable energy. A total of 5 points plus one Exemplary

Performance point can be earned based on the scenarios in Table 6.9.

MATERIAL AND RESOURCES

The material resource category (MR) addresses minimizing the impacts of the material supply chain for construction. For the first time. LEED v4.1 provides for the use of the life-cycle assessment and Environmental Product Declaration (EPD) to earn points. It also addresses additional transparency measures on the part of manufacturers to stimulate the lowering of their contribution to environmental and human health impacts see Table 6-10.

Category or Credit Name	New Constru ction	Core And Shell	Schools
Material and Resources (MR)	13	14	13
Storage & Collection of Recyclables	P	P	P
Construction/Demolition Waste Management Planning	P	P	P
Building Life Cycle Impact Reduction	5	6	5
Building Product Disclosure and Optimization Environmental Declaration	2	2	2
Building Product Disclosure and Optimization Sourcing of Raw Materials	2	2	2
Building Product Disclosure and Optimization Material Ingredients	2	2	2
Construction/Demolition Waste Management	2	2	2

Table 6-10 Points Assigned to LEED BD+C Material and Resources

P = Prerequisite for the rating system

MR Prerequisite: Storage and Collection of Recyclables

An easily accessible area must be set aside for the separation of, collection, and storage of recyclable materials, which at a minimum must include paper, corrugated cardboard, glass, plastics, and metals. The area must also have space for the proper disposal of least two of the following electrical items, mercury-containing lamps, and electronic waste.

MR prerequisite: Construction and Demolition Waste Planning

The project is required to develop and implement a construction and demolition waste plan that identifies at least five materials targeted for diversion and provides an estimate of percentage of total construction and demolition waste that these materials represent. The plan must also specify how the waste will be separated or how the mixed or comingled waste will be handled by the recycling facility. A daily progress report addressing all waste streams must be generated and include both the total waste generated and the disposal and diversion rates.

MR Credit: Building Life-cycle Impact Reduction

The project must demonstrate a strategy to minimize the environmental impacts of the building's materiality, either through reusing existing building resources or via life-cycle assessment. Four options are available:

Option 1. Historic Building Reuse. Reuse a historic building by maintaining its structure, envelope, and interior nonstructural elements of a historical building. A qualified building must be listed on the local, state, or national register of the historic places.

Option 2.Renovation of abandoned or brightened building. Renovate an abandoned or blighted by maintaining at least

50 percent of exterior and interior elements. The building must be renovated at least the degree that productive occupancy is possible.

Option 3. Building and Material Reuse. Reuse or salvage materials from on-site or off-site. Structural elements, such as floors and roof decking, enclosure materials, such as the building skin or framing, and permanently installed interior elements, such as walls and doors, are included. Table 6-11 lists the number of points earned versus percent of surface area reuse for this option.

Option 4. Whole Building Life Cycle Assessment. Perform a life cycle assessment that is compliant with International Standards Organization ISO 14044 and select materials and products that demonstrate a 10 percent reduction in impact compared with the reference building. The reference building must closely resemble the designed building in size, function, and orientation with the energy performance of both reference and designed buildings that same the stated in the EA Prerequisite, Minimum Energy Performance. There must be reduction in at least three of the six environmental impact measures (global warming potential, stratospheric ozone depletion potential, acidification, eutrophication, formation of the tropospheric ozone, and depletion of nonrenewable energy resources) by virtue of materials/product selection process.

MR Credit: Product Disclosure and Optimization – Environmental Product Declarations

Environmental Product Declarations (EPDs) are life-cycle assessments of products performed in a format specified by ISO 14044. Multi-Attribute Standards are standards that specify multiple green attributes for products. This credit provides two options that can be used to earn two points for the project.

Option 1. EPDs. At least 20 different products must (1) have an ISO 14044 product-specific declaration, or (2) have an ISO compliant EPD that discloses impact information, or (3) comply with other USGBC-approved EPD frameworks.

Option 2.Multi-Attribute Optimization. Use products that comply with one of the following criteria for a minimum of 50 percent by cost of the total value of permanently installed products: (1) third party certified products that have impacts below the industry average for three of the six environmental impacts measures (global warming potential, stratospheric ozone depletion potential, acidification, eutrophication, formation of the tropospheric ozone, and depletion of nonrenewable energy resources), or (2) comply with some other USGBC approved multi-attribute framework.

Percentage Surface Area	Points for New Construction. Schools	Points for Core and Shell
25	2	2
50	3	3
75	4	5

Table 6-11 Points Awarded versus Percentage Project Surface Area Reused

MR Credit: Product Disclosure and Optimization – Sourcing of Raw Materials

Option 1. Raw Material Source and Extraction Reporting. Use a minimum of 20 products from at least five different manufacturers that have released a report regarding their raw materials that demonstrate responsible extraction procedures. These reports can be a combination of self-declarations by the manufacturer and third-party corporate social responsibility reports (CSRs) that contain information on raw materials extraction impacts. The CSRs must use one of the acceptable CSR frameworks, such as the UN Global Compact or the Global Reporting Initiatives.

Option 2.Leadership Extraction Practices. Use products that meet the responsible extraction criteria from one of the following for at least 25 percent of the total value of permanently installed building products: extended producer responsibility, bio-based materials, wood products that are Forestry Stewardship Council certified, materials reuse, recycled content, or a USGBC approved program.

MR Credit: Product Disclosure and Optimization – Material Ingredients

Option 1. Material Ingredient Reporting. Use at least 20 materials from manufacturers that use one of the following programs that indicate the chemical inventory of the product with at least 0.1 percent or 1,000 parts per million accuracies: Manufacturer Inventory, Health Product Declaration, or Cradle to Cradle, or a USGBC-approved program.

Option 2.Material Ingredient Optimization. Use one of the following material ingredient optimization paths for at least 25 percent of the total value of permanently installed building products: GreenScreen v1.2 Benchmark, Cradle to Cradle Certified, REACH Optimization, or a USGBC-approved program.

Option 3.Product Manufacturer Supply Chain Optimization. Use products for at least 25 percent by cost, of the total value of permanently installed building products that are sourced from manufacturers engaging robust, validated safety, health, hazard, and risk programs that document at least 99 percent by weight of the ingredients in the product or material. The products also must be sourced from manufacturers with independent third-party verification that verify a variety of specified criteria are met for safety, health, hazard, and risk programs.

MR Credit: Construction and Demolition Waste Management

The project must recycle and/or salvage nonhazardous construction and demolition materials and measure the results consistently using weight or volume units. Included is wood waste converted to biofuel. Excluded are excavated soil, land-clearing debris, and alternative daily cover. The two available options for this credit are described below:

Option 1. Divert 50 Percent and Three Material streams. Divert at least 50 percent of the total construction and demolition waste stream from at least three different material streams.

Option 2.Reduction of Total Waste material. Generate no more than 2.5 pounds per $ft^2$of waste.

INDOOR ENVIRONMENTAL QUALITY

The indoor environmental quality (IEQ) category focuses heavily on indoor air quality and includes several other indoor environmental quality issues such as acoustic performance, lighting quality, day lighting, daylighting and views, and thermal comfort, see table 6-12.

EQ Prerequisite: Minimum indoor Air Quality Performance

Building or building spaces that are mechanically ventilated must follow the requirements of the ventilation rate procedure in ASHRAE Standard-2016 or stringent local equivalent. If outside of US, the ventilation system design must adhere to specific requirements of ISO Standard 17772-1:2017. Similarly, for naturally ventilated buildings, the natural ventilation procedure in ASHRAE 62.1-2016 must be followed. For mechanical ventilation, the outdoor air flow must be monitored by direct airflow measurement. For natural ventilation, the outdoor airflow measurement or measurement of carbon dioxide (CO_2) concentration is required.

EQ Prerequisite: Environmental Tobacco Smoke Control

Smoking must be prohibited inside the building and outside the building except in designated areas at least 25 feet from all building air intakes. Signage banning smoking must be posted near building entrances. For residential buildings, smoking is prohibited in all common areas, and excess leakage between units must be prevented by weather stripping and careful sealing of all penetrations between units.

Category or Credit Name	Quality New Construction	Core and Shell	Schools
Indoor Environmental Quality (EQ)	16	10	16
Minimum IAQ Performance	P	P	P
Environmental Tobacco Smoke Control	P	P	P
Enhanced IAQ Strategies	2	2	2
Low-Emitting materials	3	3	3
Construction IAQ management Plan	1	1	1
IAQ Assessment	2	n/a	2
Thermal Comfort	1	n/a	1
Interior Lighting	2	n/a	2
Daylight	3	3	3
Quality Views	1	1	1
Acoustic Performance	1	n/a	1

Table-6-12 Points Assigned to LEED (BD+C) Indoor Environmental

P= Prerequisite for the Rating

n/a=not applicable for the rating system

EQ Prerequisite: Minimum Acoustic Performance

For school projects, a maximum background noise level of 40 decibels (dBA) must be achieved for HVAC systems in classrooms and other core learning spaces. ANSI Standard S12.60-2010, Part 1. Annex A.1: the 2015 HVAC applications ASHRAE Handbook, Chapter 48, "Sound and Vibration Control"; and AHRI Standard 885-2008 provide best practices for preventing noise and sound transmission from HVAC equipment and components. For high noise sites (those with peak-hour 60 dBA sound levels during school hours), acoustic treatment and other measures to minimize noise intrusion from exterior sources and control sound transmission between classrooms and other core learning spaces must be utilized.

EQ Credit: Enhanced Indoor Air Quality Strategies

There are two options for gaining points for additional measures to improve indoor air quality:

Option 1. Measures such as entryway system, interior cross-contamination prevention, and filtration are required for this strategy. The entryway systems must be permanently installed and must be 10 ft in length to capture dirt and particulates that enter the building, generally on the shoes of the building, users or visitors. Grates, grilles and slotted systems can be used for this purpose. Interior cross-contamination prevention measures must be used where hazardous gases and chemicals may be present. These areas must be negatively pressurized, have self-closing doors, be sealed with deck-to-deck partitions, and be safely exhausted to the building exterior. Outdoor air systems supplying air to occupied spaces must be equipped with Minimum

Efficiency Reporting Value (MERV) 13 or higher filters per ASHRAE Standard 52.2-2007, which specifies how to test air filters or, if outside the United States, must have Class F7 or higher filters as defined by CEN Standard EN779-2002. The CEN Standard is the European equivalent of ASHRAE 52.2.

Option 2. An alternative set of equipment includes exterior contamination prevention, increasing ventilation, monitoring CO_2 levels, and providing additional source control and monitoring. Using advanced design strategies, such as employing computational fluid dynamics modeling, can result in entryways that inherently prevent pollutants from entering the building. Monitoring CO_2 concentrations for densely occupied spaces and providing indication of levels and altering occupants to setpoints being exceeded provide additional assurance of a quality indoor air environment.

EQ Credit: Low Emitting Materials

This credit covers volatile organic compound (VOC) emissions into indoor air, the VOC content of materials, and means for testing indoor VOC levels. Seven categories of material are covered: (1) interior paints and coatings applied on site, (2) interior adhesives and sealants applied on site, (3) flooring, (4) composite wood, (5) acoustic insulation, (6) furniture, (7) exterior products (school projects only). The dividing line between indoor and outdoor spaces is the weatherproofing barrier, for example, the waterproofing membrane and air or water-resistant barriers. For school projects, there are restrictions on the VOC content of some classes of exterior materials such as coatings and roofing.

EQ Credit: Construction Indoor Air Quality Management Plan

The air quality during construction and numerous potential contaminants resulting from sanding, painting, and finishing work can pose a threat to the health of future building's occupants. To prevent these threats from occurring, several measures are required. First, control measures to protect air-handling system components from contamination are required as specified in Sheet Metal and Air Conditioning Contractors (SMACNA) IAQ Guidelines for Occupies Buildings under Construction, 2nd edition, 2007. Second, all absorptive materials to be used in construction must be stored in a manner that protects them from moisture damage. Third, air-handling equipment must be operated only with Merv 8 or better filters installed on all return air grilles and transfer duct inlets. These filters must be replaced or removed just prior to occupancy. Finally, the use of tobacco products inside or within 25 ft of the building is prohibited.

EQ Credit: Indoor Air Quality Assessment

Option 1. Just prior to occupancy and all with interior finishes installed and withnew filtration media installed, flush out the building with 14,000 cubic feet of air per ft^2 of outdoor air. The flush-out also can occur during occupancy at specified air flow rates, interior temperatures, and duration.

Option 2. After construction is complete but prior to occupancy, conduct baseline IAQ testing for formaldehyde, particulates, ozone, total VOCs, carbon monoxide, and other specified chemicals.

EQ Credit: Thermal Comfort

Option 1. Design the project in accordance with ASHRAE Standard 55-2017, Thermal Comfort Conditions for Human Occupancy.

Option 2. Design the HVAC systems and building envelope to meet the requirements of ISO 7730:2005, Economics of the Thermal Environment, and ISO 17772-2017, energy Performance of buildings-Indoor Environmental Quality.

EQ Credit: Interior Lighting

Option 1.Lighting Control. At least 90 percent of individually occupied spaces must have lighting controls that allow the occupants to adjust lighting levels to suit their needs. For shared spaces, lighting must be able to be adjusted to meet group needs, lighting for presentation/projectors must be separately controlled, and the switches for luminaries must be in the spaces served by the luminaries.

Option 2.Lighting Quality. All spaces, with limited exceptions, must be provided with luminaries with luminance of less than 2,50 foot-candles per m^2 between 45 and 90 degrees from Nadir. Light sources must have a color rendering index of 80 or higher, with some limited exceptions. For at least 75 percent of the connected load, light sources with a rated life of at least 24,000 hours must be used.

EQ Credit: Daylight

Option 1. Use computer simulation to demonstrate daylight autonomy for 55, 75, or 90 percent of normally occupied spaces. Additionally, show that an annual sunlight exposure of no more than 10 percent is achieved.

Option 2. Using computer simulation shows that illuminance level of between 300 and 3,000 lux are achieved between 9 am and 3 pm for regularly occupied floor areas for 75 or 90 percent of normally occupied spaces.

Option 3. Use direct measurement to determine that lighting levels between 300 and 3,000 lux are achieved for 75 or 90 percent of normally occupied spaces.

EQ Credit: Quality Views

By achieving line-of-sight view to the outdoors through vision glazing for 75 percent of all regularly occupied spaces, the building occupants are connected to the exterior outdoor environment, and the project can earn points under this credit.

EQ Credit: Acoustic Performance

Strategies that promote high-quality acoustic design will address HVAC background noise, sound isolation, reverberation time, and sound reinforcement and masking. This credit requires minimum sound transmission class ratings between various types of interior spaces, minimizing HVAC noise and providing appropriate reverberation times depending on the uses for space.

INNOVATION

IN Credit: Innovation

LEED v4.1 allows the awarding of points for this credit for innovation (IN), use of pilot credit from the LEED Pilot Credit Library, or exemplary performance. In the context of LEED, innovation is the achievement of measurable environmental performance for a strategy not specifically addressed by LEED. For example, a green cleaning program implemented in the operation of the building could meet this requirement because this strategy is not addressed in any of the LEED credits. The LEED Pilot Credit Library is a set of credits that are being beta tested for eventual incorporation into future versions of LEED and that project teams can use to earn IN points. Exemplary performance provides points for specifically designed credits in other categories where the project either doubles the credit requirements or achieves the next incremental percentage threshold. For example, for the MR Credit addressing construction waste management, if the project team achieves both waste diversion and waste reduction targets, an extra point can be achieved under the IN credit (See Table 6-13). Note that the maximum that can be earned for this credit is five points from among the three listed options.

Category or Credit Name	New Construction	Core and Shell	Schools
Option 1: Innovation (refers to innovative approaches not otherwise addressed in LEED	1-3	1-3	1-3
Option 2: Pilot Credits (from the LEED Pilot Credit Library)	1-3	1-3	1-3
Option 3: Additional Strategies (e.g., Exemplary Performance)	1-2	1-2	1-2

Table 6-13. Points Assigned to LEED BD+C Innovation and Design

REGIONAL PRIORITY

For any area of world where there is a LEED project, additional credit can be earned depending on the priorities of the USGBC chapter in that region. These priorities are based on the existing credits from each LEED category and are selected based on input from the USGBC regional councils and chapters (see Table 6-14). Each region is assigned six RP credits for possible extra credit, and the project team can earn points for four of the six designated credits. The RP credits are available on the USGBC website (www.usgbc.org) to zip code.

Category or Credit Name	New Construction	Core and Shell	Schools
Regional Priority	4	4	4
Regional Priority Credits	4	4	4

Table 6-14 Points Assigned to LEED BD+C Regional Priority

CONCLUSIONS

The LEED building assessment system is a suite of rating products that can be used to guide the life-cycle design, construction, and operation of a high-performance green building. The current version of the LEED assessment system is version 4.1. The various LEED rating products provided a wide variety of options, one of which is applicable to almost any type of project. A successful certification process results in use of four levels of certification: platinum, gold, silver, and certified. Construction professionals who are engaged in green building design and construction can earn a variety of LEED credentials, ranging from LEED Green Associate to LEED AP with specialty to LEED Fellow.

SEVEN
THE GREEN BUILDING DESIGN

BACKGROUND

Designing a green building is different than designing a conventional building in terms of the nature of the building itself and choosing a type of delivery system. The major difference between the design of the conventional building and green building is the collaboration of design team at the onset of the project which can take weeks and maybe months for the project team to come up with a well-thought design parameters after collaboration between the team members. The outcome of such collaborative approach is called Charrette. This process is probably one of the most distinguishing hallmarks of contemporary green buildings.

TYPES OF DIFFERENT DELIVERY SYSTEMS

There are bsically four primary project delivery methods in construction industry as follow:

1. Design-Bid-Build

2. At-Risk-Construction Management
3. Design-Build, and
4. Integrated Project Delivery

It should be noted that the functions associated with construction management are required in all four methods. These four project delivery methods differ in five fundamental ways:

- The number of contracts the owner executes,
- The relationship and roles of each party to the contract,
- The point at which the contractor gets involved in the project,
- The ability to overlap design and construction,
- Who warrants the sufficiency of the plans and specifications.

Regardless of the project delivery method chosen, the three key players are the owner, the designer (architect and/or engineer), and the contractor that are always involved. However, the relationship and accountabilities change with the various project delivery methods.

Each of the project delivery systems has specific advantages and disadvantages, and it is the responsibility of the owners to assess the project delivery choices in relation to their project needs.

DESIGN-BID-BUILD

Design-bid-build is commonly referred to as the traditional method of project delivery, and the traditional accountability applies. In this scenario, the owner first hires an architect or engineer to design the building or structure.

The design professional prepares a design, moving through the three standard design phases: schematic design, design development and finally contract document. Under this arrangement, the design professional is usually selected on a qualification basis and then is typically paid a fee or a percentage of the building cost for their services.

After the plans and specifications are complete, the owner selects the general contractor who will provide the construction and construction management services. The most common means for selecting the general contractor under this method is by low price or low bid: several competing contractors estimate the project based on the contract documents, and the builder with the lowest price gets the contract. The general contractor typically subcontracts various sections of the work to specialty contractors. Under this method, the owner holds two separate contracts, one with the designer, and one with the contractor.

In this arrangement, all dealings between the designer and contractor go through the owner. There is no legal agreement between the designer and the contractor. Per Barbara J. Jackson (2004), this method is very linear in nature, and the contractor does not have any input regarding the design of the project. The contractor is only responsible for carrying out that work as spelled out in the plans and specs and will utilize the various construction management functions to accomplish this task.

Under the design-bid-build method, the owner warrants the sufficiency of the plans and specs to the contractor. If there are gaps between the plans and specs and owner's requirements or any error and omissions in the design, the owner is responsible for paying for those corrections. Usually, the contractor finds these errors and submits the change orders to correct the work. This is never a happy occasion for the owner. A constructability review can help preserve such surprises.

AT-RISK-CONSTRUCTION MANAGEMENT

The at-risk-construction management approach can be viewed from several different prospectives, and one of them is as a specific project delivery method. Under this approach, construction management services are provided to the owner independent of the construction work itself. The owner has two options to choose from as to how they want the Construction Manager (CM) to conduct the actual construction of the project.

Option 1: agency CM in which the Construction Management Association of America (CMAA) defines agency CM as a free-based service in which the construction manager is responsible exclusively to the owner and acts in the owner's best interests at every stage of the project. In this case, the construction manager offers advice uncolored by any conflicting interest because the construction manager does not perform any of the actual construction work and is not financially at risk for it.

The arrangement involves three contracts: one between the owner and the designer, one between the owner and contractor and one between the owner and the construction manager. The owner hires a designer and a contractor exactly as described under the design-bid-build scenario. In addition, the owner also contracts with a third party, a construction management firm that provides construction management functions but as an independent agent or representative of the owner.

The party responsible for the construction work is still an individual general contractor, and that contractor still carries out construction management functions relative to their internal requirements for managing the project to completion. In some unique situations, the owner may elect to bypass a general contractor and contract directly with several specialty contractors and allow CM firm to manage each of them. This is called "multiple prime contracts".

Option 2: at-risk CM (also called CM/GC): The CMAA defines the at-risk CM project delivery method as an option that entails a commitment by the construction manager to deliver project within a guaranteed maximum price (GMP). The construction manager acts as consultant to the owner in the development and design phases but does the work of a general contractor during the construction phase.

In this scenario, there are only two contracts, one between the owner and designer and one between the owner and the at-risk CM.

Although this might seem like a tricky relationship to manage, owners are comfortable with it because the at-risk CM is also responsible for performing the construction and carries financial liability for bringing the project in on time and within budget. By contrast, the agency CM model, which is not classified as a project delivery method, the construction manager is not at risk for the budget, the schedule, or the performance of the work.

Per Barbara Jackson (2020), one of the great things about construction management project delivery is that it provides a contractual avenue by which the owner can bring the CM function into play early in the project planning and make direct dialogue with the design team possible. In these circumstances, the construction manager, acting on behalf of the owner, can contribute by providing value engineering, life cycle cost analysis (which will be discussed fully in this book), conceptual estimating and scheduling, building information modeling (BIM), and what is called constructability reviews during the design process and head off possible incompatibilities between design and construction, which typically result in cost overruns, time delays, conflicts, and legal claims down the road.

As Brabara Jackson 2020, stated one of the most frustrating things that happens with traditional design-bid-build project delivery is when the estimates all come in too high. Then the project goes back to the drawing board for redesign and is significantly delayed, leaving everyone involved deeply disappointed. The way to avoid this common occurrence is by using a project delivery method that allows for early contractor involvement and encourages a team approach.

Design-Build: The design-build methodology continues to grove in popularity. It is often referred to as single-source project delivery and is the closest method in construction of high-performance buildings. In this arrangement, there is only one contract. The owner contracts with a design-build entity, which will be responsible for both the design and the construction of the project.

Whereas design-bid-build has linear sequencing of the work, design-build often integrates and overlaps design and construction and always for something called fast tracking. Since time is one of the pillars of construction success, then fast racking is a valuable commodity in construction business. Under the design-build method, the design-builder warrants the sufficiency of the plans and specs to the owner. The design-builder is liable for any gaps between the plans and specs and the owner's requirements for the performance of the building. If there are any shortfalls, the design-builder picks up the tab. One of the greatest advantages of the design-build is the possibility for early contractor involvement. Under this method, all the team players such as designers, contractors, material suppliers, and the manufacturers have an opportunity to be in continuous communication throughout the project. The collaboration between these major players is very important and guarantees the project success.

INTEGRATED PROJECT DELIVERY

Although the term integrated project delivery (IPD) is not new, it really gained prominence in 2007 when American Institute of Architects (AIA) formally began to articulate it as a unique approach to designing and constructing projects. The IPD is defined by AIA in 2007 as:

"The project delivery approach that integrates people, systems, business structures, and practices into a process that collaboratively harness the talents and insights of all participants to optimize project results, increase value of the owner, reduce waste, and maximize efficiency through all phases of design, fabrication, and construction".

IPD system originated in the mid-1990s when a group of companies in Orlando (FL) were attempting to increase the speed of productivity and project delivery without typical conflicts. The IPD Task Group is composed of a variety of stakeholders including owners, architects, contractors, engineers, lawyers, bankers, real estate agents, tenants, city government staff, etc.

Charles Thomsen (2011), suggested that the main ingredients of IPD are:

- A legal relationship,
- A management committee,
- An incentive pool,
- A no-fault working environment,
- Design assistance,
- Collaborative software,
- Green construction,
- Integrated leadership.

The IPD process is designed to produce shorter delivery times than other construction delivery systems, such as design-bid-build. The emergence of collaborative software provides opportunities to improve the flow of the documentation, communications, and work to ensure that all parties engaged in the project are working on the same set of documents and collaborating on the same end.

Another aspect of IPD is the relational contracts, which are the key documents becausethese contracts define the relationship between the key players. The relational contract is a single agreement signed by the owner, architect/engineer, and the contractor. Relational contracts in the IPD process have incentive clauses such that any potential savings are shared among the IPD team members and with the owner. An incentive pool for this purpose is created and put at risk depending on the collaboration of the team. Another component of the IPD process is adopting a Lean Construction strategy. Lean construction is a project delivery process that uses Lean methods of maximizing stakeholder value while reducing waste by emphasizing collaboration between teams on a project. The goal of Lean construction is to increase productivity, profits, and innovation in the industry. Regarding high performance green buildings, IPD is an untested idea because it is relatively new.

One of the main requirements of IPD process requires a resolution at the onset of project prepared with stakeholders such as architects, engineers, construction crew, real estate agencies, future building users and local government staff involved in the process. The outcome of such resolution is referred to as Charrette. Charrette is a French word means "cart" and is often used to describe the final and intense work effort expended by the project team to meet the project requirements and come up with the appropriate design parameters for the building especially if the building needs to be a green structure.

One of the long-standing criticisms of the green building has been its higher first cost, probably due to long and exhausting meetings for preparation of Charrette. IPD, together with these tools such as Building Information Modeling (BIM), and lean construction, provides the potential to deliver high-performance green building at a same or lower cost compared with conventional, code-compliant facilities.

WHICH DELIVERY SYSTEM SHOULD BE SELECTED FOR GREEN BUILDING?

Due to the challenging and adversarial nature and lack of collaboration between the design team and construction company, utilizing the design-bid-build delivery system is not suitable for construction of high-performance green buildings. The collaborative spirit needed for a successful high-performance green building project is difficult to develop in this adversarial climate. The design-build delivery system has significant potential to deliver green buildings because, like negotiated work, it is designed to minimize adversarial relationships and simplify transactions among the parties. Using the design-built delivery system seems reasonable but may not be the best delivery system due to lack of transparency between the design team and construction entity are virtually absent. Transparency is an important characteristic of green building projects. Despite this potential problem, several successful green building projects have been executed using the design-build delivery system.

IPD is a relatively new delivery system, but with its high emphasis on collaboration. It seems to be an approach that is highly compatible with green building delivery.

Steps in Executing the Green Building Project

Since the construction of a green structure is different than the conventional building in terms of choosing an appropriate project delivery system, the project team needs to be educated about the steps that need to be taken at the onset of the project. The project owner is the key team player who will direct the project team based on his/her desire and available funding to be used. The execution of high-performance green building has the following steps:

1. Selecting the project delivery system,
2. Selecting the project team,
3. Implementing an Integrated Design Process (IDP), the IDP is different that IPD.
4. Conducting Charrette to obtain input from project team, the owner and users, the community, and other stakeholders.
5. Executing the design process, consisting of schematic design, design development, construction documents, and documentation of green building measure for a project that is to be certified, all conducted using IDP.
6. Conducting Phase I and if it is required Phase II ESA as well as preparation SWPPP for controlling

stormwater runoff during construction phase of the project.
7. Construction of the building and finally,
8. Decommissioning and handover to owner.

SETTING PRIORITIES AND MAKING INITIAL CHANGES

When the owner decides to have a high-performance green building, depending upon the location and regional requirements, he/she must decide the priorities for the building. For example, if the building is in an area where shortage of water is a limiting factor, then the priorities should be focused on conservation of water rather than energy. In this case, the design team should seek different alternatives for saving indoor and outdoor water usage such as rainwater harvesting, using plumbing fixtures with less flow rates as well as choosing drought tolerant landscape plants and employing more efficient irrigation system and recycling of gray water for conservation of drinking water at the site. Or if the goal is saving energy, then the priorities will focus on promoting using passive energy such as adjusting the building orientation and using passive cooling, passive heating and utilizing daylights for reducing the energy usage at the site. By using less energy, the building will score more in LEED rating system if the project priority is gaining more score for creating a higher rated building LEED rate, such platinum or gold.

Another priority to be set and decisions to be made is the financial investment the owner is willing to make in a high-performance building. Green buildings normally involve systems not commonly used in conventional buildings. For example, rainwater harvesting systems, with their associated piping, pumps, and cisterns, entail additional design effort.

Many state governments are forced by law to operate within strict square footage cost guidelines. As a result, very simple, cost-effective measures must be considered. Other types of organizations may have revolving funds that can be used to invest in high- performance options that will pay back the fund over time. Harvard University, for example, has 12 million revolving funds that can be used for investing in higher-capital projects that are repaid out of savings. The federal government requires life cycle costing (LCC) analysis to be employed to justify building investment decisions, a requirement that works in favor of high-performance building decisions. Private-sector owners have considerably more leeway, and their decisions can be based on LCC, as is the case for the federal government which requires the use of LCC. Certified green buildings will have additional documentation requirements, for commissioning, fees for registration and certification review, and other costs that must be allocated in building's budget (Kibert, 2022).

ROLE OF THE LEED ACCREDITED PROFESSIONAL IN THE PROCESS

The inclusion of building industry professionals trained in the use of the green building certification system helps facilitate decision making and the flow of information required to navigate the relatively complex requirements of these systems. Both the US Green Building Council (USGBC) and Green Globes, have training programs and designations for these individuals.

The USGBC offers training and testing that, if passed successfully, designates the individual as a LEED Accredited Professional (LEED AP). This designation provides the building owners with a high degree of assurance that the requirements of the USGBC certification program will be understood, and that the extensive documentation required for the certification will be provided. The LEED AP examination tests the individual's knowledge of the green building principles as well as familiarity with the LEED requirements. The LEED AP examination covers the following points:

- In depth familiarity with the LEED building assessment system
- Understanding of the LEED project registration/technical support/certification process
- Demonstrated knowledge of the design and construction industry standards and process
- General understanding of the various designs referenced in LEED.
- Understanding of green and sustainable design strategies and practices, and corresponding credits in the LEED rating system
- Familiarity with key green and sustainable design resources and tools

One of the other benefits of having LEED AP on the project team is that one credit is awarded for the certification of the project. One drawback of the current system of awarding this credential is that it does not require in-depth knowledge of the building design and construction process, nor does it require professional experience. One challenge for the USGBC has been to create a rigorous accreditation process with requirements either for periodic recertification or for continuing education to maintain currency on green building issues and the LEED system.

In a similar fashion, Green Building Institution (GBI) offers training and testing designed to produce Green Globes professionals (GGPs) who have a detailed knowledge of Green Globes, the certification process, and documentation requirements. Unlike LEED AP, GGP must be qualified building industry professional prior to taking the accreditation examination to become a GPP.

INTEGRATED DESIGN PROCESS

As mentioned earlier, the level of interaction/communications between the project team in creating green structures is much higher than the creation of conventional buildings. Because the green building is somehow new idea, then it is essential that all members of design and construction team is required continuously to collaborate and cooperate in creating the plans and specs, so that no last-minute correction or revision is needed. This kind of cooperation and collaboration will serve three purposes: 1) it can fulfill its primary purpose of informing the project team about all project requirements, 2) it can familiarize the project team with the owner's priorities for the high-performance green building aspect of the project. 3) it can provide an opportunity to accomplish team building in the form of group exercises for familiarizing the group with the building, the building program, the building's green issues.

Integrated building design or integrated design is the name given to the high level of collaboration and teamwork that help differentiate a green building design from design process found in a conventional project. According to the US Department of Energy, integrated design is:

A process in which multiple disciplines and seemingly unrelated aspects of design are integrated in a manner that permits synergistic benefits to be realized. The goal is to achieve high performance and multiple benefits at a lower cost than the total for all the components combined. This process often includes integrating green design strategies into conventional design criteria for building form, Function, performance, and cost. A key to successful integrated building design is the participation of people from different specialties of design: general architecture, HVAC [heating, ventilation, and air conditioning], lighting and electrical, interior design, and landscape design. By working together at key points in the design process, these participants can often identify highly attractive solutions to design needs that would otherwise not be found. In an integrated design approach, the mechanical engineer will calculate energy use and cost very early in the design, informing designers of the energy-use implications of building orientation, configuration, fenestration, mechanical systems, and lighting options.

The IDP is characterized by early significant collaboration in the design process. In conventional design, the team members begin their joint effort at the start of schematic design; in a green building project employing integrated design, the collaboration starts at the very beginning of the project, and all team members have input on design decisions during the entire cycle of design. The earlier integrated design is implemented, the greater the benefits. Figure 7-1 demonstrates this idea.

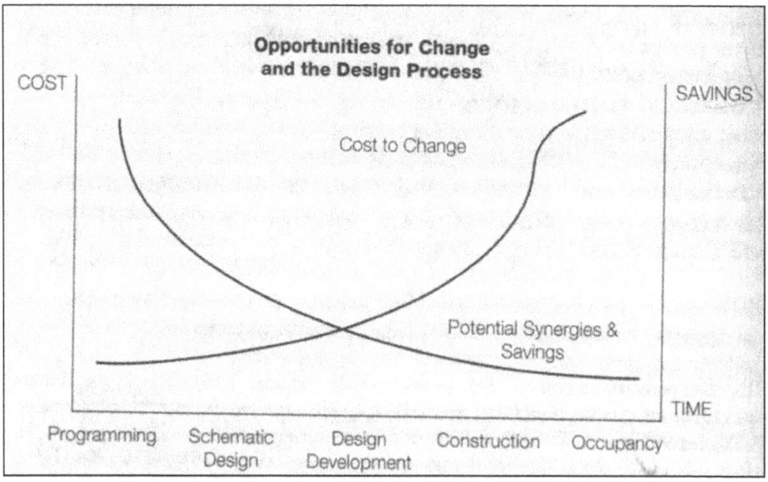

Figure 7-1 The earlier an integrated design is implemented, the greater, the potential savings and lower the cost of changes to the building design.

There are numerous potential areas for integrated design in any building project: the building envelops, the daylighting scheme, green roofs, minimization of the light pollution, indoor environmental quality, and building hydrologic cycle, to name but a few.

WHAT IS THE ROLE OF CHARRETTE IN DESIGN PROCESS

As mentioned above, in creating a green-built environment, especially at the design stage unlike the conventional structures a great deal of collaboration and brainstorming is required to come up with bullet-proof design parameters so that during the construction stage no correction or revision of the plans and spec is required. This requires a major effort by the project owner to involve a wide range of users, customers, local government, and citizens to obtain maximum input. As mentioned, the process of gathering this input is referred to as a charrette.

According to the National Charrette Institute (NCI), there are four guiding principles for a charrette. These principles are as follows:

1. Start-up. In the context of a building project, the start up for a charrette is very simple. It involves determining who the stakeholders are, engaging the stakeholders in the process, establishing the goals for the charrette, determining the time and place for holding the charrette, and notifying the participants of the details.

2. Research. education and concepts. Prior to charrette the building owner, the charrette facilitator, and the design team members should discuss the information needs for the charrette. The owner's directions, the building program, site details, utility information, and other pertinent data should be gathered and compiled for the charrette. Information on specific technologies may be useful. For example, if a fuel cell is a strong candidate technology for the project technical information about the device, issues of connecting the fuel cell to the grid, and information about fuel and emissions should be gathered for use during the process. The gathering of some of the information may need to involve some other organizations in the charrette processes. The location of the charrette is preferred to be held at Owner space equipped with good equipment for presentation and accommodation of the people involved.

3. The Charrette. Generally, the charrette should be conducted by a facilitator familiar with the green building process. A typical building charrette might take several days with different phases until complete. The first step would be the effort to educate all the participants on the owner's requirements and concept of the high-performance green building. The second step would be to review the building program, previously generated architectural schemes,

building siting, proposed construction budget, and construction schedule. The third step would be to lay out the goals of the project with respect to green high-performance aspects. The owner may desire a specific level of certification, for example a LEED gold certification that will affect many of the decisions made during the charrette (Kibert, 2016). When these steps have been completed and the project team and stakeholders understand the concept of green building, the actual charrette begins. The facilitator conducts a guided brainstorming session that draws out input from the group about every aspect of the project, with special emphasis on the sustainability of the building. During the conduct of the charrette, the team should keep a running scorecard on how the decisions made during the process are affecting the building assessment score. The economies of each decision also need to be considered, and the construction manager should ensure that enough data is available to provide a conceptual cost estimate for review by the owner.

4. Review, revise, and finalize. After the charrette is complete, the design team reviews the results with the owner, makes any appropriate adjustments and changes, and then produces a report of the charrette to guide the balance of the design process.

REFERENCES

AIA, 2007 Publication on IPD Delivery Building Delivery System

Barbara J. Jackson, 2020 “Construction Management, Jump Start, Third Edition, John Wiley and sons, Inc. 111 River Street, Hoboken, New Jersey.

Department of Energy’s View of Integrated Design Process

Kibert, J. Charles, 2022, Sustainable Construction, Green Building Design and Delivery, Fifth Edition John Wiley and Sons, Inc., New Jersey

Thomson, Charles 2007, DPR Construction. One of the First Real Collaborative Environment IPD Projects in Arizona, Great BIM and IPD Promoter.

US Department of Energy's Building Technology Program at http://energy.gov/eere/amo/toolbox-and-expertise

EIGHT
SUSTAINABLE SITES AND LANDSCAPE

BACKGROUND

Sustainable landscaping is the practice of using multiple strategies to create an environmentally friendly and climate appropriate landscape. Major goals and benefits of sustainable landscaping include water conservation, improving soil health, reducing maintenance labor and organic waste generation, carbon sequestration, and creating habitat through appropriate plant selection.

Shifting towards sustainable landscaping practices helps preserve our local ecosystems and prevent air, water, and soil pollution. In turn, this can reduce long-term costs and cut down on maintenance. Sustainable landscapes mean more than hot, dry gardens of cactus and gravel. They can incorporate beautiful flowering plants–including native plants as well as edible plants such as pomegranates, or shade trees that can cool our living environment. A well-designed sustainable yard provides immediate benefits to local communities, while also protecting the environment and providing nourishment for birds, bees, and wildlife.

ENVIRONMENTAL IMPACT OF LANDSCAPES

Urban and agriculture development results in habitat loss that directly impacts biodiversity. Removal of vegetation causes loss of protective cover, food, and disturbances in natural processes. As the construction industry grows, the building industry uses more resources and generates more waste. Historically, urban landscapes are water intensive and are chemically treated using fertilizers and pesticides, which can run off into waterways. Maintaining a typical lawn creates noise and air pollution from gas-powered equipment; small engines are among the dirtiest engines allowed in a residential neighborhood. Proper management practices will reduce resource depletion, waste generation, and pollution problems while also helping to combat a changing climate.

Role of Landscape Managers

Creating a climate appropriate landscape can be done on many scales including residential or commercial properties through thoughtful planning. Landscape contractors, site managers, and homeowners can play an active role in efficiently managing resources, reducing waste, and preventing pollution. Using sustainable landscape maintenance practices makes good business sense to help reduce costs. With forethought and planning, they can design self-sustaining landscapes that are environmentally sound and beautiful.

Recycling and Upcycling

Recycling materials from the construction, installation, or upkeep of landscape sites will reduce waste. Wood waste converts to mulch and plastic pots can be recycled into products for landscape use. Buying recycled-content landscaping products, such as plastic edging or lumber, conserves natural resources and strengthens markets for these recyclable materials. Composting clippings and organic materials on site are encouraged when available. If materials must be removed, they should be transported to local composting facilities for processing.

HOW TO LANDSCAPE SUSTAINABLY

Activities that can help support more sustainable and climate appropriate landscapes include xeriscaping, grass cycling, erosion control, water conservation, carbon sequestration, creating habitat, mulching, and composting. To learn more about these topics click on the links below:

https://calrecycle.ca.gov/organics/landscaping/

Landscape design has typically been an afterthought in the conventional building delivery system. The landscape design is given short shrift, treated apart from the building. Today, the role of landscape design in high-performance green buildings is a state of transition and is a part of the entire design and construction of the building. It is:

- Integral to the building,
- Helps in the control of stormwater,
- Treats waste and provide food,
- Contributes to biodiversity,
- Earn Credits for High Performance Green building.

The following is the definitions of important terminologies used in landscape sustainability:

Xeriscaping. Xeriscaping is landscaping with slow-growing, drought tolerant and native plants to conserve water and reduce yard trimmings. The practice of Xeriscaping will vary from region to region in California due to variation in climate throughout the state.

Grass-cycling. Grass-cycling is the natural recycling of grass by leaving clippings on the lawn when mowing. Grass clippings decompose quickly and release valuable nutrients back into the soil. This will reduce water and fertilizer usage and green waste generation as well as maintenance costs.

Create Habitat. In even the smallest space, you can create a habitat that encourages birds, pollinators, and other wildlife to thrive. Planting with California natives in your home garden can act as a "bridge," connecting wildlife to nearby wildlands.

Carbon Sequestration. Carbon occurs naturally in the environment but is also released into the atmosphere because of human activities such as tilling of soil and burning of fossil fuels. Carbon sequestration is the process of pulling carbon from the atmosphere and returning it back to the soil for storage, which can help combat a changing climate. We will discuss carbon sequestration in Chapter 12 of this book.

Erosion Control. Using compost-based erosion control products is the best management practice (BMP), as compost bonds tightly with soil, leaving no gap between soil and cover, which means less opportunity for water to run underneath and undermine the protection. For more information refer to Chapter 2.

Mulching. Application of compost and mulch enhances soils by adding organic matter. Mulch will insulate plant roots, reduce weeds, minimize water loss, and control erosion, dust, and mud problems. Decomposition of mulch helps condition the soil and adds nutrients. Adding organic matter increases the water-holding capacity of soil, reducing erosion, and conserving water. Apply 3 to 4 inches of mulch on top of soil surrounding your plants. Keep mulch at least 1 foot away from tree trunks. Replace periodically as organic mulch decomposes over time.

Composting. Many California soils are low in organic matter and high in clay; compost improves the texture of these soils so water can infiltrate more easily, reducing runoff and saving that water for plants. You can make compost at home or purchase it. Mix 1 to 2 inches of compost into clay soils and 2 to 3 inches to sandy soils, to a depth of at least 6 inches. Follow the above information for applying compost to a landscape garden and for applying compost on a lawn.

Water Conservation. Effective and sustainable landscape designs reduce water usage and cost. This can be achieved in many ways such as selecting water efficient plants, irrigation systems, or applying compost and mulch.

Farmland. Land on which crops can be produced for the least cost and with the least damage to the resource base. Prime farmland has an adequate and dependable supply of moisture (from precipitation or irrigation), nutrients and with an appropriate length of growing season (temperature). Every single minute, America loses 2 acres of farmland from 1992 to 1997 (more than 6 million acres of agricultural land has been lost). From 1992 to 1997, the prime farmland conversion rate was 30 percent faster than nonprime farmland. Food production is impacted by prime farmland conversion. US population grew by 17 percent from 1982 to 1997. Every state is losing some of their best farmlands to urban development.

Greenfields. Greenfield are the properties that have experienced little or no impact from human development activities. They can also include agricultural lands that had no activity other than farming. Industries formerly fleeing to greenfield outside urban areas are returning to former industrial sites because the economy dictates the return to the city, bringing back jobs into the city.

Land Recycling. LEED encourages reusing land that has already been impacted by human activities, rather than using greenfield.

Brownfields. Can be defined as abandoned, idled, or underused industrial and commercial facilities where expansion and redevelopment is complicated by environmental contamination (US EPA), and need for remediation for redevelopment purposes. In many US cities, brownfields are now valuable real estate properties because of their proximity to infrastructure and a potential workforce.

Grayfields. Grayfields can be defined as blighted or obsolete buildings sitting on land that is not necessarily contaminated. Boarded-up housing can be an indication of a grainfield. The Congress for the New Urbanism points out that former or declining malls can be classified as grayfields because they occupy impacted land that can be returned to productive use. Larger abandoned big stores are now referred to as "Ghost boxes".

Some of the strategies used to deal with Ghost boxes are as follows:

- Adaptive reuse, Turning ghost boxes into office space, entertainment space, etc.
- Demalling, reversing storefronts to face the street and making connection with nearby housing, etc.
- Razing and reusing older malls demolished to make room for new retail developments.
- Passing community ordinances to prevent future

grayfields and ghost boxes, example: setting maximum size for big-box stores.

Both grayfields and brownfields are becoming valuable properties. Grayfields are not explicitly addressed in either LEED-NC or Green Globes.

Blackfields. Blackfields are abandoned coal mines. They are found in coal-mining areas such as eastern Pennsylvania. They can be considered as an expanded definition of brownfields.

Sustainable Landscape Design:

Land use and landscape design are closely coupled and offer perhaps the greatest opportunity for innovation in the application of resources needed to create the built environment. Buildings, while altering the local ecosystem, can contribute to the to the ecosystem and function synergistically with nature. Carefully designed and executed work by architects, landscape architects, civil engineers, and construction managers is required to produce a building that does the following (Kibert 2022):

- Optimize the use of the site,
- Is highly integrated with the local ecosystem,
- Carefully considers the site's geology, topography, solar insolation, hydrology, and wind patterns,
- Minimizes impacts during construction and operation,
- Employs landscaping as a powerful adjunct to its technical system.

Other members of the project team must also have a voice in the decisions made about land. Buildings can become a contributory part of the ecosystem and function synergistically with nature. The location of the site, the type and color of exterior finishes, and the materials all affect the thermal load landscaping can be a powerful adjunction with technical systems.

The participation of key members is essential to make optimal use of the site and landscape. Contemporary green building approaches call for reusing of the land, its cleanup, and increasing density and at the same time for minimization of the need for greenfield use. The landscape design must integrate ecology and nature within the built environment. Appropriate use of land is a major issue in green building construction followed by minimizing the construction footprint, the heat-island effect, etc.

Buildings require multiple categories of resources for their creation and operation such as energy, materials, water, and land. The following are some approaches of land use to fit with the concept of high-performance green building:

1. Building on land that has been previously utilized,
2. Protecting and preserving wetlands
3. Using native, adapted, and drought-tolerant plants,
4. Developing brownfields and grayfields
5. Reusing existing buildings
6. Protecting key natural features
7. Minimizing the impacts of construction on the site
8. Minimizing earth moving and compaction of soil during construction
9. Fully using the sun, prevailing winds, and passive solar energy
10. Maintaining as much as possible natural hydroperiod of the site
11. Using alternative stormwater management

technologies such as green roofs, pervious pavement, bioretention, etc.
12. Minimizing heat-island effect
13. Using natural wetlands to the maximum extent possible in the stormwater management scheme.

The selection of a building site is generally the purview of the building owner, but often may be affected by input from members of the project team. One of the most important green measures in sitting a new building is to locate it where the need for automobiles is minimized, while conserving open spaces and amenities. The urban locations next to mass transit are the most desirable building sites.

The five characteristics of a sustainable landscape are as follows:

1. Sustainable landscape uses primarily renewable energy at rates that can be regenerated without ecological destabilization,
2. Sustainable landscapes maximize the recycling of resources, nutrients, and by-products, and produce minimum waste,

3. Sustainable landscapes maintain local structure and function and do not reduce the diversity of ecosystems,
4. Sustainable landscapes preserve and serve local human communities,
5. Sustainable landscapes incorporate technologies that support these goals.

J. William Thompson and Kim Sorvig (2018) provided a set of principles to guide landscape design and construction for green buildings. These principles are outlined below. In general, they are straightforward and parallel to logic of LEED, which addresses these issues (Kibert, 2022):

1. Keep healthy sites healthy. Ensure that biologically productive sites with healthy ecosystems are not harmed by the building project. Special attention must be paid to utility installation and road construction, which can be especially destructive to natural systems.
2. Heal injured soil and sites. Using grayfields, brownfields, or blackfields reduces pressure on biologically productive sites and can result in restoration of blighted properties to productive ecosystems.
3. Favor living, flexible materials. Slope erosion can be controlled with living structures rather than artificial physical structures. For instance, living materials on roofs create eco-roofs that provide additional green area and assist heating and cooling of the building.
4. Respect the waters of life. Water bodies, including wetlands, should be protected and even restored. Rainwater can be harvested from roofs, stored in cisterns, and used for nonportable applications.

Landscape irrigation should be minimized, and landscape designed to be durable and drought resistant.

5. Pave less. Paving destroys natural systems and should be minimized. Stormwater should be quickly infiltrated by use of porous concrete and asphalt paving. Heat- islands should be minimized by appropriate landscaping.
6. Consider the origin and fate of materials. Minimize the impact of landscape materials by carefully analyzing their embodied energy and other effects. Emphasize reused and recycled materials and avoid toxic materials.
7. Know the cost of energy over time. Landscape construction requires considerable energy in the form of work by machinery, embodied energy materials. The total energy consumption for all purposes, including maintenance, should be minimized.
8. Celebrate light, respect darkness. Landscape lighting should be accomplished such that plants are unaffected by lighting schemes, and lighting should be energy efficient. Lighting should not spill over to areas where it is not wanted. Low-voltage lighting, fiberoptic lighting, and solar lighting should be considered.
9. Quietly defend silence. Minimize the impact of noise pollutants by integrating special design elements.
10. Maintain to sustain. Keep account in design and follow practices for proper and regular maintenance of spaces and on-site resources such as machinery.
11. Demonstrate performance, learn from mistakes.

Some of the innovations emerging from today's high-performance green buildings include application of land-scaping directly to buildings in the form of **Green Roofs, or Living or Eco-Roofs**, and use of **Vertical Landscaping,** especially for skyscrapers. These to emerging land scape concepts are described in the next sections.

GREEN, OR LIVING ROOFS

A green roof provides a rainwater buffer, purifies the air, reduces the ambient temperature, regulates the indoor temperature, saves energy, and encourages biodiversity in the city. Green roofs are part of climate-proof construction. What's more, people are happier in a green environment than in grey surroundings.

green roof system is an addition made to the roof of an existing building for growing flora. Depending on the type of green roof you install, the plants may be modular or have drainage layers. However, all green roofs include a few important features, such as waterproofing and root repel-lent, to keep the structure safe and undamaged.

There are two primary forms of green roofing: intensive and extensive. These are differentiated by the amount of vegeta-tion utilized. While extensive green roofs can support up to 25 pounds of vegetation per square foot, intensive green roofs can hold 150 pounds per square foot. Another factor that differentiates these styles from one another is the amount of labor required to maintain them. Intensive green roofs, as the name suggests, take a lot more work to manage, as they involve keeping the various types of plants separate and maintaining an actual garden. Extensive green roofs, on the other hand, are left to grow naturally and without restriction, taking care of themselves as they would in the wild, and thus only requiring yearly weeding and fertilization.

For private homeowners, the intensive format allows you to pick and choose which flowers you would like represented, enabling you to sculpt the aesthetic you desire. Extensive green roofs are designed only to be entered for their yearly maintenance, so they become more naturally overgrown than their intensive counterpart. This makes extensive green roofs harder to navigate, meaning individuals can't walk through the space to enjoy the flora.

Both green roofs utilize a layering system, which establishes a base for the planters using various soils, mats, and other materials to retain the nutrients while forcing out any waste by-products. Through these layers, a soil profile is mimicked, thus creating a drainage process that allows liquid to be filtered through safely while also nourishing the plants.

Recent studies and advancements have resulted in a new form of green roofing, known as comprehensive roofing. This format is made to support a wide variety of plants, which is commonly found in intensive roofing, while allowing for a greater amount of vegetation to be supported, like extensive roofing.

Figure 8-1 demonstrates the vertical cross-section of the green roof including the multiple layers of protection to prevent potential water damage and penetration of rooting system into the roof during its operation.

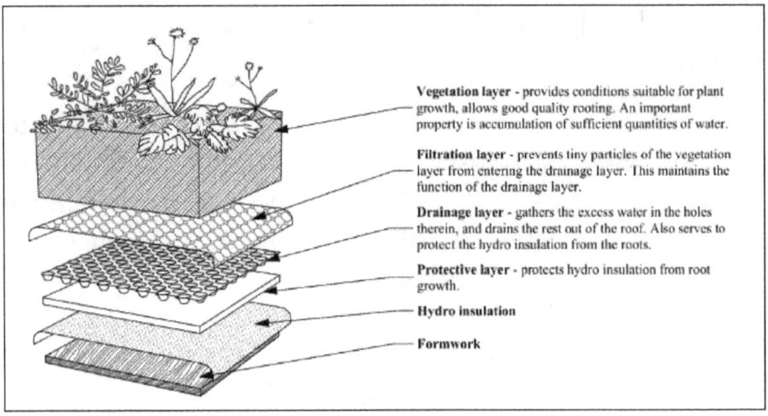

Figure 8-1 Schematic Diagram of Green Roof Structure

BENEFITS OF GREEN ROOFING

There are a variety of reasons why an individual would want to pursue green roofing, both for the betterment of the community and the home. The following is the description of these benefits:

1. Community Benefits. The installation of a green roof reduces the need to manage any stormwater that accumulates, as well as the stress put on local sewer systems. In urban settings, the inclusion of green roofs reduces the overall heat conducted by the buildings, covering many of the surfaces that would generate the most warmth. The plants prevent the distribution of smog and dust while also catching many of the pollutants found in the air. Green roofs can be public spaces, such as gardens or recreational areas, to be enjoyed by the community.
2. Due to the amount of work and money that goes into the creation and installation of green roofs,

the use of them in a neighborhood can result in an increase of jobs and revenue for local businesses.
3. Personal Benefits. There are also personal benefits to a homeowner, whether you live in an urban or rural setting. These benefits include creating a natural barrier, green roofs can reduce the amount of noise that comes into your home. In the other hand since the roof is covered it enables the roof membrane to last much longer than if unprotected.
4. A green roof reduces the amount of energy your house needs for heating, saving you money.
5. The fire retardation provided by green roofs help protect your home.
6. By adding a green roof to your house, you'll increase your property value due to the enhanced efficiency of your appliances and your home becoming more aesthetically pleasing.

GREEN ROOFS DISADVANTAGES

While green roofing can benefit you and the community, it is a costly process that can lead to some unforeseen expenses such as:

1. Insurance for green roofs is important, due to the potential of droughts or unforeseen storms, but also is expensive.
2. The weight of the green roof may result in liability issues if the pressure of it causes sagging.
3. Any damage the plants might do to the building, such as the roots growing into the shingles, will result in costly repairs.

The Development of Green Roofs

The first green roof was developed long ago, with the earliest recorded instance being the Hanging Gardens of Babylon, created in 500 B.C. At the time, the structure was built on stonework, using tar and reeds as the lower protective layers.

The modern method of building green roofs, however, was formed in Germany during the 1960s, creating the layering system that is implemented today. While common in Europe, this practice is only now starting to gain popularity in the United States.

Modern green roofs also differ from those made in ancient times because technology has greatly advanced. The layers utilized to create today's iterations include soils, three different fabrics, and drainage plates and mats. The drainage systems also make these stand out from previous versions, utilizing the soils' natural tendencies to guide the water in ways that help better maintain the system.

Design Considerations

When your green roof is being installed, three key points should be closely considered to ensure your product lasts many years without issue or extensive maintenance costs.

1. The first of these is wind uplift, which can cause severe damage to both the green roof and the structure of your house. If there is too much air pressure caught underneath the green roof when the wind blows overhead, it may cause a pressure shift, pulling the greenery off the roof. To avoid this happening RP14 Wind Design Standard for Vegetable Roofing Systems was created to guide and protect buyers from this issue.

2. While the green roofs provide fire retardation for your building, there is still a risk of the greenery catching fire due to natural events in your area. This potential disaster can be minimized through the implementation of VF-1 Fire Design Standard for Vegetative Roofs, a document made to guide the "prudent design and mandatory maintenance regiment" of each unit.
3. Lastly, one of the biggest issues green roof owners faces is the harm caused by the roots of the plants burrowing into the roof of the house, tearing up the material and compromising the overall structure. The VR-1 Procedure for Investigating Resistance to Root Prevention presents guidelines on how to safely check the efficacy of barriers protecting your home and how they can be repaired or improved. This is prominently for shorter-term effects and should not be relied on when studying the long-term effects of chemicals or barriers.

Scientists and engineers are still researching the best methods for green roofing. This is primarily focusing on how green roofs might be built in different climates and environments, as well as the scale in which the green roof can be made.

LEED CREDITS

For those unfamiliar with LEED processes, it is a program dedicated to rewarding those who build and/or maintain highly green structures. Receiving certificates from LEED will financially assist with projects that fall under their definitions and regulations of what is environmentally positive.

VERTICAL LANDSCAPING

The French designer Patrick Blanc is generally considered to have developed the notion of a **vertical garden or green wall** that takes advantage of vertical surfaces to provide buildings with some degree of ecological capacity. Skyscrapers are not normally thought of as candidates for landscaping. However, Ken Yeang, Malaysian architect has been advocating what he calls vertical landscaping to at least in part render these very large structures green. He also advocates vertical landscaping for reducing energy consumption, stating that a 10 percent increase in vegetated areas can produce annual cooling load savings of 8 percent. He described vertical landscaping as "greening the skyscraper" (Yeang 1996), which he said involves introducing plants and ecosystem components at a high level, in addition to the ground-level landscaping. Figure 8-2 shows a typical vertical landscape.

The vertical landscape creates a microclimate at the façade on each floor:

1. Can be used as a windbreak
2. Absorbs CO2 and generates O2.
3. Improves well-being of occupants.

Figure 8-2 A typical view of the vertical landscape

LOW IMPACT DEVELOPMENT

Low Impact development (LID) is a relatively new strategy that integrates ecological system with landscape design to manage stormwater runoff effectively. LID techniques minimize runoff to prevent pollutants from adversely affecting water quality and can decrease the required size of traditional retention and detention basins, resulting in cost savings over conventional stormwater control mechanisms. LID can be applied to a range of land uses, from high-density ultra-urban settings to low density development (Kibert, 2022).

By definition, the term LID refers to systems and practices of land development that use natural processes to infiltrate, evapotranspiration, or reuse stormwater runoff on the site where it is generated. LID employs principles such as preserving and recreating natural landscape features, minimizing impervious-ness creating functional and appealing site drainage, and treating stormwater as a resource rather than as a waste product. Many practices can be used to implement LID, including bioretention facilities, rain gardens, vegetated roofs, rain barrels, and permeable pavements.

By implementing LID principles and practices, water can be managed in a way that reduces the impact of built areas and promotes the natural movement of water within the system or watershed. Applied on an urban scale, LID can maintain or restore the watershed's hydrologic and ecological functions.

The Nature Conservancy described six principles for successful LID. These principles should be used in designing LID:

1. Use existing and valuable features. Identify and work with all cultural and natural features that will immediately add value to the development: hedge rows, mature trees, wildlife habitats, streams, rural architectural character, and heritage features.
2. Let natural resources work for the project. Use natural drainage by mimicking the existing systems and patterns. Minimizing construction disturbances and changes within the watershed will benefit the environment and reduce costs.
3. Increase the value of the site with open spaces. Clustering homes or buildings in the development prevents the provision of open spaces and scenic views. Connection to open spaces within a larger city network can provide natural amenity, resulting in higher property values.
4. Reduce the size of water management needs on the site. By limiting impervious surfaces, the amount of stormwater infrastructures can be greatly reduced. Buildings with smaller footprints, green roofs, permeable pavement, and

narrow roads make stormwater management easier.

5. Treat stormwater close to the source. Instead of expensive underground infrastructure, catch basins, piping, and stormwater ponds, use low-cost low-maintenance, low-tech, nonstructural rain gardens and bio-swales to infiltrate runoff.
6. Smart landscaping can save money. There is no question that good landscaping increases property values and that smart landscaping can also save money. Money is wasted on techniques, such as clear-cutting, gardening, and costly stormwater ponds that only address one problem. With a multifunctional landscape, it is impossible to manage runoff, improve water quality, reduce power bills, increase property value, and save money.

The Nature Conservancy also described 10 implementation measures that can help manage runoff while at the same time providing a landscape with natural amenity. These 10 techniques are not meaningful individually but can be highly effective as part of a larger LID strategy.

1. Impervious surface reduction. Reducing impervious surfaces include reduction of number of parking spaces, sharing parking with adjacent neighbors, reducing the length of driveways.
2. Tree preservation. Not only do trees increase the property value, but they are also excellent landscape features for the uptake of stormwater. Trees act as mini reservoirs that absorb and store large quantities of water. They are excellent for controlling runoff at the sources, reducing soil erosion, decreasing temperatures, absorbing

carbon dioxide, and providing habitat for wildlife. A 12-inch-caliper oak tree can intercept roughly 2,000 gallons of rain per year while a 30-inch-caliper maple can intercept as much as 12,000 gallons per year.

3. Reduce lawn area/increase planted areas. Lawns require a lot of watering, mowing, aerating, and chemicals and are not effective at absorbing water. Native grasses, shrubs, trees, and wildflowers are excellent species for absorbing stormwater. As is the case with trees, increasing the size of planting areas can result in higher property values while at the same time enhancing biodiversity.
4. Bioswales and vegetated swales. Bioswales assist in capturing rainwater runoff and then filter the runoff through prepared soil medium with suitable plants. They are ideal for medium strips and parking lots along streets. Prepared depth of bioswale is about 6 inches and with an overall size of 25 x 50 feet.
5. Permeable pavements. A wide range of permeable pavements are available, including interlocking pavements, grass pavements, porous asphalt, and porous concrete. These materials allow water to permeate the surface to underlying water-bearing zones or aquifers.
6. Buffers and filter strips. Buffers and filter strips are barriers between surfaces such as roads and parking lots, and waterways and sensitive aquatic environments. These buffers contain trees, bushes, wild grasses, and other natural plant species to remove particulates and other pollutants from stormwater crossing between the paved surfaces and sensitive bodies of water. Buffers can be linked to create a network of green infrastructure and

provide opportunities and benefits to wildlife corridors. These buffers are also referred to as **conservation easements** and like many other components of LID, can contribute to the present value of a property.

7. Rain gardens. A rain garden is a shallow depression planted with suitable trees, shrubs, flowers, and other species to capture stormwater from impervious areas. They can be used as a buffer to capture runoff from landscaped areas before it enters a lake, pond, or river.
8. Soil quality management. Surficial soils can create standing water if they contain fine-grained materials such as clay and silt. To prevent soils being overly compacted, driving on a wet soils beyond the parking area and over tree roots should be prevented.
9. Green roofs. Green roofs can absorb rainwater provide insulation, create wildlife habitat, and reduce the heat island effect.
10. Rain barrels and rainwater harvesting systems. A rainwater collection system can help capture and store stormwater from the roof for future use, reducing stormwater flows and decreasing drinking water dependency and costs.

LID saves money. Figures 8-3 and 8-4 illustrate a LID project in CSU East Bay Campus and City of San Jose, California.

Figure 8-3 Bioswale to Promote Rainwater Infiltration at CSU East Bay Campus

Figure 8-4 Rainwater Infiltration Gallery along the Street San Jose, California

HEAT ISLAND MITIGATION

Urban areas usually experience the urban heat island effect, that is, they are significantly warmer than surrounding rural areas. The temperature difference is usually larger at night than during the daytime.

The heat islands are formed since the urban surfaces such as roadways and rooftops absorb and emit heat to a greater extent than most natural surfaces. On a warm day with a temperature of 91°F, conventional roofing materials may reach as high as 60°F warmer than air temperatures. The heat islands raise demand for electricity in summer. Companies that supply electricity typically rely on fossil fuel power plants to meet much of this demand, which in turn leads to an increase in air pollution and greenhouse gas emissions.

Due to the heat island effect, people who live in cities are more at risk than those in suburban or rural areas. What's more, some areas within a city are often hotter than others. These neighborhood-level hotspots are called "intra-urban" heat islands.

Extreme heat events often affect our most vulnerable population first. Trees, green roofs, and vegetation can help reduce urban heat island effect by shading building surfaces, deflecting radiation from the sun, and releasing moisture into the atmosphere. Dark surfaces/roofs absorb heat more than any other colors.

The EPA launched the Urban Heat Island Pilot Project in 1998 to quantify the potential benefits of reducing heat island. For the City of Sacramento, California, the Lawrence Berkeley National Laboratory study showed:

- Citywide energy bill reduction of $26.1 million/year, assuming the high penetration of reduction measures,
- Savings of 468 million watts of peak power and reducing 92,000 tons of carbon annually,

- An improvement in air quality caused by a decrease in ozone concentration of 10 parts per billion,
- Cooling-energy savings of 46 percent and peak power savings of 20 percent by increasing roof albedo, or reflectivity, on two school buildings.

The LEED rating system gives points for mitigating heat islands in the Sustainable Sites category and also LEED gives credit for creating shades or reducing heat islands for site's impervious surfaces, such as parking lots, walkways, and plazas credits also given for providing high-albedo or vegetated roof, Kibert (2022).

The heat island effects can be reduced by several measures such as:

1. Installing highly reflective (or high albedo) roofs.
2. Planting shade trees near homes and buildings
3. Using light-colored construction materials where possible

Figure 8-5 shows the air temperature increases in and around the buildings in summertime by the heat island phenomenon at downtown area.

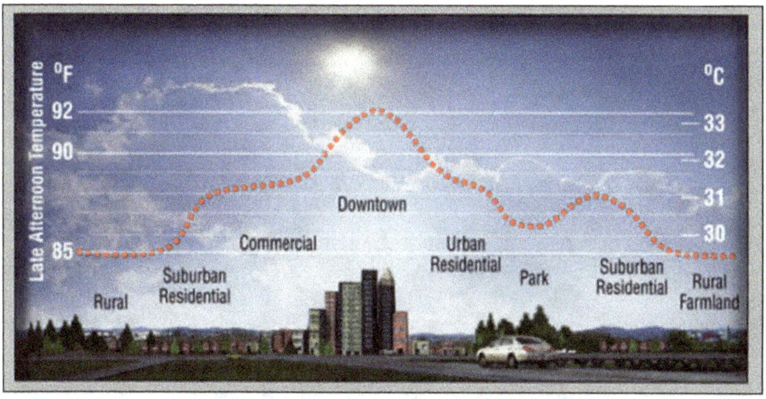

Figure 8-5 Heat Island effect on Air Temperature (Source: Courtesy of Bulge Celik)

LIGHT POLLUTION REDUCTION

Light pollution is a human-made alteration of outdoor light levels from those occurring naturally. When we over-light, fail to use timers and sensors, or use the wrong color of light, we can negatively affect many parts of our world, including migratory birds, pollinators, sea turtles, and mammals, including humans. As Figure 8-6 shows, light pollution could erase views of the stars and planets. Figure 8-6 shows the best way to adjust the light fixtures to avoid light pollution.

Figure 8-6 Right and Wrong Way to Install Light Shields

Due to light pollution, which is especially widespread in urbanized places like New Jersey, most people can't step outside and experience the true wonder of the night sky. The glare from artificial lighting is causing a glow in the sky that dims the view of all but the brightest stars and planets.

Since the beginning of human civilization, the magnificent dome of constellations and planets has mesmerized and inspired. But today, 99 percent of Americans live under light polluted skies with limited views of the night sky. Many people will live their entire lives without ever seeing the Milky Way! And, according to a new report in the Journal of Science, light pollution is doubling every eight years.

"In some areas of the world where it's still not lost, the Milky Way is so dense and so bright, you can see your own shadow cast on the ground," says Bakos in the film's opening scene. "You appreciate that you live in a galaxy – you see the shape of it, you see the dark lanes in it, you see the spiral arms projected in front of each other." The impacts of light pollution go beyond erasing views of the night sky. A growing body of research suggests that the loss of dark skies can impact on human health and the rhythms of the natural world.

For nocturnal animals, light pollution can disrupt feeding and mating patterns. It can also have harmful impacts on migrating birds, sea turtle hatchlings, and insects. For humans, excessive exposure to artificial light at night – especially blue light – has been linked to increased risks for depression, sleep disorders, obesity, and diabetes. Polluting outdoor lighting also wastes money and energy and contributes to climate change. What can be done to restore dark skies and prevent the spread of light pollution? The solutions are relatively simple: make lights dimmer, shield lights so they only shine downward, and use warm-colored light bulbs, see Figure 8-6. Shield outdoor lighting around your home or at least angle it downward to reduce light trespass beyond your property. Turn on outdoor lights only when and where needed and use motion detectors and timers.

Being able to peer into the cosmos with the naked eye and observe all its magnificence and splendor can help us all reflect on this life of ours on this "pale blue dot" called Earth. Future generations will thank us for this important movement to protect our dark night skies.

To learn more about dark skies and ways to protect them, visit the International Dark Sky Association website at https://idsw.darksky.org/. To view Dark Sacred Night, go to https://www.youtube.com/watch?v=FW0WZX75Nmo&ab_channel=OfficeofSustainabilityatPrincetonUniversity.

REFERENCES

Kibert, J., Charles, 2022 Sustainable Construction, Green Building Design and Delivery. Fifth Edition John Wiley and Sons Inc., New Jersey

Ken Yeang, 1996. The Skyscraper Bioclimatically Considered: A Design Primer. London: Academy Editors.

Dark Sky Association website at https://idsw.darksky.org/.

Dark Sacred Night, go to: https://www.youtube.com/watch?v=FW0WZX75Nmo&ab_channel=OfficeofSustainabilityatPrincetonUniversity.

William Thompson and Kim Sorvig, 2018 "Sustainable Landscape Construction, Guide to Green Building Outdoors.

NINE
BUILDING ENERGY EFFICIENCY

BACKGROUND

About 40 percent of electricity in the United States is consumed by buildings operations. About the same amount of energy in US is also used for transportation purposes. In a greater extent, for the energy production in US, we are currently dependent on using fossil fuels. During the conversion of fossil fuel to electricity a great deal of carbon in the form of carbon dioxide is being released to the environment which is believed to be a number one cause of global warming problem that our future generation is facing with. Therefore, of all challenges facing the development of high-performance green buildings, reducing the energy and carbon footprint of the built environment is perhaps the most important and the most daunting. This Chapter discusses the most important issues for reducing the energy consumption in our built environment which will help in reducing the emission of greenhouse gases to the environment.

The world's crude oil reservoir is limited, and as it is getting depleted due to its non-renewable nature. As the oil reservoir gets depleted the cost of its extraction sharply increases. For instance, in 1970, the profit margin for oil companies was \$25: \$1, in another word, for every dollar spent, the profit was \$25. Currently, this ratio has been reduced to \$7:\$1. If this trend continues the profit margin falls less than \$3:\$1, the extraction of oil will no longer be economically feasible. As Figure 9-1 shows, the point at which the cost of oil production is almost equal to its value, therefore, it will be no incentive for oil companies to be involved in oil extraction. This point is called oil rollover point.

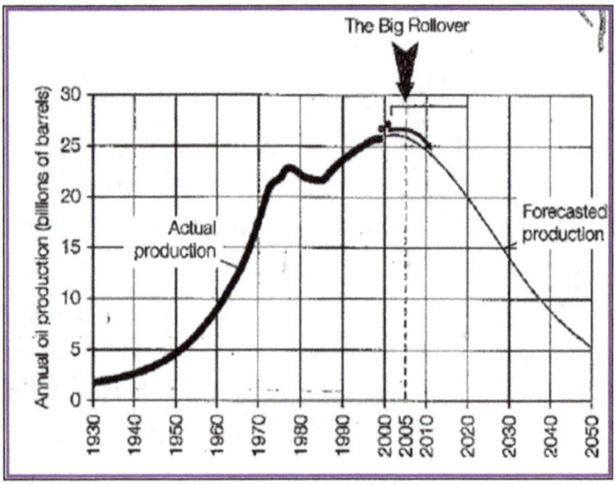

Figure 9-1 Oil Rollover Point

According to Figure 9-1, the rollover point was supposed to occur in or around 2010, however, due to the advent of fracking, the occurrence of rollover point has been delayed. We are not sure when the oil rollover will occur at this time. On the positive side, this means that renewable energies will become economically superior alternatives. On the negative side, if we wait for that day to come to make the switch, the cost of switching will impose massive economic and social hardships globally on mankind.

The green building movement and building design crew's efforts to improve building energy performance are working together to reduce the building energy requirements. As such, the energy requirements of the green buildings have been reduced from 300 KWh/m^2per year to nearly 50 KWH/m^2 per year. Today, the green buildings advocates are attempting to create carbon-neutral buildings by utilizing and focusing on integrating renewable energy generation systems into building and determining offsets for the embodied carbon of their constituent materials and products. Advancing the use of the passive heating, cooling and ventilation along with use of solar energy, ground coupling, radiant cooling, and other radical approaches may indeed enable buildings to generate at least as much energy as they consume and perhaps even export energy to offset their embodied carbon.

Per Kibert, 2022 US energy demand is staggering, with Americans consuming 100.2 quads of primary energy in 2020. One quad alone is an enormous unit of energy, equaling one quadrillion or 10^{15}BTU. The US, with 5 percent of world's population, accounts for 20 percent of the world's total primary energy consumption see Figure 9-1.

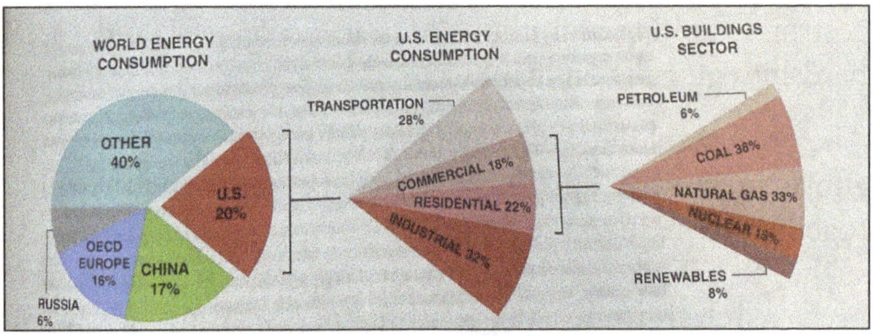

Figure 9-1 Energy Consumption Patterns Worldwide (Source: US Energy Information Administration)

Primary energy is the best measure of energy consumption because it is the energy in fuel sources such as coal, oil, and natural gas before they are converted into electricity and other forms of secondary or site energy. Total Chinese energy consumption surpassed US consumption in 2009 and reached 130 quads in 2018. It is useful to note that total US energy consumption has been flat for almost 10 years and that Chinese consumption is flattening out. In the United States, buildings consume 30 percent of generated energy or about 8 percent of global primary energy. Commercial buildings have the smallest fraction of energy at 18 percent, but this number is growing the most rapidly.

Figure 9-2, which has been adopted from US Energy Information Agency, shows how the total energy in use is generated by different fuel types. As Figure 9-2 shows, the use of non-renewable energy such as wind and solar is in their infancy but they are rising while the use of coal as the energy source is diminishing. The use of natural gas is also on the rise. The use of nuclear energy and conventional hydroelectric power stations are almost steady.

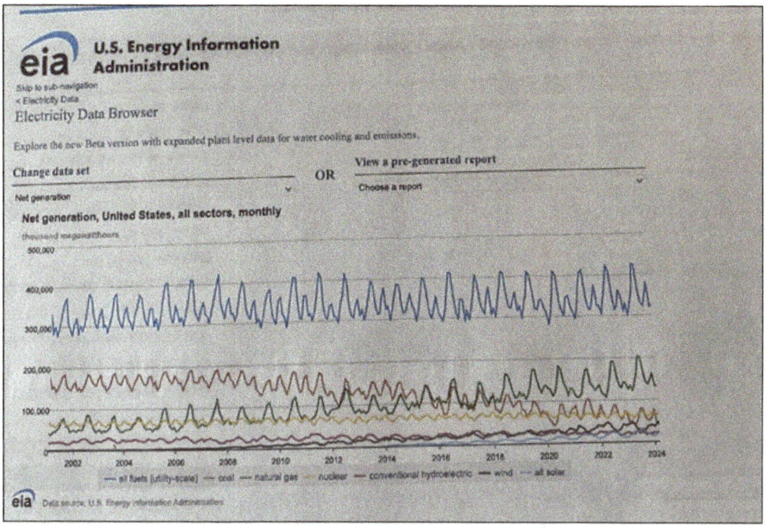

Figure 9-2 Net Energy Generation from Different Fuel Types (Source US Energy Information Administration)

Although at first glance transportation and industrial energy appear to be unrelated to building energy, they are, in fact, coupled together (Figure 9-3). The relationships between buildings and the distances between them are a major contributor to the transportation energy. In addition, a considerable amount of industrial energy is invested in building products and infrastructure materials, and it is likely that the total energy, including the embodied energy of the materials and the transportation energy attributable to buildings, is well over 60 percent of total primary energy. As such, energy consumed to support the built environment is dominated by coal at 33 percent of primary energy, but natural gas consumption is growing rapidly, and nuclear energy is a growing fraction of electrical energy generation.

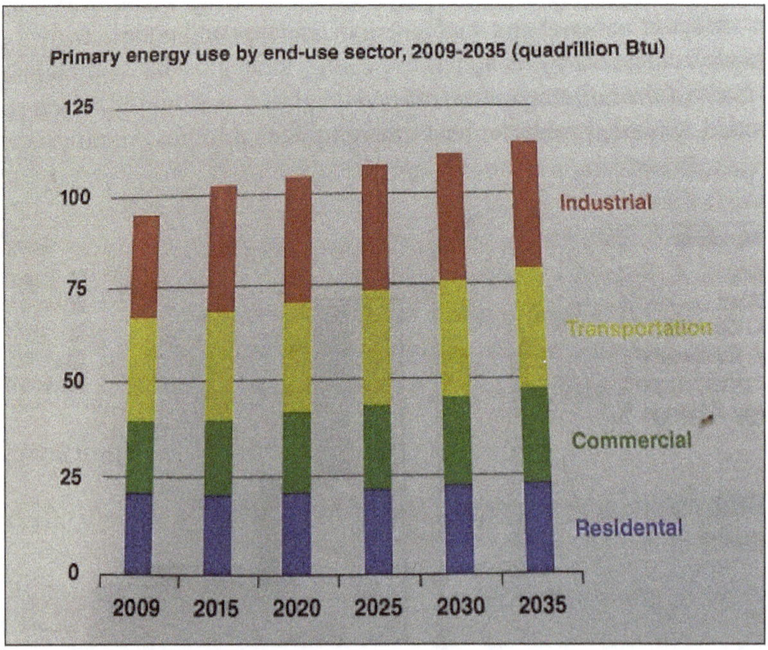

Figure 9-3 Primary Energy Use by end-use Sector, 2009-2035, in quads.

There is some good news with respect to energy, namely, energy use per capita and per unit of electric production is falling and will continue to decrease for the foreseeable future see Figure 9-4. However, the total energy consumption is still rising due to an increase in population and growing economy. Per Kibert 2022, energy prices affect economic production, and higher demand drives prices higher, thereby putting a damper on the economy. Second, most of the energy consumption is via fossil fuel

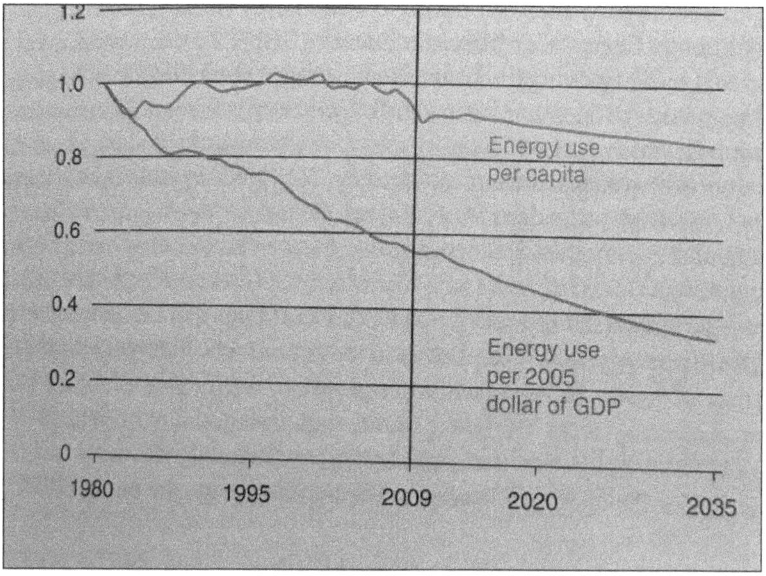

Figure 9-4 Index of energy use per capita and per dollar GDP from 1980 to 1935 (the index for 1980 is set equal to 1). Source: US Energy Information Administration

combustion, which has impact on human health. Finally, greater energy production generally means higher greenhouse gas production, contributing even more to climate change. To meet the challenges of the future per capita and per gross domestic product (GPD) energy consumption must decline far more rapidly by developing more efficient manufacturing processes, reconfiguring cities to be more compact, and designing high-performance buildings and retrofit existing buildings to consume far less energy.

In addition, most of the energy production is via fossil fuel combustion, which has human health impacts. On the other hand, energy production greatly means higher greenhouse gas emissions, contributing even more to climate change. To meet the challenges of future, per capita and per gross domestic product (GPD) energy consumption must decline far more rapidly by developing more efficient manufacturing processes, refiguring cities to be more compact, and designing high-performance buildings and retrofitting existing buildings to consume far less energy. The key factor is shifting energy production to use renewable energy sources such as solar, wind and biomass which will not generate greenhouse gases.

Building energy consumption can be reduced using better design backed up by more stringent national and state energy standards and codes. Per Kibert 2022, truly low-energy buildings are achievable by using passive energy strategies that consider the orientation and mass of building to maximize daylighting and maximize heat gain except when needed (see Figures 9-5 and 9-6). Coupled with the best emerging technologies, significant reductions in building energy use can be realized. Energy consumption in US is declining due to these very reasons. Significant additional reductions in energy consumption possible by using the most efficient building components and systems and by the US Department of Energy.

Recent programs in Germany indicate that building can designed to use far lower levels of energy that even the most ambitious US high-performance buildings. As part of 10-year demonstration program, group of 23 office buildings throughout of Germany were designed, built and monitored with a goal of using 100 kWh/m2 of primary energy per year. Primary energy is the source of energy for the energy delivered to the building (for example the energy value of coal before it is combusted to produce electricity). The efficiency of coal-fired electrical generating plants is such that only one-third of coal energy become electrical energy. Consequently, the electrical energy used in the building, referred to as the "site energy" , is multiplied by the factor of "3" to account for primary energy. A building-code compliant US office building consumes on the order of 252 kWh/m2/year of energy.

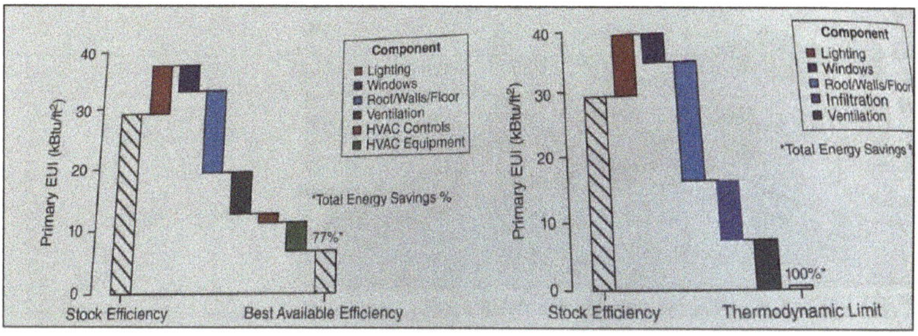

Figure 9-5 Building Heating Energy Can be Reduced by As Much as 77% Through Use of Passive Design Source: US Energy Administration

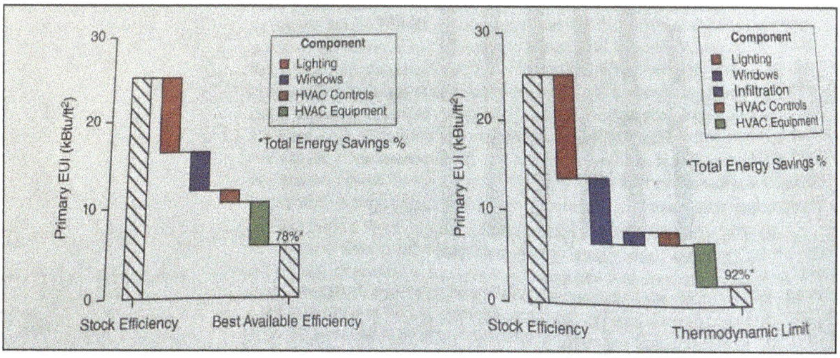

Figure 9-6 Building Cooling Energy Can be Reduced by As Much as 78% Through Use of Passive Design Source: US Energy Administration

Truly, low energy buildings are achievable by using passive solar energy depending on the orientation and mass of the building. In 2003, buildings consumed about 91,000 BTU/sf/year (287 kWh/m^2/year). Since then building consumption has been pushed lower by ever more stringent standards, the American Society of Heating Refrigeration and Air-Conditioning Engineers (ASHRAE) Standard

90.1-2019 set a ceiling on commercial building energy consumption of about 36,000 BTU/sf/year (114 kWh /m^2/year), a 60 percent reduction since 2003 (see Table 9-1).

Building Type	90.1-2016	90.1-2019	Change (%)
Office	0.79W/ft^2	0.64W/ft^2	-19
Retail	1.06W/ft^2	0.84W/ft^2	-21
School/University	0.81W/ft^2	0.72W/ft^2	-11
Hospital	1.05W/ft^2	0.96W/ft^2	-9
Manufacturing	0.90W/ft^2	0.82W/ft^2	-9
Warehouse	0.48W/ft^2	0.45W/ft^2	-6
Source: ASHRAE			

Table 9-1, Estimated Reductions in Building Energy Consumption Resulting in the Implementation of 2019 Version of ASHRAE 90.1

ENERGY DESIGN STRATEGIES IN HIGH-PERFORMANCE GREEN BUILDINGS

Since the appearance of building assessment systems about 30 years ago, a process for designing low energy buildings has emerged that can produce 100 kWh/m^2/year primary energy buildings. Listed next are the 10 steps involved in designing energy systems with low energy and low-carbon footprints:

1. Use building simulation tools throughout the design process (BIM)
2. Optimize passive solar design of the building
3. Maximize the thermal performance of the building envelope
4. Minimize internal building loads
5. Maximize daylighting and high-efficiency lighting systems

6. Design a hyper-efficient HVAC systems that minimizes energy use
7. Select high-efficiency appliances and motors
8. Maximize the use of renewable energy systems
9. Harvest and use waste energy
10. Incorporate innovative emerging strategies, such as ground coupling and radiant cooling.

The design of an energy-efficient building is a complex undertaking, and these steps cannot just be performed in sequence; they are, in effect, part of an iterative process that starts with passive design. Trade-offs inevitably must be made, often because of the client's requirements and budget. Designed properly, a building with low energy and a low-carbon footprint should provide greatly reduced operational costs for minimal or no increase in capital costs. In some cases, a well-executed passive design strategy can markedly reduce the cost of HVAC equipment due to the reduction in heating, cooling, ventilation, and lighting loads that can occur.

BUILDING ENERGY AND DAYLIGHTING SIMULATION

In designing the high-performance green buildings, the required energy simulation is a very important task using computer-generated data. Using computer modeling, the simulated energy requirements reveal how the project owner will prepare in providing the necessary power requirement and evaluate the operating costs of the building ahead of time. Such design parameters can also be submitted to the LEED authorities for further assessment if the building is supposed to be LEED-certified building. Today's simulation tools allow the integration of active and passive building systems and can easily examine the interplay and trade-offs among heating, cooling systems, walls, and roof choices, insulations, lighting windows and door, exterior and interior shading, and skylights.

Perhaps the best-known whole lighting energy simulation tool is DOE-2.2 which now has user-friendly interfaces and wizards to speed the energy simulation process (see www.doe2.com). Daylighting is a key component of an energy-efficient building and performing simulations that optimize daylighting, which is important to understand the trade-offs among fenestration, envelope thermal resistance, and energy use for artificial lighting.

Some building energy simulation tools, such as Energy-10, allow the integrated evaluation of daylighting, passive solar heating, low-energy heating, and cooling strategies, and envelope design. Daylighting also can be evaluated with sophisticated software such as Radiance (radiance.lbl.gov/radiance), developed by Lawrence Berkeley National Laboratory (LBNL). Radiance contains libraries of materials, glazing, electric lighting luminaries, and furniture to facilitate the daylighting analysis. The simulation provides a quantitative check on the intuitive guesswork of the design team about the interrelationship of the building system.

Typical software for whole-building energy simulation includes: EnergyPlus, DesignBuilder, TRNSYS, and eQuest. BIMi plug-ins such as Autodesk Ecotect are also available for energy modeling. One of the challenges of using energy modeling software is that they can have a wide range of results when comparing the model input to actual building energy performance. In some cases, the modeling is quite accurate while in others it is far off. The real value of modeling is to find the relative importance of the changes to the building's envelope and energy systems; providing an accurate prediction of building energy performance is less important. The modeling of plug loads (computers, printers, fax machines, copiers, appliances) is notoriously inaccurate because the behavior of building users is unpredictable. Actual plug loads often are substantially higher than those simulated in the energy model. Additionally, with the continual addition of new electricity powered devices in office buildings, plug loads tend to increase over time. The issues of plug loads and techniques to reducing them will be addressed later.

PASSIVE DESIGN STRATEGIES

Passive Design is the design of the building's heating, cooling, lighting, and ventilation systems, relying on sunlight, wind, vegetation, and other naturally occurring resources on the building site. Passive design includes the use of all possible measures to reduce energy consumption prior to the consideration of any external energy source other than the sun and wind. Thus, it defines the energy character of the building prior to the consideration of active or powered systems (chillers, boilers, air handlers, pumps, and other powered equipment). Randy Croxton, is the Managing Director at Stancold Plc, one of the pioneers of contemporary ecological design, describes a good passive design as one that allows a building to "default to nature". A building that has been well designed in a passive sense could be disconnected from its active energy sources and still reasonably functional because daylighting, adequate passive heating and cooling, and ventilation are provided by the chimney effect, cross-ventilation, operable windows, and prevailing winds. A successful passive design scheme creates a truly climate-responsive, energy-conserving building with a wide range of benefits (Kibert, 2022).

Passive design has two major aspects: (1) the use of building's location and site to reduce the building's energy profile and (2) the design of building itself, its orientation, aspect ratio, massing, fenestration, ventilation paths, and other measures. Passive design is complex, as it depends on many factors, including latitude, altitude, solar insolation (incoming solar radiation), heating degrees days (HDDs), and cooling degrees days (CDDs), humidity patterns, annual wind strength and direction, the presence of trees and vegetation, and the presence of other buildings. An optimized passive design can generally reduce the energy costs of heating, cooling, ventilation, and lighting.

Some of the factors that should be included in the development of a passive design strategy are listed below:

1. Local climate, sun angles and solar insolation, wind velocity and direction, air temperature, and humidity throughout the year
2. Site conditions. Terrain, vegetation, soil conditions, water-table, microclimate, relationships to other buildings
3. Building aspect ratio, the ratio of building's length to its width
4. Building orientation, long axis orientated east-west, room layout, glazing
5. Building massing. Energy storage potentials of materials, fenestration, and color
6. Building use. Occupancy schedule and use profile
7. Daylighting strategy. Fenestration, daylighting devices (light shelves, skylights, internal and external louvers)
8. Building envelope. Geometry, insulation, fenestration, door, air leakage, ventilation shading, thermal mass and color

9. Internal loads. Lighting, equipment, appliances, people
10. Ventilation strategy. Cross ventilation potential, paths for routine ventilation, chimney effect potential.

Like any concept, passive design can be applied improperly to building design. Its success is highly dependent on the wide range of factors just listed, and its application differs widely from New York to California, Colorado, or Florida. For example, using thermal mass as a passive design strategy, an excellent choice in the high desert altitudes found in New Mexico, with its abundant sunlight and wide daily temperature swings, would not be an appropriate choice in a hot, humid climate with generally narrow daily temperature differences, as would be found in Tampa, Florida. The optimum building orientation which will be discussed further in this chapter, the location and type of windows, the use of daylighting, and many other decisions must be based on a careful examination of the situation found in each locale.

SHAPE, ORIENTATION AND MASSING

Depending upon the location of the building and climatological conditions building orientation plays a major role in reducing energy demand. The classic passive design approach to orienting a building on its site is to locate the long side on a true east-west axis to minimize solar loads on the east and west surfaces, particularly during the summer. By definition the aspect ratio is the ratio of a building's length to its width, which is an indicator of the general shape of a building. The skin factor of a building is the total surface area of the surrounding walls. There is a direct relationship between the aspect ratio and skin factor. As the aspect ratio increases the skin factor also increases. Figure 9-6 shows the direct relationship between the aspect ratio and skin factor.

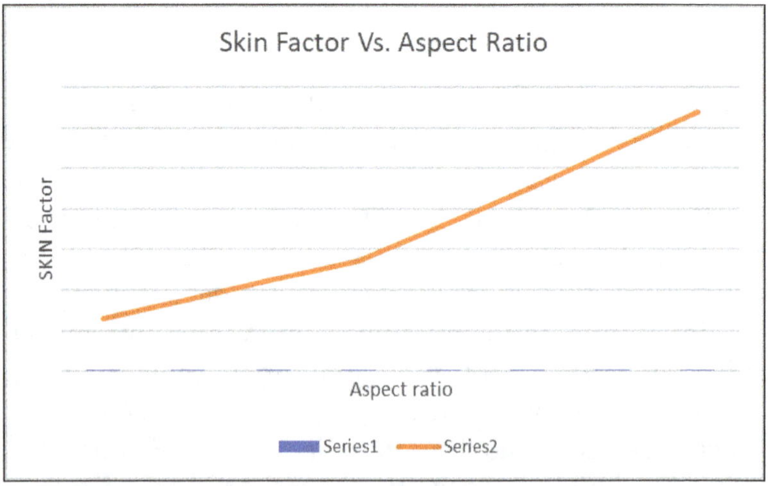

Figure 9-6 Relationship Between Skin Factor and Aspect Ratio

Passive design indicates that a building in the northern United States, where climate is cold should have an aspect ratio close to one; that is, it should be virtually square in shape. For buildings in the warmer southerly latitudes, the aspect ratio increases, with the building becoming longer and narrower. The reasoning behind this shift in aspect ratio is that a square building will have the minimum skin surface area compared to its volume. In summary the designer should consider the following:

In warmer climates:

- Design buildings with an aspect ratio of greater than one

- The higher aspect ratio, the higher skin area of building
- Place the long side on a true east – west axis
- Windows on east and west are usually minimized to mitigate potential solar loads during the morning and afternoon sunlight.

Figure 9-7 shows the orientation of the building in a warmer climate:

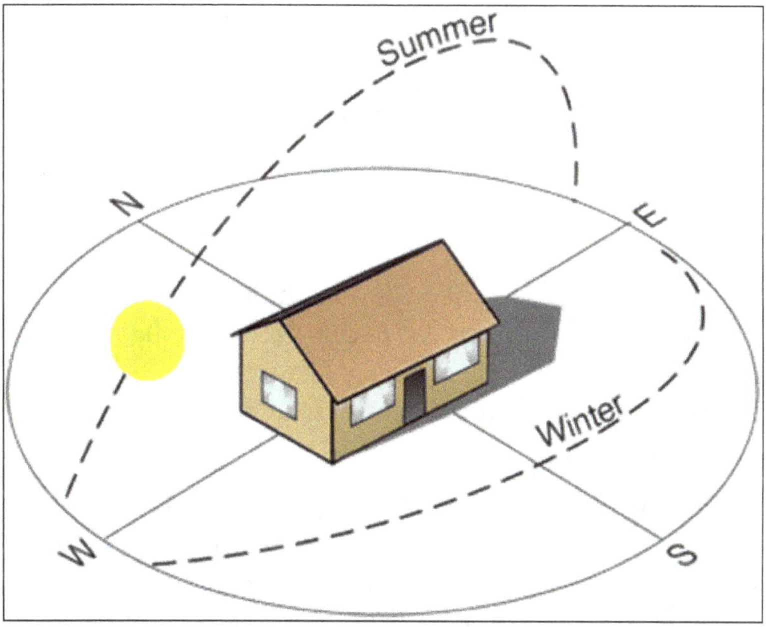

Figure 9-7 The Building Layout in Warmer Climate

In colder climate:

- The aspect ratio should be kept closer to one to lower skin area of the building. Lower skin area will minimize energy demand during winter month

- Temperature differentials for heating are generally much greater than cooling. As such, energy demand for heating is greater than cooling
- The number of windows in the east and west side of the building may be increased for solar gain during winter.

Thermal Mass

Thermal mass is an important aspect of passive design. In colder climates, passive solar heating is desired. The orientation of building should be arranged to allow materials with high heat capacity and significant mass to store solar energy during the day. Materials such as brick, concrete, concrete masonry and adobe, etc., used for floors and walls can absorb solar energy during the day and release in the evening, when internal temperature begins to drop.

In Warmer Climate, like Florida, buildings should have minimal mass for storing less energy and generally should be light weight and well insulated. Preventing solar energy transmission into the structure is a desired strategy for passive cooling.

The ideal passive heating/cooling design would include careful consideration of building orientation, fenestration, shading and massing. Shielding the windows during the summer and allowing solar energy during the winter could be an effective operation and maintenance strategy for saving energy within the built environment. In the Germany, passive cooling is the interaction of all measures that reduce heat gains and increasing daytime and night ventilation either by natural or forced ventilation.

DAYLIGHTING

Using natural light or daylight for illumination is one of the hallmarks of a high-performance building. In addition to the benefits of supplying substantial light for free, natural lighting has been shown to provide great physical and psychological benefits to building occupants. Some of the benefits include improvements to human health, and increased productivity and cognitive function (Kibert, 2022).

In late 1990, Pacific Gas and Electric Company (PG&E) of California conducted the first comprehensive scientific studies of benefits of daylighting on two general types of buildings: retail stores and schools. Daylighting in stores was shown to increase sales per square unit area of retail space from 30 to 50 percent ("Skylighting and Retail Sales", 1999), while the learning rate of students was 20 to 26 percent higher in classrooms with daylighting compared to those with only artificial lighting ("Daylighting in Schools", 1999). More recent research shows that natural light boosts vitamin D, wards off seasonal depression, improves sleep, and reduces health risks of fluorescent lighting (Garone, 2020). Clearly, daylighting produces a win-win situation, marked by lower energy costs and better school performances. Most likely, the same is true in offices. Although not yet proven by scientific methods, it is thought that a 10 to 15 percent increase in office worker productivity can be expected because of daylighting. If the connection between daylighting and human health could be proven with a high degree of certainty, this alone would cause an enormous transformation in the way buildings are designed and built.

Developing an effective daylight strategy can, however, be a complex undertaking due to trade- offs that must occur between admitting light and cooling the building. The cost of windows, skylight, light shelves, and other features that function to transmit light versus conventional construction, where daylighting is not much of an issue, must be factored in. Fortunately, experience with daylighting is growing at an exceptional rate, along with the green building movement itself; consequently, the information from these efforts is becoming available to a wider audience of designers and owners. A list of key ideas for assessing daylight feasibility from LBNL is shown below:

Key ideas for daylighting feasibility

- Windows must see the light of the day. A high-density urban site may make daylighting difficult if windows will not see much sky.
- Glazing must transmit light. A strong desire for very dark glazing generally diminishes the capacity to daylight in all but very sunny climate
- Install daylight-activated controls. To save energy, lights are dimmed or turned off with controls. Automated lighting controls in a daylit building can have other cost-saving applications and benefits
- Design daylight for the task. If the occupants require very bright light, darkness, or a highly controllable lighting environment, tailor the design to meet their needs
- Assess daylight feasibility for each portion of the building. Spaces with similar orientation, sky views, ground reflectance, and design can be treated together. Within a single building, the feasibility and cost effectiveness of daylighting may vary greatly.

An excellent checklist for daylighting from Environmental Building News (EBN) is shown below:

General Daylighting:

- Provide a daylighting scheme that will work under the range of sky conditions expected at that location.
- Orient the building in an east-west axis
- Brighten interior surfaces
- Organize electric lighting to complement daylighting
- Provide daylight controls on electric lighting
- Commission the daylight controls

PERIMETER WALL DAYLIGHTING:

- Extend windows high on perimeter walls
- Provide perimeter daylight zones
- Provide light shelves on south facing windows
- Minimize direct-beam sunlight penetration into workspaces
- Choose the right glazing
- Arrange interior spaces to optimize the use of daylighting

Roof Daylighting

- Provide roof apertures for daylighting
- Optimize skylight spacing
- Consider extending skylight performance with trackers
- Use reflective roofing on sawtooth clerestories
- Diffuse daylight entering the building through roof apertures

Light shelf is a horizontal surface that reflects daylight deep into a building. Light shelves placed above eye-level and have a high reflectance upper surface, which reflects daylight into the ceiling and deeper into the space. It can help save money on artificial lighting and provide occupants with the benefit of daylighting.

DAYLIGHT STRATEGY

The daylighting strategy is controlled through occupancy sensors, passive infrared technology, and a manual light switch. In addition, installing clerestory windows provides ample lighting into the built environment. Figure 9-8 shows the clerestory windows.

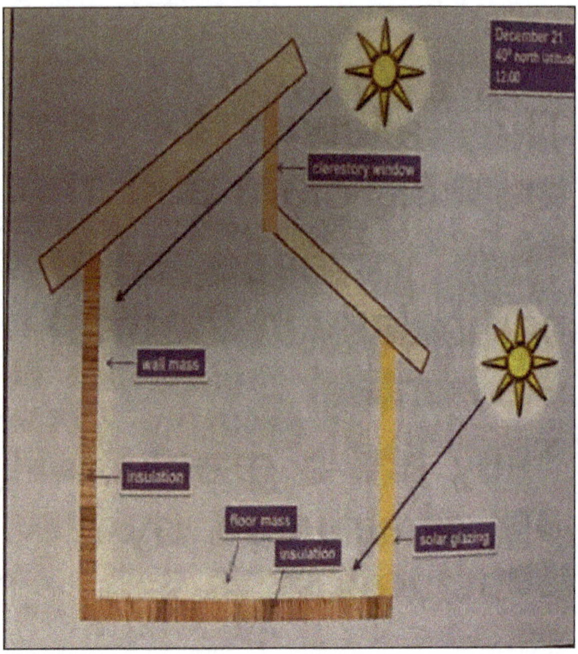

Figure 9-8

Clerestory Windows

Clerestory windows have many benefits:

1. They allow more lights into the building, while reducing glare and improving privacy.
2. They increase solar gain in the living space and particularly if thermal mass is placed on the north side of the home.
3. They are a great choice if the bottom story windows are obstructed by trees, neighboring homes or other structures.
4. They add to the overall aesthetic of the home.

Clerestory windows have some disadvantages including:

- The extra light extends into the night, as clerestories capture street lighting, ambient light from neighbors' homes and even moonlight. This can make them a bad choice in sleeping areas.
- They provide another exit for the home's stored heat and since they are high up, they are closer to the warmest air in space.
- They are extremely difficult to shade, which means they generally always remain uncovered, thus increasing heat loss. You may need to install expensive powered shutters to reduce heat loss and nuisance light.
- They require very high ceilings compared to the most homes, which means the living space contains a lot more air to heat. Vaulted ceilings always make home heating more difficult.

PASSIVE VENTILATION

Providing ventilation to building occupants normally is accomplished by using fans, dampers, and controls to move outside air into the building while at the same time removing an equal amount of interior air to the outside. In more advanced designs, an economizer cycle uses outside air for cooling, providing significant savings. Figure 9-9 shows different type of ventilation schemes.

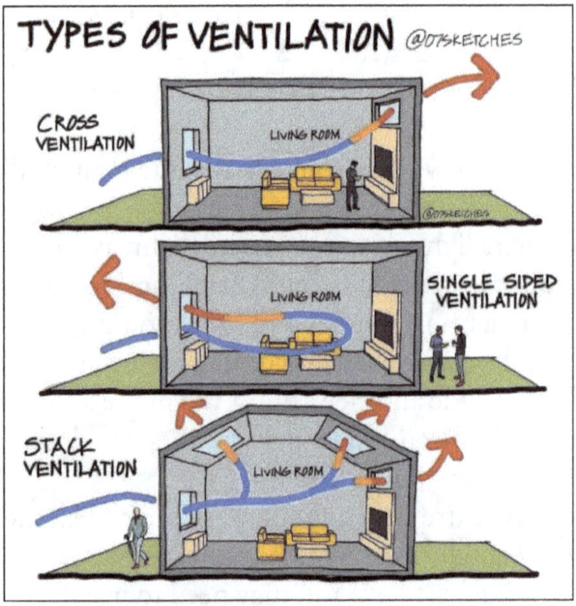

Figure 9-9 Schematic Diagram of Building Ventilation Concept

Ventilation air using natural forces to move the air, rather than mechanical systems, also can be provided, gently reducing the energy needed to move air. Passive ventilation can be accomplished by using either a thermal chimney effect, whereby air normally rises due to heating, inducing airflow in a generally vertical direction, or a Venturi effect, whereby air movement is induced by the development of a low-pressure zone created by wind flow.

Passive ventilation is accomplished by "Thermal chimney effect or Venturi effect without using fans or dampers. Heated air rises due to its low density, upon upward air movement, a low-pressure zone is created which induce fresh air from outside. Figure 9-10 shows the principle of thermal chimney effect or venturi effect.

Figure 9-10 The hot air moves upward and induces the ambient air from the bottom. This is the principle of passive ventilation

The Jubilee Campus of the University of Nottingham in the United Kingdom, designed by the Sir Michael Hopkins and partners and built in 1999, still has one of the most advanced passive ventilation strategies among modern buildings. Wind catchers are used to position the air exhaust stacks for optimal ventilation. The wind catchers automatically turn in the direction of wind, creating suction behind them and driving ventilation system for the building. Cool, clean air is brought in at a high level and funnel down to the floor levels, where it starts to rise with the sunlight, body and equipment heat. This intricate pattern of environmental cause and effect is echoed throughout the building's staircases and corridors. Thermal wheels are used in conjunction with the wind catchers to exchange energy between existing exhaust air and incoming fresh air. The innovations in this design resulted in Jubilee Campus winning the Royal Institute of British Architects sustainability award in 2001. Figures 9-11 and 9-12 show the passively ventilated building on the Jubilee Campus.

Figure 9-11 Wind catchers (upper right) on the Jubilee Campus of the University of Nottingham in the United Kingdom. The wind catcher pivots in the wind, with the vane indicating the direction of airflow. Wind flowing past the vane induces the convection of air through the structure. (Source: photograph courtesy of Hopkins Architects and Ian Lawson)

In a typical European passive ventilation design, the first determinant is the quantity of air required for ventilation. In England, the Chartered Institution of Building Services Engineers published standard Guidelines requiring these ventilation rates.

- Classrooms: 2 to 4 air changes per hour
- Offices: 4 to 6 air changes per hour
- Theaters: 6 to 10 changes per hour
- Storage areas: 1 to 2 changes per hour

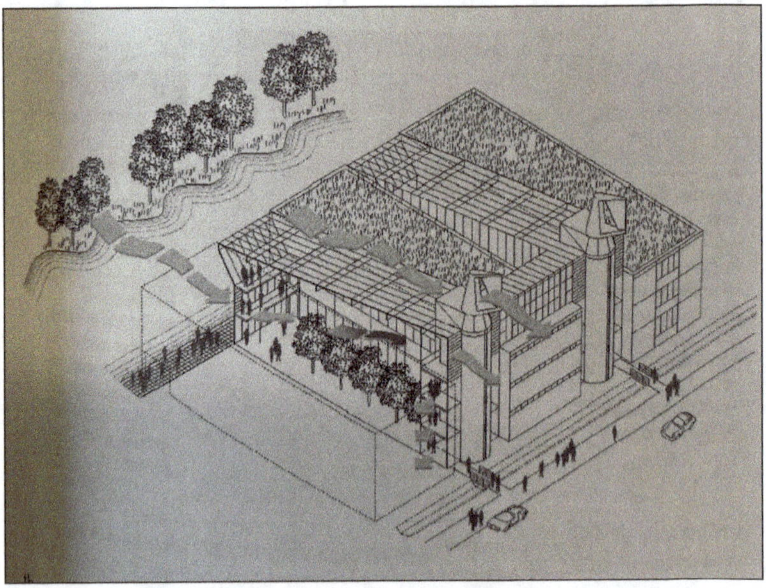

Figure 9-12 Schematic of natural ventilation strategy for the Jubilee Campus of the University of Nottingham, United Kingdom Air flows from the low level at the rear of the building, moves gradually upward and then exits through the pivoting wind catchers on the front of the building. A Venturi effect is induced in the wind catcher by the wind following past the vanes. (Source: Illustration courtesy of Hopkins Architects)

The outside wind speed, which is generally in the range of 3 to 19 ft per second in England, is factored into the design, and the number of passive ventilation stacks required to move the calculated amount of ventilation air are designed into the structure. In the base of the stack, dampers are connected to the building's energy management system (EMS). And possibly to carbon dioxide (CO_2), humidity, and/or temperature sensors, control the rate of ventilation. Diffusers at ceiling level introduce the ventilation air into the occupied spaces. Solar tubes that bring in light as well as air are incorporated into some passive ventilation stacks.

In contrast to Europe, which has a wide range of examples of passive ventilation systems, the concept has not had much success in the United States.

PASSIVE COOLING AND HEATING

Designers now assume moderate interactive by means of natural ventilation, daylighting and passive cooling or heating.

Lean building is the building that uses smaller HVAC and ventilation system rather than using conventional equipment. Heat loads are transferred to the surrounding environment with some time delays, and heat storage in the building itself is substantial. Cooling is accomplished by using night ventilation in which the building mass is cooled off by using earth-to-air heat exchangers. Figure 9-13 thru 9-16 show the schematic diagram of passive/cooling and heating system. During summer months the hot air from the ambient air enters the underground piping system at about 10 feet below the ground surface, the hot air cools off due to high coefficient of performance (COPs) between the air inside the pipe and geologic formation and as a result the cool and fresh air enters the building. In wintertime the cold air from the ambient air enters the piping system and upon heat exchange process between the pipe and surrounding medium warms up and conditioned air with pleasant temperature enters the building.

COP of a heat pump, refrigerator or air conditioning system is a ratio of useful heating or cooling provided to work required. Higher COPs equate to lower operating costs.

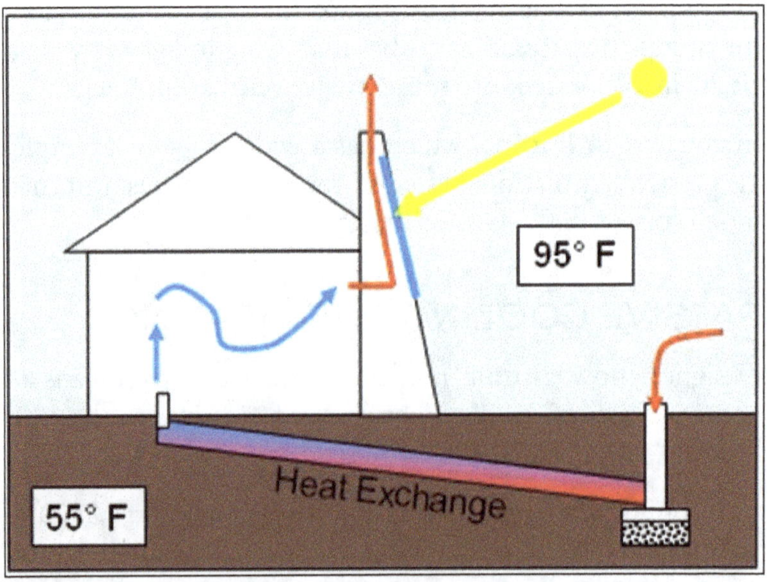

Figure 9-13 shows the schematic diagram of heat exchange process in underground

The following figures show construction activities of installation underground piping system for installation of passive cooling/heating systems.

Figure 9-14 Subterranean piping system

Figure 9-15 Subterranean piping system

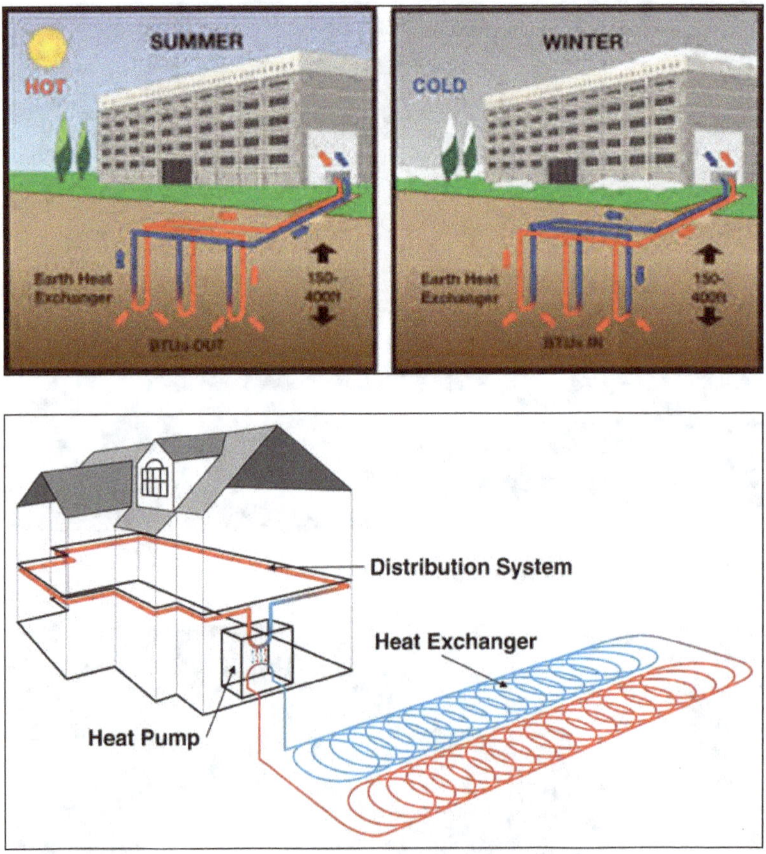

9-16 Examples of heat exchange systems for achieving passive heating and cooling process

Passive cooling is a building design approach that focuses on heat gain control and heat dissipation in a building to improve the indoor thermal comfort with low or no energy consumption. This approach works either by preventing heat from entering the interior or by removing heat from the building. Therefore, passive cooling means using design choices to reduce heat gain and increase heat loss. Buildings in warmer climates require some form of cooling at some time of the year to be comfortable. Passive cooling is especially useful in hot and humid or hot and dry climates.

Passive cooling can be established using shade and insulation to keep heat off the building in summer or using breeze and air movement inside the building to keep it cool.

The advantage of passive cooling technique lies in energy efficiency and lower cost, making it an astute systems design choice for the thermal management of both building and electronic products. A passively cooled building uses the building's materials and design to transfer heat to heat sinks, such as the air around the building, in the same way air conditioning does, but without using electricity.

Passive cooling has also some disadvantages. For instance, it has a lower heat transfer rate and higher thermal resistance which can limit the performance which can limit the performance or scalability of this technique. As such, relying on passive cooling strategy may not be sufficient to survive against extreme weather conditions which may occur occasionally.

BUILDING ENVELOPE

The building envelope or building enclosure is a physical separator between the conditioned and unconditioned environment of a building, including the resistance to air, water, heat, light, and noise transfer. The building envelope which includes walls, windows, roof, and foundation, forms the primary thermal barrier between the interior and exterior environments. Therefore, the building envelope must control solar, heat gain, conduction or direct transmission, and infiltration or leakage heat transmission. The three major building envelop issues that need to be addressed are:

- Thermal resistance of the walls,
- Window selection, and
- Roof strategy

WALL SYSTEM

The thermal conductance of U-value building is an important factor in building energy efficiency because walls are generally the dominant component of the envelop. U-values are measured in the units of BTU/hr-ft^2-F^o(W/m^2-^{o}C). The lower the value of an assembly, the greater is its resistance to heat transfer.

Maximum U-values are set by state building energy codes and by ASHRAE 90.1. The maximum U-values is the function of the number of HDDs and CDDs for the various climate zones in the United States, see Figure 9-17 and Table 9-2. Both CDD and HDD are measures of how much heating or cooling probably will be required in a given climatic zone.

In general, wall thermal resistance becomes more important the farther north where the building is located in the United States. Two other considerations in selecting wall system are the thermal mass of the exterior surface that receives direct sunlight during the day and placement of insulation with respect to the building façade. Placing insulation closer to the exterior nearest to outdoor conditions and having the thermal mass closest to interior provides ideal conditions for using the mass beneficially and for minimizing the thermal loads transmitted into building's interior that must be removed by air conditioning. In southern climates, it is generally important to design energy-shading façades that reflect energy or are ventilated to carry away energy that is absorbed on the building's skin.

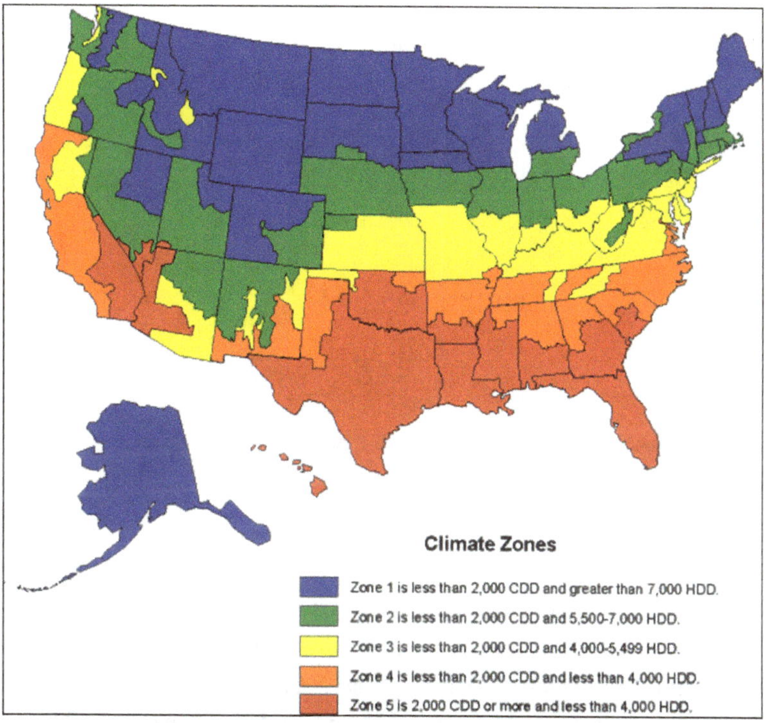

Figure 9-17 Climate zones by County in the United States based on International Energy Conservation Code, (Source: Pacific Northwest National Lab)

Zone Number	Thermal Criteria	
	IP Unite	SI Units
1	9000<CDD50F	5000<CDD10C
2	6300 CDD50F <9000	3500<CDD10C < 5000
3A and 3B	4500 CDD50F <6300 and HDD65F <5400	2500<CDD10C < 3500 and HDD 18C < 3000
4A and 4B	CDD50F <4500 and HDD65F <5400	CDD 10C < 2500 and HDD 18C < 3000
3C	HDD65F <3600	HDD 18C < 2000
4C	3600<HDD65F <5400	2000<HDD18C<3000
5	5400<HDD65F <9000	3000<HDD18C<4000
6	7200<HDD65F <9000	4000<HDD18C<5000
7	9000<HDD65F <12600	5000<HDD18C<7000
8	12600<HDD65F	7000<HDD18C

Table 9-2 US Climate Zones Defined by HDDs and CDDs

Note: Heating degree days (HDDs) and cooling degree days (CDDs) are defined as the differences of daily average temperature From a base temperature, summed over the entire year. CDD50°F means the baseline for calculation CDD is 50°F. when the Mean daily temperature is above 50°F, CDDs are calculated. A day with a mean temperature of 75°F would have (75°F 50°F) = 25 CDDs. These are totaled for the full year in each climate zone to establish the annual CDDs for that zone. HDDs are based on a baseline of 65°F.

WINDOW SELECTION

Windows play a variety of roles in building envelopes. They allow (1) light into the room space, (2) permit occupants to admit air into the space (in the case of operable windows), and (3) provide thermally resistant layer to energy movement. Windows must be installed to balance the amount of light admitted into the structure with the control of solar heat gain and conduction of energy through the window assembly.

Window performance is a combination of several factors: the solar heat control coefficient (SHGC), the visible transmittance (VT) of the glass, thermal conductance (U-value), and the infiltration or leakiness character of the window assembly.

Solar heat gain is largely a function of where windows are placed in the building and the type of glass used. SHGC and VT are used to express the radiation performance of windows in the building envelope. SHGC, with a value between 0 and 1, is the fraction of solar heat that enters the window and becomes heat; it includes both directly transmitted and absorbed solar radiation. The lower the SHGC, the less solar heat the window transmits through the glazing from the exterior to the interior and the greater its shading capacity. In general, south-facing windows in buildings designed for passive solar heating should have windows with a high SHBC to allow beneficial solar heat to gain in the winter. East- or west-facing windows encounter high levels of solar energy in the morning and afternoon and generally should have lower SHGC assemblies.

The VT, ranging in value from 0 to 1, refers to the percentage of the visible spectrum (380-720 nanometers) that is transmitted through the glazing. When daylight in a space is desirable high- VT glazing would be the logical choice. However, lower-VT glazing may be more applicable for office buildings or where reduced interior glare is desirable. A typical clear, single-pane window has a VT of 0.90, meaning that it admits 90 percent of the visible light.

The ratio of VT to SHGC, known as the light to solar gain (LSG) ratio, provides a gauge of the relative efficiency of different glass types in transmitting daylight while blocking heat gains. The higher the LSG, the brighter the room is without adding excessive amounts of heat. Table 9-3 shows average values of SHGC, VT, and LSG for typical windows. Figure 9-18 is the diagram of the characteristics of a contemporary high-performance window that is optimized for hot climates.

windows that are filled with argon or krypton gas have more thermal resistance than those filled with air. Argon is inert, relatively abundant, and less costly than krypton, which provides higher thermal resistance but at a higher cost. Windows with a low SHGC were used in the relatively high-temperature climate of northern Florida, as shown in Figure 9-18.

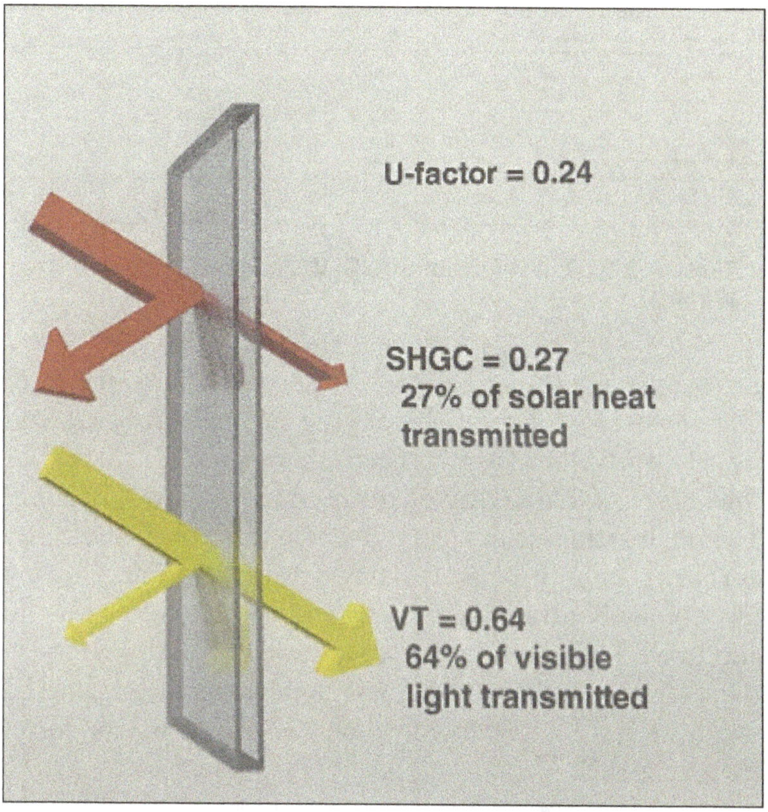

Figure 9-18 Characteristics of a typical double-glazed window with low SHGC, low-E glass, filled with argon gas. These windows are often referred to as spectrally selected low-E due to their ability to reduce solar heat gain while retaining high visible transmittance such coatings reduce heat loss and transmit less solar heat gain, making them suitable for climates with both heating and cooling concerns. (Source: Courtesy of Efficient Windows Collaborative)

Window Type	Glazing	VT	SHGC	LSG
Single-glazed	Clear	0.69	0.79	0.87
Double-glazed	Clear	0.57	0.58	0.97
Double-glazed	Bronze	0.43	0.48	0.89
Double-glazed	Spectrally selective	0.51	0.31	1.65
Triple-glazed	Low-E	0.48	0.37	1.29

Source: From "Solar Heat Gain Control for Windows" 2006

Table-9-3 Typical values of SHGC, VT, and LSG Total Window

Low-emissivity (low-E) and effective coatings are applied to glazing to control the light passing through the glass and usually consist of a layer of metal a few molecules thick. The thickness and reflectivity of the metal layer (low-E coating) and the location of the glass to which it is attached directly affect the amount of solar heat gain in the room. Coating technology is advancing rapidly, and there are new low-E2 and low-E3 windows, with two three silver coatings, respectively, that greatly improve glass performance. The low-E3 windows have a remarkedly low SHGC of 0.3 or lower (Kibert, 2022).

Any low-E coating is roughly equivalent to adding an additional pane of glass to a window. Low-E coating reduces long-wave radiation heat transfer by 5 to 10 times. The lower the emissivity value (the measure of amount of heat transmission through the glazing), the better the material reduces the heat transfer from the inside to the outside. Most low-E coatings also slightly reduce the amount of visible light transmitted through the glazing relative to clear glass. Representative emissivity values for different types of glass are listed next.

- Clear glass, uncoated: 0.84
- Glass with single hard-coat low-E: 0.15
- Glass with single soft-coat low-E2: 0.10

Increasing the window area to maximize daylight has the effect of replacing a highly thermally resistant wall with far less thermally resistant glass, creating an opportunity for the infrared or heating component of light to enter the envelope and creating the potential for infiltration around the window frame. In trading off daylighting to optimize the thermal envelope, controlling solar heat gain through windows is critically important. Prior to development of today's window glazing and film technologies, 75 to 85 percent of infrared energy could pass through typical single- or double-paned glass, (Kibert, 2022).

ROOF SELECTION

The roof of high-performance buildings is especially important because it is a major area for heat transmission due to its large area and exposure to the sun.

According to Cool Communities (www.coolcommuni ties.org), a nonprofit organization that advocates for measures that prevent urban heat islands and based in Rome, Georgia, the roofs of structures such as shopping malls, warehouses, and office buildings can reach 150°F in summer, enough to affect whole neighborhoods. Using surfaces with high albedo (a measure of reflectivity of solar radiation) for roofing can reduce the ambient air temperature so that the entire area is cooler. Light-colored roofs have high albedo, or high reflectivity, which helps reduce the thermal load on the building as well as the surrounding neighborhood. Both the LBNL and the Florida Solar Energy Center estimate that buildings with high-colored, reflective roofs use 40 percent less energy than similar buildings with dark roofs (Parker et al. 2000).

The Solar Reflective Index (SRI), which measures how hot materials are in the sun, is used to easily describe the amount of solar energy reflected by roofing materials. A building with light-colored shingles and an SRI of 54would reflect 54 percent of incident solar energy and would be very cool relative to a building with conventional dark shingles. Manufacturers have developed clean, "self-washing" white shingles with an even higher SRI, up to 62 percent. this is useful because the labor costs of maintaining the white color and high reflectivity of a conventional roof may exceed the worth the energy being saved; consequently, a self-washing roof system significantly reduces maintenance costs and improves energy performance. The reflectance of commonly used roofing materials is shown in Table 9-4. As can be seen, dark-colored roofs tend to absorb solar radiation, and they can be as much as 90°F hotter than the air just above the roof. Because heat transmission is a function of temperature difference, a dark-colored, hot roof will have proportionately more heat reduction than a light-colored, relatively cool roof. Figure 9-19 shows highly reflective, high-SRI roof used on a building in Hollywood, California.

Material	Solar Reflectance (%)	Temperature of Roof over Air Temperature (F°/C°)
Bright white coating (ceramic, elastomeric) on smooth surface	80	15°/8°
White membrane	70-80	15°-25°/8°-14°
White metal	60-70	25°-36°/14°/20°
Bright white coating (ceramic, elastomeric) on rough surface	60	36°/20°
Bright aluminum coating	55	51°/28°
Premium white shingle	35	60°/33°
Generic white shingle	25	70°/39°
Light brown/gray shingle	20	75°/42°
Dark red tile	18-33	62°-77°/34°/43°
Dark shingle	8-19	76°-87°/42°-48°
Black shingle or materials	5	90°/50°

Source: Parker et al. (2000)

Table 9-4 Reflectance of Roof Materials and Air Temperatures above Roof

Figure 9-19 In a renovation project, the roof of this art studio in Hollywood California, was coated for waterproofing with SureCoat, a white finish, which could be applied to the mechanical system as well giving it a very high reflectivity, or albedo. Research by the LBNL and Florida Energy Center has demonstrated that high-colored roofs with a high SRI use 40 percent less energy than dark roofs. (Source: SurCoat Systems)

INTERNAL LOAD REDUCTION

The internal heat load reduction in designing the high-performance green building is quite important in reducing energy requirements of the built environment. This is achieved by using a good daylight strategy that has a dual benefit of further reduction of energy demand.

The building occupants constitute a major fraction of the building's internal heat load, and we generally can assume that reducing the number of people in a building is not a viable/feasible alternative. Reducing loads due to computers, peripherals, copiers, and other miscellaneous equipment is a promising strategy because it has been found that these loads constitute a substantial fraction of a building's energy consumption. Increasing wiring sizes beyond those required by code has the benefit of reducing energy losses in the wiring system and proportionately reducing the impact of these heat losses on the building's cooling system.

INNOVATING ENERGY OPTIMIZATION STRATEGIES

A wide variety of innovations in building systems are emerging. Four of the more innovative approaches are:

1. Radiant cooling
2. Ground coupling
3. Renewable energy systems, and
4. Fuel cells

Each of these is a cutting edge of strategy that can have a marked effect on energy consumption if used properly in a building.

RADIANT COOLING

In general, cooling is delivered to conditioned spaces using air that is pressurized by fans and delivered via duck-work to the various spaces. because air has a very low heat capacity, rather large quantities of air must be delivered to space to provide the cooling effect needed. Additionally, air, a compressible medium, is a relatively energy intensive to move, compared to water, which is compressible, has very high heat capacity and can be moved comparatively cheaply via pumping. That's why, in Europe, radiant cooling is often used for cooling spaces. Radiant cooling systems are water, which has 3,000 times the energy transport capacity of air, as the medium for delivering cooling to space. In Germany, radiant cooling system have become a new standard.

Radiant cooling system circulates cool water through tubes in ceiling, wall, or floor elements or panels. The water temperature does not differ noticeably from the room temperature, so care must be exercised to ensure that the temperature of the circulated water did not reach the dew point of air in the space. Otherwise, condensation will occur, resulting in moisture problems. The cost of the radiant cooling systems is approximately the same as that of Variable Air Volume (VAV) system, but the life-cycle savings are 25 percent higher to those of VAV system. Moreover, the energy required for circulating water is only about 5 percent of the energy needed to circulate a compressible capacity of air.

There are three types of radiant cooling systems:

1. Concrete core. Plastic tubes are buried in concrete floor and ceiling slabs.
2. Metal panels. Metal tubes are connected to aluminum panels.
3. Cooling grids. Plastic tubes are embedded in plaster or gypsum.

The metal panel system is the most used radiant cooling system. Due to its metal construction, it has a relatively fast response time to changing conditions. Cooling grids are generally the choice for retrofit projects because the grid of plastic cooling can be placed easily in plaster of gypsum in existing walls. As a guide to system sizing, the total heat transfer rate (combined radiation and convection) is about 11 $W/m^2/^oC$ temperature difference for cooled ceilings. Five design guidelines for radiant cooling are listed below:

1. The building should be well sealed.
2. In humid area, the intake fresh air should be dehumidified prior to its entry into conditioned spaces.
3. Radiant cooling requires a large surface area due to the relatively small temperature difference between the cooling surface and the room air.
4. The set points for cooling and heating must be considered carefully to deliver maximum conditioning without causing moisture problems. For instance, for a typical system in Germany, during the cooling season, the room temperature set point is about 80°F, with cold or chilled water entering the radiant cooling panels at 61°F and leaves at 66°F. for heating, the room set point is 68°F, with heated water delivered at 95°F and leaving the radiant panels at 88°F.
5. Humidity sensors should be used to detect when the temperature of the supply water is approaching the dew point to activate valves that will prevent condensation from occurring.

GROUND COUPLING

A ground-coupled heat exchanger is an underground heat exchanger that can capture heat from and/or dissipate heat to the ground. Therefore, the Earth's near constant subterranean temperature is used to warm or cool air or other fluids for residential, commercial/industrial use. For example, it is feasible to use air or groundwater in 60°F range in a radiant cooling system for a building and virtually eliminate the need for a chilled-water plant. For more information see Figures 9-12 through 9-15 and the related narratives.

RENEWABLE ENERGY SYSTEMS

Renewable energy systems can be generated on-site by three different techniques: PVs, wind energy. And biomass. The following is descriptions of PVs, wind energy and biomass in using as the renewable energy source:

Photovoltaics (PVs)

A solar cell or photovoltaic (PV) cell is an electronic device that converts the energy of light directly into electricity. Most of us have utilized solar panels. When the sun shines onto a solar panel, energy from the sunlight is absorbed by PVcells in the panel. This energy creates electricity charges that move in response to an internal electrical field in the cell, causing electricity to flow. PVs have no moving parts and are efficient and require relatively low maintenance. Energy storage is needed, if provided by batteries. PV modules are currently electricity at hundreds of thousands of installations throughout the world. Especially exciting are building-integrated photovoltaics (BIPV) technologies that integrate PV cells directly into building materials, such as semitransparent insulated glass windows, skylights, spandrel panels, flexible shingles, and raised-seam metal roofing. PV elements can be fabricated in different forms. They can be used on or be integrated into roofs and facades as part of the outer building cladding, or they can be used as part of a window, skylight or shading device. PV laminates provide long-lasting weather protection. Their expected life span is more than 30 years. Warranties currently are available for 20-year periods. They usually can be added, removed, and reused in other applications.

A variety of attractive BIPV products are available that allow building surfaces, such as the roof, walls, skylights, sunshades, to double as solar collectors. Integrating these products into the building envelope creates a large solar collection area, enabling solar power to displace more of the electricity used in the building.

Figure 9-20 BIPVs in the Solaire Building in New York City's Battery Park. The BIPV are the specked surfaces between The windows. (Source: Courtesy of the Albanese Development Corporation)

Depending upon the type of collection medium, BIPV can generate approximately 5 to 10 W of power per $ft^2$of collector area in full sunlight. It means that a collector area of 100 to 200 sq ft typically is needed per one kilowatt of capacity. Off course this is also the function of latitude of the building location as well as the orientation of the building. Figure 9-20 shows the BIPV system installed on the façade of Solaire, a 27-story luxury residential tower in New York City's Battery Park.

Wind Energy

Wind energy or wind power, is created using a wind turbine, a device that channels the power of wind to generate electricity. The wind blows the blades of the turbine, which are attached to a rotor. The rotor then spins a generator to create electricity. Wind energy is the fastest-growing form of energy production, with an estimated yearly growth of 25 percent (Kibert, 2022). According to the DOE's National Research Energy Laboratory (2001), the cost of wind energy to a utility has declined from \$0.40/kWh in the 1980s to less than \$0.02kWh today. There are now over 109,000 gigawatts of wind energy in the US with over 60,000 operating wind turbines in 41 US states and two US territories, see Figure 9-21. This represents an increase of over 270 percent over the past decade. Texas has the highest wind energy capacity in the United States, about 27,000 megawatts in 2020. The reason for this high wind power capacity was the Texas Renewable Portfolio Standard, which was available between 2005 and 20015, and which mandated installation of wind energy by utilities, and federal tax credits, which provide a \$0.015/kWh write off for the first 10 years of a project's operation.

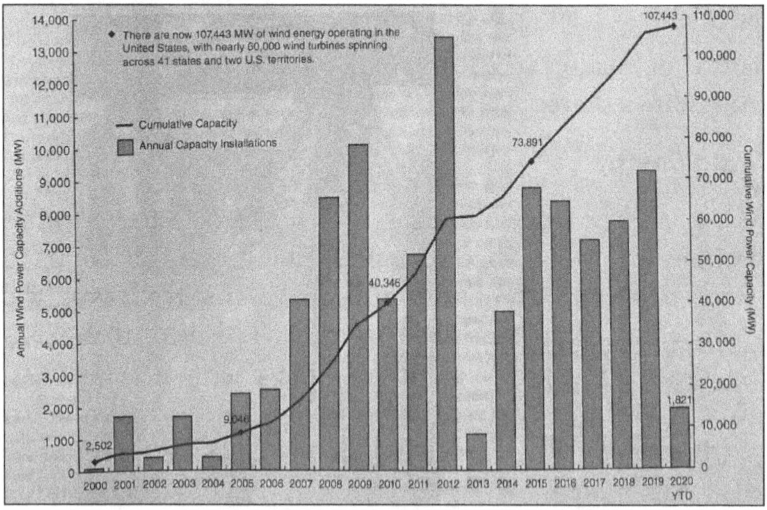

Figure 9-21 Utility-scale wind capacity includes installations of wind turbines larger than 100-kW for purpose of AWEA US Wind Industry Quarterly Market Reports (Source: American Wind Energy Association, 1st Quarter 2020 Report)

Small wind turbines (those with less than 100 kW output) suitable for building-scale applications are available, and there are innovative programs that can make their incorporation into a building project financially feasible.

Biomass Energy

The term biomass refers to any plant-derived organic matter available on a renewable basis, including dedicated energy crops and trees, agricultural food and feed crops, agricultural crop wastes and residues, wood wastes and residues, aquatic plants, animal wastes, municipal wastes, and other waste materials. Handling technologies, collecting logistics, and infrastructure are important aspects of the biomass resource supply chain.

Shifting of the $50 billion now spent for oil imports and other petroleum products to rural areas would have a profoundly positive effect on the economy in terms of jobs created (for production, harvesting, and use) and industrial growth (facilities for conversion into fuels and power). David Morris of the Institute for Local Self-Reliance refers to using biomass as moving partway from a "hydrocarbon economy to a carbohydrate economy".

OZONE-DEPLETING CHEMICALS IN HVAC SYSTEMS

The building industry as well as fire protection systems employ the largest quantities of ozone-depleting chemicals. Removing these chemicals from building inventories and replacing system with new products that use non-ozone-depleting chemicals are priorities. This section discusses the replacement of refrigerants in air-conditioning systems with newer technologies that do not impact the ozone layer.

Before 1986, chemicals known as chlorofluorocarbons (CFC) were commonly used as refrigerants in chillers, mechanical devices that are used to generate cooling; CFC-11 and CFC-12 were the two most common CFCs used. Then, in 1986, their release into the atmosphere was found to be a major cause of destruction of the ozone layer, and international treaties soon called for their phase-out. The impact of CFCs on the ozone layer is indicated in terms of a quantity called the ozone-depleting potential (ODP). The ODP is defined as 1.0 for CFC-11, meaning that a substance with an ODP of 10.0 depletes ozone at 10 times the rate of CFC-11. Other typical CFCs have a value of 1.0. For hydrochlorofluorocarbons (HCFCs), the ODP ranges from about 0.02 to 0.11(or about 10 to 50 times less impact than caused by CFCs).

Several families of chemicals have been used to replace CFCs, among them HCFCs and hydrofluorocarbons (HFCs). Although HCFCs are a great improvement over CFCs, they still have relatively high ODPs. HFCs, by contrast, have zero ODP and, as a result have no impact on the ozone layer. HFC-containing equipment is available from all major manufacturers. HFC-134a has become the dominant refrigerant, replacing HCFC-123 and HCFC-22 in most chillers designed for building use. HCFC-22 currently is used in large proportion of positive-displacement compressor-based chillers and in some larger-tonnage centrifugal chillers. These uses predate the 1987 Montreal Protocol, which banned the use of ozone depleting chemicals, but will be phased out as part of the overall HCFC phase-out. In the United States, HCFC-22 was phased out for use in new equipment as of January 2010.

According to the Carrier Corporation, HFC-134a has been proven to be an optimal refrigerant in chiller application because of its no chlorine molecules and does not contribute to ozone layer depletion. HFC-134a is a highly efficient thermodynamic refrigerant in application. Current centrifugal chillers using HCF-134a are 21 percent more efficient than chillers sold just 6 years ago and 35 percent more efficient than the chillers installed during the 1970s and 1980s. Because HFC134a is a positive pressure refrigerant, pressure vessels using it must conform to the American Society of Mechanical Engineers pressure code., and every step in their construction must be inspected by third-party insurance companies. As a result of the stringent testing and applied technology, chiller leak rates can be lowered to less than 0.1 percent annually. Existing chillers have a leak rate of 2 to 15 percent. HFC-134a has also a smaller molecular mass than the past CFCs and HCFCs. This is an important feature, as it results in an overall product size that is 35 to 40 percent smaller, a size reduction that helps offset the cost of construction and facilitates the use of smaller interconnecting pipes. This advantage has led to the addition of isolation valves to the chiller piping connection so that the HFC-134a can be stored in the chiller during service. This feature gives the end user the option of never having to remove the refrigerant from the vessels once charged, a real "no emission" feature. An additional advantage of HCF134a chillers is their smaller size, requiring much less plant space than the CFC-11 chillers they replaced.

ADVANTAGE AND DISADVANTAGE OF RENEWABLE ENERGY SYSTEMS

Finally, this chapter concludes by discussing the advantages and disadvantages of renewable energy systems. The following are the advantages of renewable energy systems:

1. Reduces the consumption of fossil fuels. The consumption of fossil fuels is a major source of greenhouse gases including carbon dioxide, and air pollution. The generation of greenhouse gases is the major contributor to climate change.
2. It is more feasible and economical to be used in remote areas where there are no power plants to produce electrical energy such as rural areas, developing countries, hospitals, telecommunication devices, etc.
3. Since renewable energy is produced on-site, and it is not transmitted from faraway locations, no conveyance loss will be involved.
4. Unlike the power plants, which require water for power generation, renewable energy does not require any water for its generation.

5. It does not produce thermal pollution of waterways.
6. PV cells last about 30 years and the cost of their replacement is minimal.
7. No hidden costs involved in generation of renewable energy

The following are the disadvantage of renewable energy systems:

1. Currently, electricity using PV systems is more expensive than electricity production from fossil fuel or nuclear power plants.
2. Even more expensive to purchase.
3. Requires engineering expertise to design and install.
4. Sunlight is not always available. During rainy and cloudy days its production rate drops.
5. Energy panels and BIPVs take a big surface area to produce enough power in a multi-story building.
6. Manufacturing, repairing and maintenance of PVs requires nonrenewable energy, also axillary equipment such as batteries and inverters also would need some form of nonrenewable energy.

REFERENCES

American Wind Energy Association (AWEA). 2020. "Wind Powers America: Second Quarter 2020 Report." American Wind Energy Association, and Wind IQ.

Daylighting: Energy and Productivity Benefits, 1999. Environmental Building News 20(2): 1-14.

Daylighting in Schools: An investigation into the Relationship between Daylighting and Human

Performance: A Condensed Report., 1999. Conducted by the Heschong Mahone Group for the Pacific Gas and Electric Company. Available at www.nrel.gov/docs/fy01 osti/29267.pdf

Department of Energy. 2001. "Greening of Federal Facilities, 2nd ed. DOE/GO-102001-1165. Washington, DC: Department of Energy. Available at www.nrel.gov/docs/fy01osti/29267.pdf

Energy Management Systems. 1998.energy Design Brief. Energy Design Resources. Available at www.energydesignre sources.com

Garone, Sarah 2020. "The Health Benefits of Natural Light (and Seven Ways to Get More of it). Healthline. Available at htpp://www.healthline.com/health/natural-light-benefits.

Haves, P. et al. 2010 "Model Predictive Control of HAVAC Systems: Implementation and testing at the University of California, Merced campus. "Lawrence Berkeley National Laboratory Report. Available at http://escholarship.org/uc/item/3pt2d32h.

Kibert, J., Charles, 2022 'Sustainable Construction, Green Building Design and Delivery, Fifth Edition, John Wiley ad Son, Inc, Hoboken, NJ

Parker, D. S., J.E.R. McIlvaine, S. F. Barkaszi, D. J. Beal and M. T. Anello, 2000. "Laboratory Testing of the Reflectance Properties of Roofing Material." "FSEC-CR-670-00, Florida Solar Energy Center, Cocoa, FL., Available at www.fsec.ucf.edu/en/publications/html/FSEC-CR-670-00/.

Powering Texas. 2020. "Texas Wind: Powering the Future of Electricity Demand for Texas." Available at https;//poweringtexas.com/

TEN
CONSTRUCTION AND WATER ISSUES

BACKGROUND

Among the most vital resources that every individual needs for continuation of his life on Earth is water. Water is a very precious commodity which unfortunately has gone unnoticed for many nations who live next to lakes, streams, oceans and large water bodies. Due to the occurrence of extreme drought conditions for several decades in many parts of the world and mismanagement of water resources has its vital role in our lives and well-being, as well as economy, our built environment has been felt across the world. Despite water's symbolic and practical values, water resources throughout the world are badly over-harvested. The disappearance of lakes, estuaries, rivers as well as subsurface drinking water resources (aquifers) are live examples of critical water shortages around the globe.

On July 28, 2010, United Nations General Assembly agreed on Sustainable Development Goal 6 (SDG 6) one of 17 goals whose purpose is to eradicate global poverty SDG 6 calls on the UN member states to "Ensure the availability and sustainable management of water and sanitation for all. At present time, over 2 billion people live in water stressed areas of the world, and 3 billion people have no running water within about 0.6 miles of their homes. Over 4 billion people experience severe water shortage during at least one month of the year. Every eight seconds, a child dies of water-borne disease, which would be preventable if families had adequate financial resources. According to World Health Organization (WHO, 2019), over 297,000 children under the age of 5 died from diarrhea in 2017.

The world is running out of water, and the future will likely be grim for populations that cannot afford the technology and energy to produce clean water from seawater or polluted water. For bare survival, the WHO suggests that 0.5 to 1 gallon of water is needed per person for drinking and another 1 gallon for cooking and food preparation. The US Agency for International Development (USAID) states that 26.4 gallons of water a day per person are required to maintain a reasonably good quality of life. In the United States, direct per capita daily water use is approximately four times higher about 100 gallons; and if agricultural and industrial water use is included, the amount per person per day is approximately 1,800 gallons, an enormous quantity of a limited precious resource.

In addition to shortages of water across the globe, public health and hygiene are important issues. Waterborne diseases, including diarrhea, typhoid, and cholera, are responsible for 80 percent of the illnesses and health in developing countries. Some 15 million children die from these diseases per year. Raw sewage and toxic materials, including industrial and chemical wastes, human wastes and agricultural wastes, are dumped into water systems at the rate of 2 million tons per day. About 300,000 gallons of raw sewage are dumped every minute into the Ganges River in

India, which is also a primary source of water for many Indians. Wastewater treatment lags in most of the world, only 35 percent is treated in Asia and approximately 14 percent in Latin America. According to the UN (2017) high-income countries treat about 70 percent of their municipal and industrial wastewater; upper middle-income countries treat about 38 percent of their wastewater; lower middle-income countries treat about 28 percent of their wastewater; and low-income countries treat just 8 percent. These estimates support the often-cited approximation that, globally, over 80 percent of all wastewaters is discharged without treatment.

GLOBAL WATER RESOURCES DEPLETION

Water is the foundation of all sustainable development. It impacts every aspect of life: hygiene and sanitation, mental and social well-being, justice issues for women and children, education, the economy, and the environment. Yet 703 million people lack basic access to water, trapped in a cycle of poverty when their most essential human need is inaccessible.

Water is the most essential resource on earth. As the world experiences the impact of our changing climate, water scarcity is becoming more prevalent in nearly every region of the globe. One-quarter of the global population lives in countries that face water shortages due to weak governance and human capacities and a lack of infrastructure to transport water from rivers and aquifers. Implementing innovative, sustainable, and environmentally friendly water access, sanitation, and hygiene (WASH) services are crucial to navigating the future of the global water crisis. Overuse, water pollution, lack of infrastructure, and changing weather patterns due to climate change are some of the drivers of water scarcity.

The global urban population facing water scarcity is projected to double from 930 million in 2016 to 1.7–2.4 billion people in 2050. The growing incidence of extreme and prolonged droughts is also stressing ecosystems, with dire consequences for both plant and animal species (www.unesco.org). Of all Earth's water, only 2.75 percent is freshwater, and of that, three-quarters, is sequestered, or locked up, in glaciers and permanent snow cover. Only a tiny fraction of planetary water, about 0.01 percent, is surface water found in rivers and lakes and thus readily accessible, see Table 10.1.

Reservoir	Volume (km^{3}x 1,000,000)	Percentage of Total
Oceans	1,370	97.25
Ice caps and glaciers	29	2.05
Groundwater	9.5	0.68
Lakes	0.125	0.01
Soil moisture	0.065	0.005
Atmosphere	0.013	0.001
Streams and rivers	0.0017	0.0001
Biosphere	0.0006	0.00004

Source: Fundamentals eBook. www.physicalgeography.net.

Table 10.1 Inventory of Water on Earth's Surface

In much of the world, freshwater removed from both ground and surface sources is being used up far faster than it is being replenished. Western Asia has the most severe water supply problem in the world, with over 90 percent of its population experiencing severe water stress. In Spain, over half of its approximately 100 aquifers are overexploited. In the United States, the situation is better but not significantly and perhaps not for long. In Arizona alone, more than 520 million cubic yards of water are removed from aquifers each year, double the replenishment rate from rainwater.

Perhaps the best-known case of water supply depletion is the Aral Sea, which in the 1960s began supplying water to Soviet collective farms to produce cotton. Formerly, it was a source of large fish; by the early 1980s, they had been virtually eliminated. By the 1990s, the Aral Sea occupied half of its original area, and it had shrunk in volume by 75 percent. By 2019 it had shrunk by 89 percent. A once-beautiful, large, rich, and deep lake with complex ecosystem had been largely destroyed in about 55 years due to human activities.

Another example of water body depletion was Lake Urmia, in Azerbaijan Province in Iran. In 1995, Lake Urmia reached a high-water mark; then in the ensuing two decades, the lake level dropped more than 7 meters (23 feet) and lost approximately 90 percent of its area. Consecutive droughts, agricultural water use, and dam construction on rivers feeding the lake have contributed to the decline. (https://earthobservatory.nasa.gov). see Figure 10.1.

Figure 10.1 Satellite Images of Lake Urmia in 2005 and a Decade Later in 2015

In the united Staes, water crises are occurring almost everywhere.

The rate of water consumption is over 40 times that of gasoline, and some argue that one day in the not-too-distant future, water may be more expensive than gasoline. In fact, the equivalent price of bottled water in a convenience store is already at least $5.80 per gallon ($1.50/L). The bright side of this picture is that in 1980 US water use was even higher, about 450 billion gallons a day; meaning that total and per capita water use dropped in spite of an additional 100 million people and a more than tripling of US gross domestic product. Thus, less water was used in an economy of 21 trillion than in an economy of $6 billion. Although direct consumption by people in buildings is not a large fraction of total water use in the United States, water shortages in many areas of the country are having an impact on development and construction. Figure 10-2 shows the impact of excess groundwater extraction on the land development by creation of sinkholes.

Technically, in LEED rating system more credits should be allocated to water efficiency category due to its vital role in our life. Without adequate source of water, no construction activity can be initiated, as such, conservation of water and efficient use of this precious commodity should be acknowledged more in the upcoming version of the LEED rating system.

Figure 10.2 Sinkhole are an example of land subsidence due to groundwater extraction. Since groundwater serves partly as a structural component to the rock, its depletion results in voids and eventual collapse, sometimes sudden and unpredictable, creating a substantial hazard to people and infra structure. (Source: Courtesy of Ildar Sagdejev)

Agriculture is the cause of serious water supply problems because it is responsible for over 80 percent of water consumption, and 60 percent of irrigation water is wasted because of leaky canals, evaporation, and mismanagement. Similar problems occur in the cities of many developing countries, with about 40 percent of the water in large cities being lost to leaky systems.

Buildings account for about 12 percent of freshwater withdrawals. The built environment hydrologic cycle, characterized by the input of high-quality potable water and the release of used, contaminated water, is inefficient, wasteful, and illogical. In its more extended context, the built environment hydrologic cycle also includes the irrigation of landscaping and the handling of stormwater. As pointed out by Hawken, Lovins, and Lovins (1999), the invention of water closet or W/C by Thomas Crapper was perhaps the start of an unfortunate trend in decision-making with respect to building water use. In order to dispense with the human waste generated in buildings, water closets dilute high-quality potable water with disease-ridden feces and relatively clean urine. Consequently, enormous quantities of water are wasted, and a potentially useful source of fertilizer is released into sanitary sewer system to combine industrial waste. The result is a complex, chemically intense, energy-consuming, pollution-producing system of wastewater treatment plants (WWTPs). A major re-thinking of the built environment hydrologic system is clearly needed to make better use of increasingly scarce and expensive potable water and to reduce the impact and cost of treating effluent from buildings.

In this chapter, we address how high-performance buildings can help contribute to reducing pressure on increasingly scarce water resources and to improving the health of local ecosystems. We also discuss strategies for selecting water sources, employing recent technological improvements in plumbing fixtures, evaluating alternative wastewater strategies, implementing sustainable stormwater management, and optimizing landscape water consumption.

HYDROLOGIC CYCLE TERMINOLOGY

Before discussing a high-performance building hydrologic strategy, it is important to define the common terms used in the context of the built-environment hydrologic water cycle. The following is the definitions of the most common terms:

Hydrologic Cycle:The continuous cycle of water between planetary reservoirs, such as the ground, water bodies, and the atmosphere. The hydrologic cycle is also referred to as the "water cycle" Table 10.1 shows the distribution of water on Earth's surface and in the atmosphere. The residence time for water on Earth's surface varies from as little as 1 month for soil moisture to as much as 10,000 years for deep groundwater (see Table 10.2).

Reservoir	Average Residence Time
Groundwater:	
Deep (fossil)	10,000 years
Shallow	100-200 years
Lakes	50 – 100 years
Glaciers	20-100 years
Rivers	2 – 6 months
Seasonal snow cover	2 – 6 months
Soil moisture	1 – 2 months

Source: Fundamentals eBook, www.physicalgeography.net/fundamentals/contents.html

Table 10.2 Typical Residence times of water Found in Various Reservoirs

Built Environment Hydrologic Cycle:

The flow and storage of all types of water on sites altered from their natural state for the purpose of building and infrastructure. Water types include potable water, rainwater, stormwater, graywater, blackwater, and reclaimed water that are used, processed, stored, and moved by employing a variety of technologies that, in the case of high-performance buildings, are coupled with natural systems.

Potable Water: Water that is safe for human consumption i.e., has high quality and low risk of harm). Potable water generally is obtained from groundwater or surface water sources and then processed to increase its quality to drinking water standards.

Groundwater: Water that is found underground in rock formations, such as aquifers and in soils. Groundwater is extracted for human consumption using shallow wells or deep, artesian wells. Water that seeps into the ground to add to the supply of groundwater is referred to as "recharge water".

Surface Water: Water that collects on Earth's surface in rivers, streams, lakes and ponds, and which serves as a source of recharge to groundwater.

Fossil Water: Deep groundwaterthat has a long residence time, sometimes on the order of thousands of years, although this water source has existed for a long time in underground aquifers, it is being depleted rapidly because it is not readily replenished and is essentially a nonrenewable resource. In the United States, US Department of Agriculture reported that in parts of three leading grain-producing states that draw water from the Ogallala Aquifer, Texas, Oklahoma and Kansas, the underground water table has dropped by more than 100 feet, as a result wells have gone dry on thousands of farms in the southern great plane.

Stormwater: Water that does not infiltrate into the ground and either runs off into bodies of water or enters the stormwater system. Includes water from the precipitation of rain and snow, water from melting snow, and water from overwatering.

Rainwater: Water from liquid precipitation, excluding water from snow, hail, and sleet, that has not entered a stream, lake, or other body of water.

Rainwater harvesting: The collection, storage, and use of rainwater. Most systems use the roof surface as the collection area and a large, galvanized steel, fiberglass, polyethylene, or ferrocement tank as the storage cistern. When the water is to be used just for landscape irrigation, typically only sediment filtration is required. When water is being collected and stored for potable uses, additional measures are required to purify it and ensure its safety. Rainwater harvesting offers several important benefits, including reducing pressure on limited water supplies and reducing stormwater runoff and flooding. It also can be a better-quality source of water than conventional sources. After purification, rainwater is usually very safe and of high quality.

Reclaimed Water: Water from a WWTP that has been treated and can be used for non-potable purposes, such as landscape irrigation, cooling towers, industrial process uses, toilet flushing, and fire protection. In some areas of United States, reclaimed water may be referred to as irrigation quality water, but potential uses can extend well beyond irrigation.

Blackwater: Water containing human waste. Wastewater from kitchen sinks and dishwashers is considered blackwater because it contains oil, grease, and food scraps, which can burden the treatment and disposal processes.

Graywater: Water from showers, bathtubs, bathroom sinks, washing machines, and drinking fountains. Graywater also includes condensation water from refrigeration equipment and air conditioners, hot tub drain water, pond and fountain drain water, and cistern drain water. Graywater contains a minimum amount of contamination and can be reused for certain landscape applications. Although graywater reuse still is being debated by public health officials, no cause of illness has ever been traced to such reuse. Both graywater and blackwater contain pathogens, humans should avoid contact with either, but blackwater is considered a much higher risk medium for the transmission of waterborne diseases. Although they are not blackwater, these water sources should not be included in graywater that is to be used for irrigation: garden and greenhouse sinks, water softeners backflush, floor drains, and swimming pool water. In building served exclusively by composting toilets and thus producing no true blackwater, it may be useful to include kitchen wastewater in graywater by taking special precautions to eliminate organic matter.

Xeriscaping: A landscaping strategy that focusses on using drought-resistant native and adapted species that require minimal to no water for their maintenance. The term is derived from the Greek word "xeri", meaning "dry" the strategy also referred to as "enviroscaping".

Living Machine: A trademark and brand name for a form of ecological wastewater treatment designed to mimic the cleaning functions of wetlands. The system is an intensive bioremediation system that can also produce beneficial by-products, such as reuse quality water, ornamental plants, and plant products useable for building materials, energy, or animal feed.

BENEFITS OF WATER EFFICIENCY

Reducing building water consumption and rethinking the wastewater strategy employed for the built environment can dramatically extend the available supply of water, improve human health, and reduce threats to ecological systems. In addition to these benefits, the Rocky Mountain Institute (www.rmi.org), a non-profit organization that provides a wide range of services on energy, water, development, and green building issues, suggests that water efficiency can have these tangible and calculable benefits:

1. Energy savings. More money can be saved by reducing the energy needed to move, process, and treat water than the actual value of the water saved.
2. Reduced wastewater production. Reducing the water consumption also reduces wastewater generation, lowering costs for building owners. Wastewater costs are significantly higher than the cost of portable water.
3. Lower facilities services investments. Designing water-efficient buildings reduces the costs of water and wastewater infrastructure.
4. Industrial processes. Innovative in water use in production systems can result in new processes and approaches.
5. Higher worker productivity. Facilities that incorporate resource efficiency measures are known to have a more productive workforce.
6. Reduced financial risk. Implementing water efficiency can be accomplished as needed, thus reducing costs and risks for large facilities.
7. Environmental benefits. Lowering water consumption results in reduced impact on natural systems.
8. Public relations value. Protecting the environment is looked upon favorably by the general public and clients.

The building hydrological cycle and energy use are tightly coupled, with very little impact being apparent to the building owner. Complex and expensive systems extract potable water from surface water and groundwater sources, then pump it for treatment and distribution, requiring large quantities of energy that are generally subsidized by the low-cost sanitary sewer and lift stations to central WWTPs, consuming relatively large amounts of energy. (The term "watergy" sometimes is used to describe the tightly intertwined relationship of water and energy). The good news is that reducing water consumption reaps numerous positive benefits, not only by reducing flows through the system but also by lowering overall energy consumption and associated pollution from energy sources, (Kibert,2022).

HIGH-PERFORMANCE BUILDING HYDROLOGIC STRATEGY

The built environment hydrologic cycle involves the handling and use of water indoors and outdoors. Drinking water is imported to the indoor for consumption and exported as wastewater. The imported water into the building could be drinking water from the municipal water system or from wells; rainwater from cisterns; or, when permitted, graywater recycled within the building. For the outdoors, the build environment hydrologic cycle can be extraordinarily complex due to the challenges of handling sometimes large volumes of rainwater and providing it for landscape irrigation. Rainwater falling in the building site can have different fate. On a greenfield, or previously undeveloped site, most rainwater infiltrates into the ground, and the remainder flows off into streams or other bodies of water. On a developed site, the situation can be different, with relatively little water infiltrating into the ground. Buildings and landscapes, such as sidewalks, and roads, preventing infiltration of rainwater and inducing water flow across parking lots and roads. Rainwater must be either collected and discharged into the municipal stormwater system that receives and processes this water or stored onsite in retention or detention ponds. Designers of high-performance buildings have developed novel-built environment hydrologic strategies that are having significant positive impacts on water consumption. The focus of these approaches is threefold: (1) to minimize the consumption of potable or drinking-quality water, (3) to reduce generation of wastewater, (3) to maximize rainwater infiltration into the ground.

These strategies together with emergence of several key technologies, are resulting in high-performance buildings with enormous reductions in their water consumption and wastewater generations profiles. These innovative strategies are discussed below:

STEPS IN DEVELOPING A HIGH-PERFORMANCE HYDROLOGIC STRATEGY

The next eight logical steps can be used to develop a hydrologic strategy for high-performance buildings:

1. Select the appropriate water sources for each consumption purpose. Potable water must be used for human consumption or ingestion. In addition to potable water, other water sources include rainwater, graywater, and reclaimed water. These alternative sources of water can be used for landscape irrigation, fire protection, cooling towers, chilled and hot water, toilet and urinal flushing, and other applications for which valuable potable water can be minimized. In each case, the availability of each alternative water source should be analyzed to determine which mix is optimum for a particular project and its forecasted water use profile.

2. For each purpose, employ technologies that minimize water consumption. This category can include a combination of low-flow fixtures (toilets, urinal, faucets, and showerheads), no-flow fixtures (composting toilets, waterless urinals), and controls (infrared sensors). For cooling towers,

chemical free electromagnetic technology can reduce scaling caused by biological contaminants, both of which can reduce the system performance. For landscaping, highly efficient drip irrigation systems use far less water and deliver the water to the plant roots with more than 90 percent efficiency. Additionally, drought- tolerant native and adapted species can be employed in the landscape scheme, an approach that often can eliminate the need for an irrigation system.

3. Evaluate potential for a dual wastewater system. Such a system separates lightly contaminated water from sinks, drinking fountains, showers, dishwashers, and washing machines from human waste-contaminated sources such as toilets and urinals. This dual piping system separates graywater from blackwater, thus providing the capability for recycling water within the building.

4. Analyze the potential for innovative wastewater treatment strategies. For example, construct wetlands or Living Machines can be employed to process effluent. These approaches are rapidly evolving and beginning to appear in more high-performance building projects each year as the practice of using nature in symbiosis with the building process becomes more defined.

5. Apply life-cycle costing (LCC) to analyze the costs and benefits of adopting practices that reduce water flow through building and its landscape beyond the levels mandated by the Energy Policy Act Of 1992 (EPAct 1992). A simple LCC that examines nothing more than the cost of potable water generally provides long payback times, perhaps in the 10- to 20-year range. Including reductions in wastewater generation and the costs associated with its treatment provides an accelerated payback. A more liberal interpretation of costs, such as the actual energy cost of moving

water and wastewater, emissions associated with energy generation, worker productivity improvements, and general environmental benefits, shortens the payback time of the initial investment. Finally, it can be reasonably expected that the price of potable water in most regions will increase at a greater rate than the general inflation rate and perhaps dramatically faster. Including this factor in the LCC evaluation, along with other indirect cost factors, should bring the paybacks into the same range as those for good energy conservation measures (i.e., seven years or less)

6. Design landscaping to use minimal water for its maintenance and upkeep and consider the restoration of ecological systems as an important part of the building design.

7. Design parking, paving, roads, and landscaping in such a way to maximize infiltrationof stormwater. Prior to buildings being present on the site, a natural hydrological cycle functioned to move water among the atmosphere, ground, bodies of water, and ecological systems. Restoring, the natural hydrologic cycle can benefit natural systems and reduce the need for complex and expensive stormwater infrastructure.

8. Incorporate green roofs into buildings to store and naturally process stormwater and contribute to the regeneration of the ecology of the building location, (Kibert, 2022).

ESTABLISHING WATER CONSUMPTION TARGETS

Limits of water consumption in buildings are set by building codes, which are, in turn, based on legislation. Table 10.3 shows the progress in setting maximum water consumption levels for typical building plumbing fixtures. One of the landmark pieces of legislation concerning potable water consumption is the Energy Policy Act of 1992. EPAct 1992 requires all plumbing fixtures used in the United States to meet ambitious targets for reducing water consumption; as a result, building codes mandate these dramatically lower levels of water consumption. Additional requirements for water efficiency for prerinse spray valves used in commercial kitchen were set by the Energy Policy Act of 2005 (EPAct 2005). In 2007, California passed legislation that established even more stringent requirements for toilets and urinals, reducing toilet water consumption from 1.6 gallons per flush to 1.28 gallons per flush and urinal water consumption from 1.0 gallon per flush to 0.5 gallon per flush.

Beyond legislation and code requirements, the US Environmental Protection Agency (EPA) created the voluntary WaterSense label in 2006, which required that WaterSense-certified fixtures use at least 20 percent less water than the requirements of EPAct 1992. The Water Sense label is awarded based on the third-party certification that the fixture meets EPA requirements see Figure 10.3. WaterSense is now the de facto standard in many jurisdictions and green building assessment systems such as Green Globes award points in the water category only for performance that exceeds the WaterSense fixture requirements.

Figure 10.3 WaterSense label to stimulate the development of technologies that improve on the EPAct 1992 requitements by at least 20 percent Source: US Environmental Protection Agency

Setting goals for building water consumption that exceeds code requirements is an important first step in designing a strategy that makes sense. If the factor 10 concept is applied to the issue of water consumption, potable water and, by inference, wastewater, it should be reduced by 90 percent for the purpose of producing a sustainable future. This means that typical per capita household consumption of potable water in this country must be reduced from 100 gallons per day to about 10 gallons per day.

Fixture Type	Units	EPAct 1992	WaterSense	Best Technology
Water Closet, flushing	gpf	1.6	1.28	0.8*
Urinal	gpf	1.0	0.80	0.0/0.125+
Showerhead	gpm at 80 psi gpm at 60 psi	2.5/2.2	2.0/1.8	0.57/0.49
Faucet	gpm at 80 psi gpm at 60 psi	2.5/2.0	2.0/1.6	0.50
Replacement Aerator	gpm	2.5		0.50
Metering faucet	gpc	0.25		0.09

Key:
gpf = gallons per flush
gpm = gallons per minute
psi = pounds per square inch
*** = The best technology in this case is the water closet with the lowest water flush rate. Composing toilets use no water but generally have limited application.**
+ = For urinals, the best technology in the waterless urinals, which has no water use. Ultra-low-flow use about one-eighth of a gallon per flush and can be selected in cases where a waterless urinal is not appropriate or desirable.

Table 10.3 Maximum Water Consumption Levels for Typical Building Plumbing Fixtures

To accomplish this remarkable reduction requires that water be reused and recycled at high rates. For example, per capita consumption of water is almost evenly divided between outdoor and indoor uses. If any recycled water were used outdoors for irrigation landscaping, per capita consumption of potable water drops to 50 gallons per day. Indoors, almost half of the water consumed is for toilet and urinal flushing and using only recycled water for this purpose would further reduce potable water consumption to 25 gallons per day. This relatively straightforward measures produce an immediate Factor 4 reduction. Additional measures that incorporate low-flow fixtures and electronic controls can nearly produce the desired Factor 10 reduction.

As an alternative to using Factor 4 or Factor 10 strategies to set targets for building water consumption, a more recent approach known as the "net zero built environment" is emerging, Kibert, 2022. In addition to addressing energy, the net zero strategy addresses water consumption by setting limits to water usage based on annual precipitation and water recycled within the building. This is referred to as "net zero water" and is required for certification under the Living Building Challenge. a number of US military installations such as the Aberdeen Proving Grounds in Maryland and Fort Hood in Texas are participating in net zero water pilot programs. As an example, if a net zero water target were established as a criterion for a building in Gainesville, Florida, where the average annual rainfall is 36 inches, each square foot of roof would provide 3 cubic feet, or about 22.5 gallons of water. For Rinker Hall, LEED gold certified building at the University of Florida in Gainesville, with three stories and a 15,000 ft^2roof, the water budget would be about 330,000 gallons per year.

WATER SUPPLY ALLOCATION

The major objective of water supply allocation for a high-performance building is to minimize the use of drinking water consumption by utilizing the other sources of water available at the site. The other available sources of water may include rainwater, graywater and reclaim water which are considered non-potable sources of water. When the feasibility of using each of these non-potable sources has been assessed, the next step is to ensure that consumption of both potable and non-potable water is minimized. A wide range of high-efficiency fixtures are available that provide flow rates well below the EPAct 1992 requirements. Waterless plumbing fixtures are becoming more widely available and price competitive as manufacturers begin offering more alternatives. EPAct 1992 set relatively ambitious limits on water use for water fixtures. However, water use by high-performance green buildings normally exceeds the EPAct 1992 requirements. For example, LEED requires at least a reduction of 20 percent in potable water consumption over EPAct 1992 requirements.

There are many types water fixtures available in the marketplace and everyday new types of fixtures are being presented to customers. In this book we are not discussing different types of water fixtures such as toiles, urinal, showerheads, faucets, drinking water fountains, etc. However, as mentioned earlier the general objective is to minimize the flow rates as much as possible to take pressure off of the potable water in our built environment.

NON-POTABLE WATER SOURCES

Among the all-non-potable water sources, rainwater is considered a crucial source of water for survival for all human existence. Rainwater usually is collected from the rooftops of homes and buildings and stored inside a cistern or storage tank. Rainwater harvesting, which once was popular in Europe, was abandoned for some time due to the emergence of central water system in almost every municipality. However, by emergence of the high-performance building it became popular again. Currently, rainwater harvesting is the only way of providing drinking water to residents within Caribbean islands due to poor groundwater quality and high salinity levels. The "Texas Guide for Rainwater Harvesting (2005)" provides an excellent overview of rainwater harvesting, cites three factors that are propelling rainwater back into the picture as a viable water source:

1. The escalating environmental and economic costs of providing water by centralized water systems or by well drilling
2. Health concerns regarding the source and treatment of polluted waters
3. The perception that there are cost efficiencies associated with reliance on rainwater

Rainwater systems are appropriate when one or more of the following factors is present:

- Groundwater water supplies are limited or fragile. Fragile aquifer systems are those that, when pumped, can threaten ecologically valuable surface waters and springs
- Groundwater supplies are polluted or significantly mineralized, requiring expensive treatment
- Stormwater runoff is a major concern.

A rainwater harvesting system generally has seven key components:

1. Catchment area. With most rainwater harvesting systems, the catchment area is the building's roof. The best roof surface for rainwater harvesting does not support biological growth (e.g., algae, mold, moss), is smooth so that pollutants deposited on the roof are quickly removed by the roof wash system and has a minimal number of overhanging tree branches above it. Galvanized metal is the roofing material used most for rainwater harvesting.
2. Roof wash system. This is a system for keeping dust and pollutants that have settled on the roof. It is necessary for systems used as a source of potable water but also recommended for other systems, as it keeps potential contaminants out of the tank. A roof wash system is designed to purge the initial water flowing off a roof during rainfall.
3. Pre-storage filtration. To keep large particulates, leaves, and other debris out of the cistern, a domed stainless-steel screen should be secured over each inlet leading to the cistern. Leaf guards over gutters

can be added in areas with significant windblown debris or overhanging trees.

4. Rainwater conveyance. This is the system of gutters, downspouts, and piping used to carry water from the roof to cistern.
5. Cistern. This is usually largest single investment required for a rainwater harvesting system. Typical materials used include galvanized steel, concrete, ferrocement, fiberglass, polyethylene, and durable wood (e.g., redwood or cypress). Costs and expected lifetimes vary considerably among these options. Tanks may be in a basement, buried outdoors, or located aboveground outdoors. Light should be kept out to prevent algae growth. Cistern capacity should be sized to meet the expected demand. Particularly for systems designed as the sole water supply, sizing should be modeled because of 30-year precipitation records, with sufficient storage to meet the demand during times of the year having little or no rainfall.
6. Water delivery. A pump is generally required to deliver water from the cistern to the point of use, although occasionally gravity-fed systems are possible with appropriate placement of system components.
7. Water treatment system. To protect plumbing and irrigation lines (especially with drip irrigation), water should be filtered through sediment cartridges to remove particulates, preferably down to 5 micrometers. For systems providing potable water, additional treatment is required to ensure a safe water supply. This can be provided with microfiltration, ultraviolet sterilization, reverse osmosis, or ozonation (or a combination of these methods). With some systems, higher levels of

treatment are provided only at a single faucet
where potable water is drawn.

Rainwater harvesting system has immense potential for reducing potable water consumption by introducing a water source that is readily obtainable in many regions of the United States, see Figures 10.4 and 10.5.

Figure 10.4 Rainwater is harvested in a cistern system consisting of these 11 components: (1) nontoxic, noncorrosive roofing material; (2) nontoxic, noncorrosive gutters and downspouts; (3) first-flush divertor with clean out trap and bleed valve; (4) debris trap and sediment filter: (5) easily accessible but locked passageway; (6) engineered cistern that avoids direct sunlight exposure to the collected rainwater; (7) automatic water refill with air gap supplied from the building; (8) cistern; (9) pump electrical supply; (10) pump stop/start relay; (11) backflow prevention valve. The submersible pump is located behind the cistern. (source: H_2Options, Inc.)

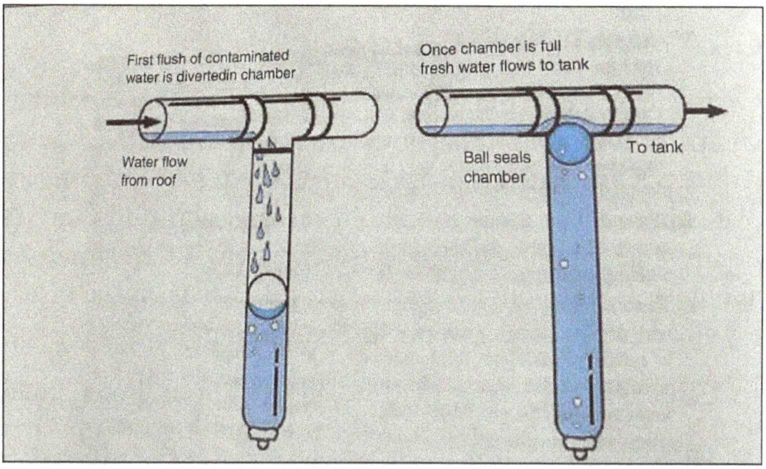

Figure 10.5 The simplest first flush diverter is a stand pipe that captures and diverts contaminants were washed from the roof. Rainwater fills the standpipe, backs up, and then allows water to flow into the main collection piping after the contaminants have been flushed out. (Source: Texas Water Development Board)

Despite this advantage, there are no standard designs or approaches to designing a rainwater harvesting system; hence, currently, each system designed for a building is unique. Factors included in the design include the roof material and slope, rainfall intensity, airborne pollutants (such as smoke, dust, and automobile exhaust), and debris generated from trees and other nearby vegetations. Because of the wide range of factors affecting the design of the rainwater harvesting system, they can be prone to failure and unreliable, resulting in a potential erosion of interest in rainwater as a substitute for potable water. The creation of clear standards, designs, and standard components would go a long way toward resolving this problem and making the implementation of these systems standard practice. Graywater is generally used for landscape irrigation, but it also can be used to flush toilets and urinals.

GRAYWATER SYSTEMS

Graywater is water from showers, bathtubs, washing machines and bathroom sinks. Using greywater is an innovative way to reduce potable water. It provides an efficient means to irrigate plants in outdoor areas and also it recharges groundwater. A homeowner who installs a greywater system, in average saves about 40,000 gallons of water per year, resulting in much lower water bills (http://waterwisegroup.com/)

In California, greywater shall be directed to and contained within an irrigation or disposal field. Ponding or runoff is prohibited and shall be considered a nuisance. Greywater may be released above the ground surface provided at least two inches of mulch, rock or soil, or a solid shield covers the shield area. A greywater shall not be used in spray irrigation (sprinkler system), allow to pond or runoff and shall not be discharged directly into or reach any storm manhole or any surface body of water.

Root crops which are eaten uncooked should not be irrigated with greywater. Plants that thrive only in acid soil should be watered with greywater, which is alkaline. Use greywater only on well-established plants, not seedlings or young plants (http://pubs.nmsu.edu). Potential hazard associated with greywater includes:

- All greywater has the potential to harbor dangerous bacteria and viruses.
- It is never potable
- Micro-organisms present in untreated greywater can cause damage to foliage
- Untreated greywater should not be used for lawn sprinklers, this could spread dangerous airborne bacteria

- Long-term irrigation of arid loess soil with greywater may result in accumulation of salts, surfactants and boron in soil, causing change in soil properties and toxicity to plants.

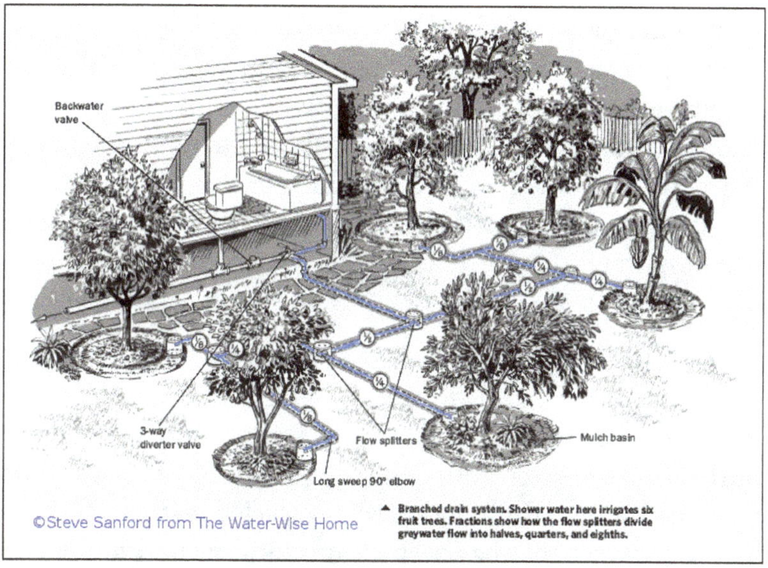

Figure 10.6 Schematic diagram of Laundry-to-Landscape (L2L)

As a rule, greywater can be used for subsurface irrigation of lawns, flowers, trees and shrubs, but it should not be used for vegetable gardens. People like the idea of reusing laundry water to irrigate their yards. It is called laundry-to-landscape (L2L) greywater system, see Figure 10.6. Installing a greywater system is part of a growing movement for homeowners eager to reduce their use of drinking water. Greywater for toilet flushing is more appropriate in commercial or other high-density properties where dedicated staff can maintain it over time. In general, toilet-flushing greywater systems usually require frequent maintenance, manual cleaning of filters, and chemical disinfectants to prevent odors in the bathroom.

RECLAIMED WATER

Reclaimed or recycled water starts out as domestic wastewater (sewage) but is then treated and tested to use for specific purposes. Reclaimed water is colorless and may have a slight chlorine smell. It is impossible to tell recycled water apart from tap water with a naked eye. Recycled water does not contain any constituents that exceed federal and state drinking water standards and is safe for permitted uses.

The disadvantages of reusing water are that there is a potential health risk posed by the bacteria it may contain such as E. Coli or other harmful bacteria which are transported to wherever the water is finally used. This might lead to the outbreak of water-born and food-borne diseases. The reclaimed water, also referred to as recycled water, is municipal wastewater that has been cleaned and treated to remove pollutants and contaminants so that the water can be safely reused for a variety of uses. Generally, reclaimed water has a little odor. The water is disinfected, so the only smell is a slight chlorine odor. Studies on the effects of recycled water on plants have found that most plants are not greatly affected, especially if the recycled water is supplemented with fresh water. However, reclaimed water cannot be sprayed directly onto any fruit or vegetable that can be eaten raw with the skin or peel still on such as lettuce, strawberries or tomatoes.

Recycled water is safe to swim, since it is regularly tested by the Laguna Environmental Laboratory, the Environmental Protection Agency, the Regional Water Quality Control Board, and the Department of Public Health.

Per Kibert 2022, to consider using the reclaimed water for drinking in a building, one or more of these conditions should be present:

1. High cost of water for the need to extend the drinking water supply.
2. Local public agency encouraging or mandating drinking water conservation
3. Availability of high-quality effluent from a WWTP, or
4. Recognizing by the building owner of environmental benefits of water reuse.

A modern WWTP has three stages of treatment primary, secondary, and tertiary with each succeeding stage requiring more energy and chemicals than previous stage. Virus inactivation is attained by granular carbon adsorption plus chlorination or by reverse osmosis, ozonation, or UV exposure. One additional precaution is to run reclaimed water mains at least 12-inches lower in elevation than potable water mains and to separate them from potable or sewer mains by a minimum of 10 psi lower pressure than potable water mains to prevent backflow and siphonage in the event of accidental cross-connection.

Piping and valves used in reclaimed water systems should be color-coded with purple tags or tape to demonstrate piping identification and cross-connection problems when installing systems, see Figure 10-7. Potable water mains usually are color-coded blue, while sanitary sewers are green.

Figure 10.7 Reclaimed water is former wastewater that is cleaned and distributed through a clearly coded system of bright purple pipes

WASTEWATER STRATEGIES

Once the wastewater was collected it is either treated by directly using WWTP, or by using indirect approach which we will discuss in this section. In the indirect approach nature is brought into the building and enclosed in tanks and vats through which wastewater is passed and cleaned up by plants, light, and bacteria. We will discuss two natural-based systems to wastewater treatment, constructed wetlands and the Living Machine concept.

Constructed wetlands are treatment systems that use natural processes involving wetlands vegetation, soil, and their associated microbial assemblages to improve water quality (http://epa.gov>wetlands). The main disadvantages associated with constructed wetland are: (1) the large land is required (2) the unpleasant odor that can be associated with certain types of influents are being treated.

Wetlands can improve water quality by removing pollutants from surface waters. Three pollutant removal processes provided by wetlands are particularly important: (1) sediment trapping, (2) nutrient removal and (3) chemical detoxification. Constructed wetlands represent an alternative treatment system to conventional treatment systems such as activated sludge process. All types of constructed wetlands exhibit high treatment efficiency for organics and suspended solids. Removal of these parameters is comparable with conventional systems. Constructed wetlands consist of water-loving plants, soils and pools of water that can help filter stormwater before it is slowly released downstream. As stormwater flows through a constructed wetland, wetland microbes, plants, and soil help filter pollutants and treat the water through natural ecological processes. Constructed wetlands using nature to perform tasks that would otherwise be accomplished by energy-intense mechanical, chemical, electrical systems that has four distinct advantages (Campbell and Ogden, 1999):

1. Nature is self-maintaining, self-regulating, and self-organizing.
2. Nature is powered by solar energy and chemical energy stored in organic materials.
3. Nature systems can degrade and absorb undesirable toxic and metal compounds, converting them into stable compounds.
4. Natural systems are easy to build and operate.

The use of wetlands to treat wastewater from buildings provides precisely this type of opportunity because these ecological systems can break down organic waste, minimizing the need for complex infrastructure and creating nutrients that benefit the species performing these services, Kibert, 2022. Constructed wetlands can be characterized as passive systems for wastewater treatment. They mimic natural wetlands by using the same filtration processes to remove contaminants from wastewater, see Figure 10-8. In addition, to removing organic nutrients, constructed wetlands have the ability to remove inorganic substances; they can be used to treat industrial wastewater, landfill leachate, agricultural wastewater, acid mine drainage, and airport runoff.

Wetlands remove contaminants from water by several mechanisms including nutrient removal and recycling, sedimentation, biological oxygen demand, metals precipitation, pathogen removal, and toxic compound degradation. When considering the use of construction wetlands for wastewater treatment, a number of site-specific factors must be considered. These factors include hydrology (groundwater, surface water, permeability of the ground), native plant species, climate, seasonal temperature fluctuations, local soils, site topography, and available area.

Figure 10.8 Per Kibert 2022, wetlands sometimes referred to as nature's kidneys, are natural habitats with distinct characteristics of soil percolation, vegetation, and wildlife habitat. They play an important role in the ecosystem while providing a filtering process whereby contaminants in stormwater runoff are degraded before entering the groundwater. Wastewater can be purposely directed into natural wetland for the highest cost benefit in terms of both conventional economic and natural capital (Source: Courtesy of T. Wyman)

Constructed wetlands are built for either surface or subsurface flow. Surface flow systems, see Figure 10 9, consist of shallow basins with wetland plants that are able to tolerate saturated soil and aerobic conditions.

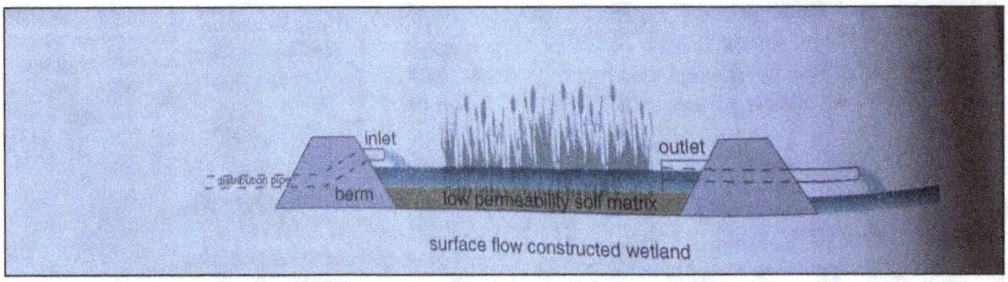

Figure 10-9 Surface flow is constructed wetlands mimics natural wetlands because the water flows aboveground as sheet flow. Wetland plants are selected to provide attachment areas for microbes, which are essential for water quality improvement. The outlet receives water from the wetland cell and directs it either to downstream wetland cells or to natural water systems (Source: Courtesy of T. Wayman)

The wastewater entering the surface system slowly moves via sheet flow through the basin and is released as clean water.

Subsurface systems, see Figure 10.10, where the wastewater flows through a substrate such as gravel, have the advantage of higher rates of contaminant removal, compared to surface flow systems, and limited contact for humans and animals. They also work especially well in cold climates due to the ground 's insulating properties. Cost is an important factor in deciding which approach is best for a particular situation. The good news about the constructed wetlands is that both the capital and operating costs are far lower than conventional WWTPs, with the added benefit of reduced direct and indirect environmental impacts associated with materials extraction, processing, and manufacturing.

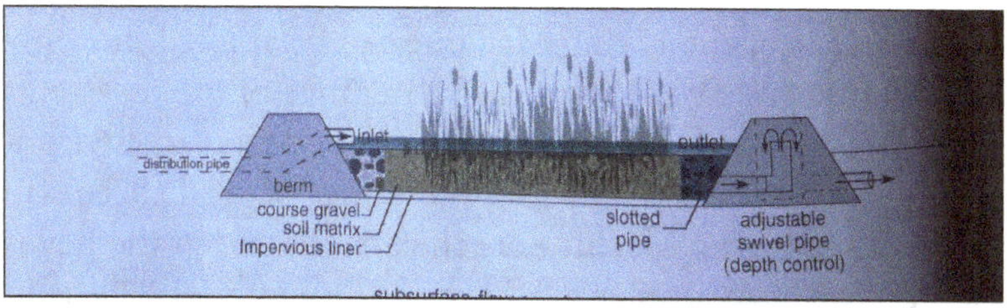

Figure 10.10 Subsurface flow constructed wetlands closely resemble WWTPs and must initiate and maintain all surface flow through the bed media to the outlet where water is collected from the base of media. (Source: Courtesy of T. Wayman)

LIVING MACHINE

Living Machine is an emerging wastewater treatment technology that utilizes a series of tanks, which supports vegetation and variety of other living organisms which naturally break down the contaminants molecule and clean up the wastewater. The best-known Living Machine is the one created by John Todd in 1999, pioneer in the development of natural wastewater processing system. The Living Machine differs from a conventional WWTP in four basic respects (Kibert, 2022):

1. The vast majority of Living Machine's working parts are live organisms, including hundreds of species of bacteria, plants, vertebrates such as fish and reptiles.
2. The Living Machin's could design its internal ecology in relation to the energy and nutrient streams to which it is exposed.
3. The Living Machin's can self-replicate through reproduction of the organisms in the system.

The concept of the Living Machine can be applied not only to an alternative WWTP but also to a range of other systems that can generate fuel, grow food, restore degraded environments, and even heat and cool buildings. Several successful examples of Living Machine are the one located in the Lewis Center for Environmental Studies at Oberlin College in Oberlin Ohio, which processes wastewater from the occupants of this 14,000-square-foot building.

Designing The High-Performance Building Hydrologic Cycle

As it was mentioned earlier, the objective of designing a hydrologic cycle for a high-performance building is to reduce the drinking water consumptions by introducing and utilizing alternative source(s) of water. To determine how well a given strategy in reducing drinking water demand is working and to meet the requirements for most green building certification schemes, a baseline model of building water and wastewater system is created to allow comparisons. Both the types of plumbing fixtures and alternative sources of water (rainwater, greywater and reclaimed water) can be varied in the baseline model to determine how much potable water has been saved.

Table 10.4 shows the flow rates for flush and flow fixtures. Flush fixtures as the name implies, are plumbing fixtures that use a fixed quantity of water for their function, while flow fixtures use a quantity of water that depends on the length of time the fixture is used. The water use and flow rates shown in the table are the starting point for determining the water use of a building.

Flush Fixture Type	Water Use (gpf)
Conventional low-flow water Closet*	1.60
High efficiency toilet (HET), single flush gravity	1.28
HET, single-flush pressure assist**	1.00
HET, dual-flush (full-flush)**	1.28
HET, dual-flush (low-flush)**	1.00
Waterless toilet	0.00
Composting toilet	0.00
Conventional low-flow urinal*	1.00
EPA WaterSense Compliant urinal	0.50
High-efficiency urinal (HEU)	0.125
Waterless urinal	0.00
Flow Fixture Type	**Water Use (gpm)**
Commercial low-flow lavatory faucet*	2.20
High-efficiency lavatory faucet***	1.50
Conventional low-flow kitchen sink faucet	2.20
High-efficiency kitchen sink faucet***	1.50
Conventional low-flow showerhead*	2.50
High-efficiency showerhead**	2.00

*Conventional low-flow fixtures meet the EPAct 1992 requirements
**HET fixtures meet the EPA WaterSense requirements; HEU fixtures exceed the 0.50 gpf WaterSense requirement and use just one pint or 0.125 gpf.
***Faucet flow requirements are based on a water pressure of 80 PSI.

Table 10.4 Water Use by Various Types of plumbing Fixtures

Establishing the population and occupancy type is the next step in identifying a baseline. Along with the population, the female-to-male ratio must be known to quantify water consumption through fixture types. For projects that have either unknown ratios or ratios that are relatively the same, it is best to model even gender distribution. Occupancy type classifies the people who use the facility's plumbing system. The largest occupant type difference is between a full-time equivalent (FTE) and a transient or someone temporarily visiting the facility. An FTE refers to a person who occupies the building in the equivalent of an eight-hour day. Table 10.5 shows the typical daily use patterns of plumbing fixtures based on gender and type of occupant.

Fixture Type	FTE	Student Visitor	Retail Customer	Resident
Water closet:				
Female	3.0	0.50	2.0	5.0
Male	1.0	0.10	0.1	5.0
Urinal	2.0	0.40	0.1	n/a
Lavatory faucet	3.0	0.50	0.2	5.0
Commercial at 15 sec, 12 sec, with auto control; residential at 60 sec				
Shower	0.1	0.00	0.0	1.0
Shower duration: commercial at 30 seconds; residential at 480 seconds				
Kitchen sink	1.0	0.00	0.0	4.0
Kitchen sink duration: commercial at 1 sec; residential at 60 sec				

Table-10.5 Uses per Day for Plumbing Fixtures by Gender and Type of Building Occupants

BASELINE WATER MODEL EXAMPLE

To best understand how to generate a baseline water model, we start with a few pieces of information. As an example, we will use a governmental building designed to have a total of 60 full-time male and 40 full-time female employees. The building is assumed to have 250 transient male visitors and 150 transient female visitors per day.

Occupant Type	Flush Fixture	Daily use	Potable Water (gpf)	No. of Occupants	Water Use (gal)
FTE	Conventional low-flow water closet (male)	1.0	1.6	60	96
FTE	Conventional low-flow water closet (female)	3.0	1.6	40	192
FTE	Conventional low-flow urinal	2.0	1.0	60	120
Transient	Conventional low-flow water closet (male)	0.1	1.6	250	40
Transient	Conventional low-flow water closet (female)	0.5	1.6	150	90
Transient	Conventional low-flow urinal	0.4	1.0	250	100
	Total Flush Fixture Potable Water Use (gal)				638

Table 10.6a Example of baseline water Model

Occupant Type	Flow Fixture	Daily Uses	Potable Water (gpm)	Duration (sec)	No. of Occupants	Water Use (gal)
FTE	Conventional low-flow lavatory faucet	3.0	2.2	15	100	165
FTE	Conventional low-flow kitchen sink faucet	1.0	2.2	15	100	55
FTE	Conventional low-flow showerhead	0.1	2.5	300	100	125
Transient	Conventional low-flow lavatory faucet	0.5	2.2	15	400	110
Transient	Conventional low-flow kitchen sink faucet	0.0	2.2	15	400	0
Transient	Conventional low-flow showerhead	0.0	2.5	300	400	0
	Total flow fixture potable water use (gal)					**455**
	Total Daily Potable water use (gal)					**1093**
	Annual workdays					**260**
	Total annual potable water use (gal)					**284,180**
	Total annual wastewater generation (gal)					**284,180**

Table 10.6a Example of baseline water Model

The baseline water model assumes that the fixtures used in the building meet the EPAct 1992 requirements for maximum flow rates for plumbing fixtures. Fixture performances are selected from Table 10.4, and number of uses per day can be found in Table 10.5. Each fixture type must be modeled to identify a total water use (in gallons) per day. This value is determined by identifying the product of multiplying the appropriate occupant type by the number of daily uses per person and by the total water consumed per fixture use. This calculation can be found in Table 10.6. The rightmost column indicates the estimated water use per day, which is then summed up for both flush and flow fixtures.

To model the annual water consumption from the facility accurately, the number of workdays must be multiplied by the daily total water use; in this case, 260 days was determined. In this example, the total annual potable water use is predicted to be 284,180 gallons per year. Note that the estimated water quantity for a conventional water system can be used to estimate the wastewater quantity for the building.

USE OF LOW-FLOW FIXTURE STRATEGY

The most straightforward strategy for reducing potable water consumption in buildings is to incorporate plumbing fixtures that use significantly less water than code-compliant fixtures. For example, while a code-compliant urinal has a maximum water use of 1 gallon per flush, a HET is required to have a maximum potable water use of 0.5 gallon per flush; furthermore, waterless urinal uses no water at all. Table 10.7 shows the same baseline water consumption calculations with the modification of installing HETs and HEUs instead of their conventional counterparts. The results of this modification indicate a water use reduction of just almost 50%. It is now possible to determine the feasibility of such a retrofit by associating costs savings in both water consumption and wastewater treatment.

USE OF ALTERNATIVE WATER SOURCES STRATEGY

Further significant savings in potable water consumption can be achieved by substituting other suitable water sources for potable water. Table 10.8 shows the impact of incorporating rainwater catchment and graywater systems. In this case, we are assuming that the size of the rainwater harvesting system is sufficient to supply graywater to be used for flushing toilets and urinals. By incorporating the potable water consumption in this scenario into the baseline model, a water use reduction of over 78 percent can be achieved. This is nearly to the potable water reduction to Factor 10.

WATER BUDGET RULES OF THUMB

Based on the example cases which were covered above, it is possible to develop some rules if thumb, sometimes called "Heuristics", set targets for potable water consumption. For the low-flow fixture strategy, high efficiency fixtures are used aggressively, and result is about a 50 percent, or Factor 2, reduction in potable water consumption compared to code requirements. For a combination of low flow fixtures and alternative water strategies, an 80 percent reduction in potable water consumption was achieved.

Occupant Type	Flush Fixture	Daily use	Potable Water (gpf)		No. of Occupants	Water Use (gal)
FTE	HET, single-flush gravity (male)	1.0	1.28		60	77
FTE	HET, single-flush gravity (female)	3.0	1.28		40	154
FTE	Waterless Urinal	2.0	0.00		60	0.
Transient	HET, single-flush gravity (male)	0.1	1.28		250	32
Transient	HET, single-flush gravity (female)	0.5	1.28		150	96
Transient	Waterless urinal	0.4	0.00		250	0.
	Total Flush Fixture Potable Water Use (gal)					**359**
Occupant Type	**Flow Fixture**	**Daily Uses**	**Potable Water (gpm)**	**Duration (sec)**	**No. of Occupants**	**Water Use (gal)**
FTE	High efficiency lavatory faucet	3.0	1.50	15	60	68
FTE	High efficiency kitchen sink faucet	1.0	1.50	15	40	15
FTE	High efficiency showerhead	0.1	2.00	300	60	60
Transient	High efficiency lavatory faucet	0.5	1.50	15	250	47
Transient	High efficiency kitchen sink faucet	0.0	1.50	15	150	0.
Transient	High efficiency showerhead	0.0	2.00	300	250	0.
	Total flow fixture potable water use					**190**
	Total Daily Potable water use (gal)					**549**
	Annual workdays					**260**
	Total annual potable water use (gal)					**142,740**
	Total annual wastewater generation (gal)					**142,740**
	Potable water saving compared to baseline model					**49.8%**

Table 10.7 Water Model for Low-Flow Fixture Scenario

Occupant Type	Flush Fixture	Daily use	Potable Water (gpf)		No. of Occupants	Water Use (gal)
FTE	HET, single-flush gravity (male)	1.0	0.		60	0.0
FTE	HET, single-flush gravity (female)	3.0	0.		40	0.0
FTE	Waterless Urinal	2.0	0.		60	0.0
Transient	HET, single-flush gravity (male)	0.1	0.		250	0.0
Transient	HET, single-flush gravity (female)	0.5	0.		150	0.0
Transient	Waterless urinal	0.4	0.		250	0.0
	Total Flush Fixture Potable Water Use (gal)					**0.0**
Occupant Type	**Flow Fixture**	**Daily Uses**	**Potable Water (gpm)**	**Duration (sec)**	**No. of Occupants**	**Water Use (gal)**
FTE	High efficiency lavatory faucet	3.0	1.50	15	60	68
FTE	High efficiency kitchen sink faucet	1.0	1.50	15	40	15
FTE	High efficiency showerhead	0.1	2.00	300	60	60
Transient	High efficiency lavatory faucet	0.5	1.50	15	250	94
Transient	High efficiency kitchen sink faucet	0.0	1.50	15	150	0.
Transient	High efficiency showerhead	0.0	2.00	300	250	0.
	Total flow fixture potable water use					**237**
	Total Daily Potable water use (gal)					**237**
	Annual workdays					**260**
	Total annual potable water use (gal)					**61,620**
	Total annual wastewater generation (gal)					**61,620**
	Potable water saving compared to Baseline Model					**78.3%**

Table 10.8 Water Model for a Combination of Alternative Water and Low-Flow Fixture Strategy

Consequently, it is possible to develop water reduction strategies that are more than Factor 4for an aggressive strategy that includes alternative water sources and low-flow fixtures and at least Factor 2 for a less aggressive strategy that uses a simple low-flow fixtures strategy.

As mentioned earlier, the concept of net zero water is receiving serious consideration, and it provides a sensible approach based on one of the core concepts of sustainability; that is, resource use should be constrained to what nature provides. If we assume the building in the three water models in the preceding sections were in a climate zone with 24 inches of annual rainfall and that the facility had a roof area of 15,000 ft^2, then 30,000 ft^3, or 224,000 gallons of water would be available for all uses. The baseline model shows that 284,140 are required, and the result is that a net zero water strategy would require about 60,000 gallons or 22% reduction in water consumption, which matches up to using high-efficiency WaterSense fixture throughout the facility.

The following sections are dealing with stormwater management which will include preventing stormwater runoff by reducing construction footprints and installing porous concrete to promote groundwater recharge. In addition, we will discuss how we should reduce consumptive use of landscaping by choosing native and drought-resistant plants and using more efficient irrigation methods to save outdoor water use in high-performance buildings.

SUSTAINABLE STORMWATER MANAGEMENT

Stormwater management has long been a challenging issue for built environment development. Replacing plants and trees that naturally uptake large quantities of water with buildings and hard-escapes such as roads and parking lots generates large volumes of stormwater run-off which carries debris, particles and chemicals along the way. This excess and chemically affected stormwater is an enormous headache for municipalities that then must build large stormwater management facilities at high cost to taxpayers and with additional costs to the environment. The excess stormwater usually causes flooding conditions which threatens the roads, infrastructures, farmlands and health of ecosystems.

Sustainable stormwater management is an innovative and effective approach that attempts to maintain the natural waterways and hydrology of the area intact during the site development. This idea was discussed earlier in Chapter 4, Ecological Design. Sustainable stormwater management recognizes that there is a relationship between the natural and built environments and treats them as integrated components of the watershed. Instead of the traditional approach of using piping and extensive collection system, it focuses on on-site collection and conveyance of stormwater from roofs, parking lots, streets to promote the infiltration of water into the ground. Vegetated natural systems slow and filter the water and enhance the interception and evaporation of rainfall through their leaves and roots. Vegetation also reduces stormwater runoff and removes pollutants in the process. Studies have shown that this approach can reduce stormwater runoff volume by as much as 65 percent. It also can remove 80 percent suspended solids and heavy metals and as much as 70 percent of nutrients, such as phosphate and nitrogen. Sustainable stormwater management integrates natural components, such as landscape swales and infiltration basins, with structural devices, such as cisterns, planters, pervious pavers, and pervious concrete and asphalt surfaces. Pervious asphalt and concrete are particularly interesting materials because they allow stormwater to infiltrate rapidly through the hard exterior surface and then into the ground. Using sustainable stormwater management reduces runoff and release stresses on stormwater collection system as well as the WWTPs. Figures 10.11 through 10.14 shows some of the components that may be part of a sustainable stormwater management system.

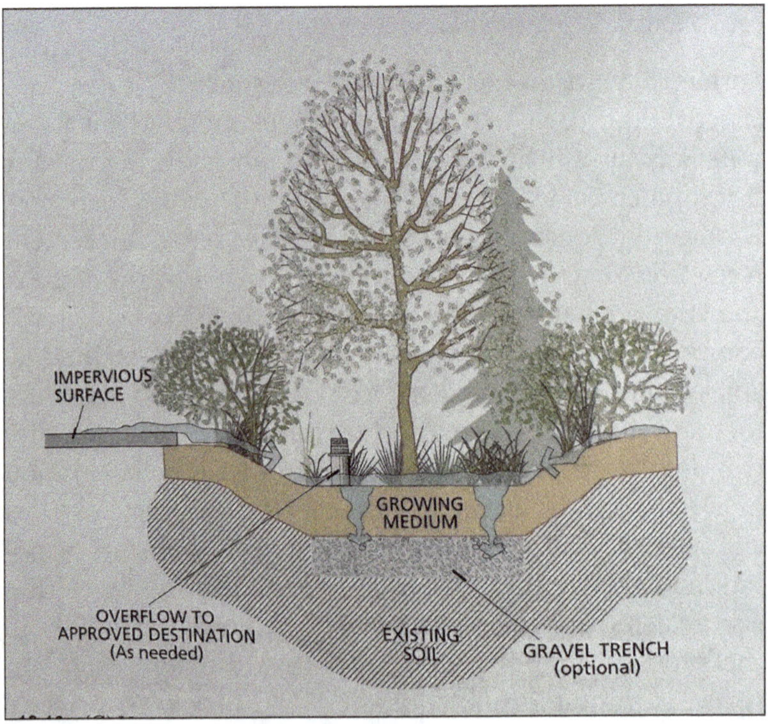

Figure 10-11 Vegetated Infiltration basins are also known as rain gardens or detention ponds. The basin is either excavated or created with barns then landscaped to temporarily store runoff until it infiltrates into the ground. These designs temporarily detain water during a large storm and usually incorporate an overflow system for safety purposes. (Source: City of Portland, Oregon)

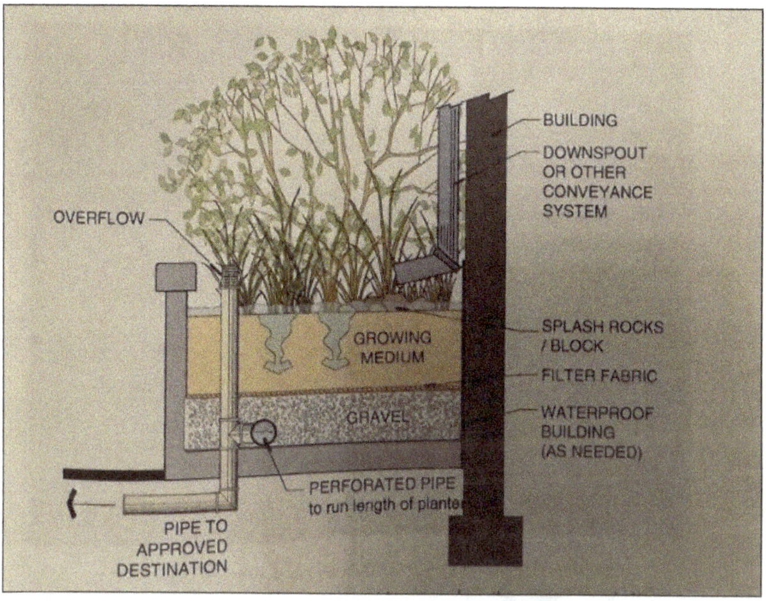

Figure 10-12 Flow-through planters are used in areas where the water cannot infiltrate into the ground. They are sealed and filled with gravel, soil, and vegetation to absorb and filter the rainwater. Excess water escapes through a perforated pipe located at the bottom of the planter or to an overflow system. (Source: City of Portland, Oregon)

Figure 10-13 Pervious paver or unit pavers replace impervious surfaces and allow stormwater to soak into the ground. They typically are made out of precast concrete, brick, stone, or cobbles and from interlocking patterns with the gaps filled with either sand or gravel. (Source: courtesy of Holly Piza)

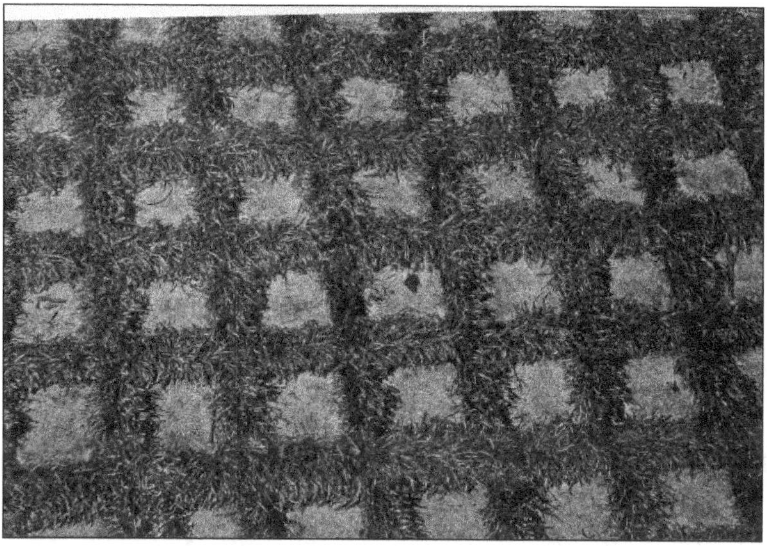

Figure 10-14 Turf block, also known as grass grid or open-cell unit pavers, has gaps filled with soil and grasses allowing water to pass through. This option accepts only precipitate and not storm water runoff; thus, suited for low traffic and infrequent car parking. Applications include patios, walkways, emergency access roads, street shoulders, and residential driveways. (Source: Courtesy of Western Interlock, Inc.)

REFERENCES

Campbell, Craig S., and Michael H. Ogden. 1999. Constructed Wetlands in the Sustainable Landscape. New York: John Wiley & Sons.

Hawken, Paul, Amory Lovins, and L. Hunter Lovins. 1999. Natural Capitalism: Creating the Next Industrial Revolution. Boston: Little, Brown.

Kibert, J. Charles 2022 "Sustainable Construction, Green Building Design and Delivery, Fifth Edition. John Wiley and Sons, Inc. Hoboken, NJ

World Health Organization. 2019. "Drinking Water Fact Sheet." Available at http://www.who.int/news-room/fact-sheets/detail/drinking-water.

UN. 2017. "Wastewater: Untapped Resource: 2017 UN World Water Development Report." Available online at http://reliefweb.int/report/wold/2017-un-world-water-development-report-wastewater-untapped-resource.

The Texas Guide for Rainwater Harvesting, Third Edition 2005. Austin, Texas: Texas Water Development Board, in Collaboration with the Center for Maximum Potential Building System. Available at www.twdb.texas.gov/publications/brochures/conservations/doc/RainwaterHarvestingManual

ELEVEN
SELECTING CONSTRUCTIOM MATERIAL FOR HIGH-PERFOMANCE BUILT ENVIRONMENT

BACKGROUND

For the project team, it is a challenging task to select environmentally friendly building material in the construction of high-performance green buildings. Because of the looming disaster of climate change reducing the carbon footprint of built environment has the highest priority. Reducing carbon footprint also calls for reduction of embodied energy in manufacturing of the building material which includes reducing the energy consumption for extraction of the material, transportation, installation, maintenance and eventually disposing them upon demolition in a certain disposal facility. The attributes that make materials and products acceptable for application in a high-performance green building include:

1. High recycled content
2. Reused building materials
3. Locally and regionally available materials
4. Certified wood products, and

5. Wood products are made from rapidly renewable resources.

However, combining these characteristics into decision making process and looking into the pros and cons of these characteristics still needs to be considered and streamlined. In light of on-going research and development by the manufacturers and stakeholders, significant progress is being made in crafting a widely accepted approach for determining the environmental efficacy of materials and products used in construction. The advent of green product certification such as Environmental Product Declarations (EPDs) and multi-attribute standard (MASs) promises to ease past problems of determining the impacts of both products and eventually the whole buildings based on universally accepted approaches. Some of the green building assessment tools such as Green Globes have incorporated both EPDs and MASs into their schemes to direct the project teams to more closely evaluate the environmental impacts of products being specified for their projects.

Lifecycle assessment (LCA) is the tool to scientifically assess and define the EPD of different available products in the market. By assessing the EPDs of the product used in a building one can define the entire Environmental Building Declaration (EBD), since EBD is the sum of all EPDs:

$$\text{EBDs} = \Sigma\text{EPDs}$$

Another approach to define EPD of a product uses the results of unbiased and independent third- party assessments. A realistic evaluation of EBDs will be possible when EPDs is completely available. At present time, EPDs simply allow a comparison between products being used for the same purpose. In old days, the manufacturers were unwilling to provide information about the ingredients of their products, however, nowadays they are getting forced to share their information as part of the overall green building process. In addition to the advent of EPDs and EBDs, independent organizations that produce standards are significantly emerging. Third-party organizations such as Green Seal, GreenGuard, the Forest Stewardship Council, etc. are providing realistic assessments of products being used. Industry-related organizations such as Carpet and Rug Institute (CRI) and the Sustainable Forestry Initiative (SFI) also are contributing in addressing the suitability of building material for green buildings.

Another issue that is highly important when discussing the environmental impact of the buildings and the construction materials and products is their fate at the end of their service. Building assessment systems address the recycling and reuse of materials from building demolition and reduction of waste in the construction process. Deconstruction, or the ability to disassemble a building for the purpose of reusing its components, also is finding its way into the mix of high-priority materials-related requirements.

This chapter addresses the issue of green building materials and products, the criteria for defining environmentally friendly products, the application of LCAs, EPD, and MASs in decision- making for materials selection and their role in high-performance green building design and construction. The role of third-party standard organizations in high performance building assessment and certification is addressed. Finaly, this chapter provides information about specific materials and product groups where new technologies and approaches are beginning to take hold in support of the green building movement.

THE CHALLENGE OF MATERIALS AND PRODUCT SELECTION

The selection of building materials and products for a high-performance building project has historically been very challenging. At the present time, there is no clear consensus about criteria for materials characterized as "environmentally preferable, environmentally responsible materials". Ross Spiegel and Dru Meadows (2010) defined green building materials as "those that use the Earth's resources in an environmentally responsible way." Currently, there is not a clear rule or criteria for selecting materials and products to be environmentally preferrable, environmentally responsible or green. Consequently, there are many different schools of thought, and multiple approaches, and abundant controversy. It is not obvious, for example, that building products made from post-commercial, postindustrial, or post-agricultural waste are, in fact, green. Many of the current green building products contain recycled content from these various sources.

To shed light on this subject, we will discuss three philosophies or points of view about what constitutes a green building product:

1. The Natural Step
2. Cardinal Rules for a Closed-Loop Building Material Strategy, and
3. A Pragmatic Approach suggested by Environmental Building News.

THE NATURAL STEP PHILOSOPHY

According to Natural Step, the concept of sustainable development is about finding ways to satisfy our basic needs in a way that does not undermine the balance of social and ecological systems. The Natural Step which is based on four scientifically based "system condition" was developed in 2002 by Dr. Karl-Henrick Robert, Swedish oncologist. The four conditions are listed next:

1. In order the society to be sustainable, nature's functions and diversity are not systematically subjected to increasing concentrations of substances extracted from the Earth's crust.

In a sustainable society, human activities such as burning fossil fuels and the mining of metals and minerals must not occur at a rate that causes them to systematically increase in the ecosphere. The ecosphere is a self-sustaining ecosystem, so one never has to feed the life within. There are thresholds beyond which living organisms and ecosystems are adversely affected by an increase in substances from the Earth's crust. Problems may include an increase in greenhouse gases leading to global climate change, contamination of surface water and groundwater, and metal toxicity which can cause functional disturbances in animals.

In practical terms, this condition requires society to implement comprehensive metal and mineral recycling programs and decrease economic dependence on fossil fuels.

2. In order for a society to be sustainable, nature's function and diversity are not systematically subjected to increasing concentration of substances produced by society.

In a sustainable society, humans will avoid generating systematic increase in persistent substances such as DDT, PCBs, Lead and CFCs. Synthetic organic compounds such as DDT, PCBs, and VOCs remain in the environment for many years, accumulating in the tissue of plants and animals, causing profound effects on predators in the upper level of the food chain. Freon and other ozone-depleting compounds may increase the risk of cancer due to added risk of UV radiation in the troposphere. Society needs to find ways to reduce economic dependence on persistent human-made substances.

3. In order the society to be sustainable, nature's functions and diversity must not be systematically impoverished by physical displacement, overharvesting or other forms of ecosystem manipulation.

In a sustainable society, humans will avoid taking more from biosphere than can be replenished by natural systems. In addition, people will avoid systematically encroaching upon nature by destroying the habitat of other species. Biodiversity, which includes the great variety of animals and plants found in nature, provides the foundation for ecosystem species, which are necessary to sustain life on this planet. Society's health and prosperity depends on enduring capacity of nature to renew itself and rebuild waste into resources.

4. In a sustainable society, resources are used fairly and efficiently in order to meet basic human needs globally.

Meeting the fourth system condition is a way to avoid violating the first three system conditions for sustainability. Considering human enterprise as a whole, we need to be efficient with regard to resource use and waste generation in order to be sustainable. If one billion people lack adequate nutrition, while another one billion have more than they need, there is a lack of fairness with regard to meeting basic human needs. Achieving greater fairness is essential for social stability and cooperation needed for making large-scale changes within framework laid out by the first three conditions. To achieve this fourth condition, humanity must strive to improve technical and organizational efficiency around the world, and to live using fewer resources, especially in affluent areas.

This condition implies an improved means of addressing human population growth. If the total resource throughput of the global human population continues to increase, it will be increasingly difficult to meet basic human needs, as human-driven processes intend to fulfill human needs and wants are systematically degrading the collective capacity of the Earth's ecosystem to meet these demands (Kibert, 2022).

Applying the system conditions to new building construction, with a particular focus on building materials, produces a matrix, as shown in Table 11.1. The matrix indicates the relationship between the conditions and the various major types of materials used or generated in construction: durables, consumables, and solid waste.

Area	Item Violation Examples	1	2	3	4
Material	Use of less abundant mined metals and minerals (copper, chromium, titanium)	x		x	
	Use of heavy metals (mercury, lead, cadmium)	x			
Durables	Use of persistent synthetic materials like polyvinyl chloride (PVC), HCFC, formaldehyde		x		
	Use of wood from rainforests and old-growth timber that is harvested unsustainably			x	
	Use of petroleum-based products, solvents, oil and plastic film	x	x	x	x
	Use of persistent synthetic materials for cleaning and building		x		
Consumables	Excessive packaging and other disposables		x	x	x
Solid Waste	Landfill disposal of construction and demolition waste including toxic components, such as lead and asbestos	x	x	x	x

(Source: Adapted from the Oregon Natural Step Construction Industry Group 2004.

Tabe-11.1 Violation of Natural Step Conditions in the Application of Construction Materials

In practical terms, applying the Natural Step to the employment of building materials would result in these materials practices, adapted from Oregon Natural Step Construction Group (2004):

1. All materials are nonpersistent and nontoxic and procured either from reused, recycled, renewable, or abundant (in nature) source:
   a. Reused means reused or remanufactured in the same form, such as re-milled lumber, in a sustainable way.
   b. Recycled means the product is 100% recycled and can be recycled again in a closed loop in a sustainable way.
   c. Renewable means able to regenerate in the same form at a rate greater than the rate of consumption.
   d. Abundant means human flows are small compared to natural flows, i.e., aluminum, silica, iron, etc.
   e. In addition, the extraction of renewable or abundant materials has been accomplished in a sustainable way, efficiently using renewable energy and protecting the productivity of nature and the diversity of species.
2. Design and use of materials in the building will meet the following in order of priority:
   a. Material selection and design favor deconstruction, reuse, and durability appropriate to the service life of the structure.
   b. Solid waste is eliminated by being as efficient as possible, or
   c. Where waste does occur, reuses are found for it on-site, or
   d. For what is left, reuses are found off-site.

e. Any solid waste that cannot be reused is recycled or composted

Natural Step calls for the ultimate elimination of synthetic materials whose concentration in the biosphere is compromising not only human health but also the very health of biosphere in which we reside. The Natural Step also cautions against the degradation of biosphere by human activities because it is the very source of the resources needed to sustain life. And, finally, it addresses the social aspects of sustainability by noting that human needs in all parts of the world must be met. In summary, the message of Natural Step is to reduce resource extraction, increase reuse and recycling, and minimize emissions that affect both ecosystem and human life.

CARDINAL RULES FOR A CLOSED-LOOP BUILDING MATERIAL STRATEGY

Ideally, the building material system should follow the Cardinal Rules for a closed loop building materials strategy listed in Table 11.2.

Table 11.2 Cardinal Rule for a Closed-Loop Building Material Strategy

1) Building must be deconstructable
2) Products must be dis-assemblable
3) Materials must be recyclable
4) Products/materials must be harmless in production and use
5) Materials dissipated from recycling must be harmless

According to the Cardinal Rules, complete dismantling of the building and all its components is required so that materials input at the time of the building's construction can be recovered and returned to productive use at the end of the building's useful life. These rules also establish the ideal conditions for materials and products used in the building. It is, however, important to point out that very few materials and products today adhere to these five rules, meaning that the behavior of materials is far from its ideal state. Devising a system of materials, products, and buildings to support closed-loop behavior is in the distant future (Kibert, 2022). Nonetheless, this thought process can be used as a touchstone for making decisions about the development of new products, materials, and technologies that support the high-performance green building movement.

A PRAGMATIC APPROACH SUGGESTED BY ENVIRONMENTAL BUILDING NEWS

To take pragmatic view of green building materials, it is useful to first examine contemporary efforts to wrestle more directly with these issues based on our current understanding, capabilities, and technologies. The foremost US publisher of well-reasoned approaches to most matters concerning high-performance buildings is BuildingGreen, Inc., an excellent source, especially about green building materials and products selection. In one of their Spotlight Reports they start off by describing the criteria that make products green (BuildingGreen, 2020):

1. **Efficient use of materials**
   a. Durable or low maintenance
   b. Reduces material use
   c. Salvage content

2. **Responsible sourcing**
    a. Postconsumer recycled content
    b. Pre-consumer recycles content
    c. Biobased and sustainability sourced
    d. Certified wood
    e. Uses agricultural waste
3. **Energy and water conservation**
    a. Embodied carbon
    b. Reduces heating and colling loads
    c. Energy-conserving equipment
    d. Conserves water
    e. Renewable energy equipment
4. **Low-impact operation**
    a. Reduces construction impacts
    b. Reduces pollution or waste
    c. Reduces stormwater pollution
    d. Reduces pesticide treatment
5. **Healthy Buildings**
    a. Improves light quality
    b. Improves acoustical performance
    c. Prevents moisture or air-quality problems
    d. Low-emitting
    e. Avoids hazardous ingredients
    f. Improves air quality
    g. Monitors indoor air quality
6. **Sustainable industry and resilient, just communities**
    a. Information transparency
    b. Reduced manufacturing impacts
    c. Responsible corporate practices
    d. Provides connection to nature
    e. Enhance community well-being
    f. Increase resilience

This pragmatic view of building materials and products is a useful starting point because it leads with the contemporary supply chain and with today's technologies and practices. The question, then, is: how do we evolve closer to the ideal of green building materials and products espoused by the natural Step and Cardinal Rules for a closed loop building materials strategy?

SELECTING BUILDING MATERIALS AND PRODUCTS

There are three priorities in selecting building materials for a project, sometimes it is referred to as 3R rules:

1. Reduce the quantity of materials needed for construction as is done with energy and water resources.
2. Reuse materials and products from existing building; this is a relatively new strategy called "deconstruction". Deconstruction is the whole or partial dismantling of existing buildings for the purpose of recovering components for reuse.
3. Use materials or products that contain recycled content and that are themselves recyclable or use products and materials made from renewable resources.

RECYCLING TECHNIQUES

There are two kinds of recycling routes for recycling, (1), technical and (2) organic. The technical recycling route is associated with synthetic materials (i.e., materials that do not exist in pure form in nature or are invented by humans). These include metals, plastics, concrete, and non-wood composites, to name a few. As noted earlier, only metals and plastics are fully recyclable; hence, they potentially can retain their engineering properties through numerous cycles of reprocessing. Materials in the technical or synthetic category require major investment of energy, materials, and chemicals for their recycling.

Materials recyclable through organic recycling route, are the one that derived from renewable sources. For example, composting is the best-known organic recycling route. This route is designed to allow nature to recycle building materials and turn them back into nutrients for ecosystems. Although feasible in theory, organic recycling has not been attempted on a large scale in the United States. For the organic route to work, it would have to incorporate products from a wide range of applications, including agricultural wastes and landscape clearing debris as well as organic waste from construction.

GENERAL MATERIALS STRATEGY

Minimizing environmental impacts of the building suggests the following strategies:

1. Reuse existing structures. By modifying an existing building and reusing as much as its structure and system as possible, one can minimize the use of new materials, with their accompanying impacts of resource extraction; transportation, and processing energy, waste, and other effects. Clearly, trade-offs must be made when considering a building for reuse. For example, a building that historically has been inefficient and would need significant changes to its envelope and mechanical/electrical systems might incur significant waste and require enormous quantities

of new materials, in order for the original structure to be retrofit for its new use.

2. Reduce material use. Using the minimal number of materials required for a building project can also lower the environmental impact of introducing products manufactured from virgin resources. In a typical building, however, the opportunities for dematerialization are few. And center on the possible elimination of systems that are not absolutely necessary. rejecting floor finishes in favor of finished concrete is an example of reducing materials use, but probably at the cost of aesthetic appeal. Materials waste caused by following the conventional construction processes also contributes to unnecessary materials use. In general, dematerializing a building is difficult because of building code provisions, the desires of users, and sometimes the need for new systems that are becoming standard in high-performance green buildings.
3. Use materials created from renewable resources. Materials created from renewable resources offer the opportunity to close materials loops via an organic recycling process. The organic route involves recycling by biodegradation, the is, by composting or aerobic /anaerobic digestion, either by nature itself or by processes that mimic the decomposing action of nature. This approach applies to all products made of wood or other organic materials, such as jute, hemp, sisal, wool, cotton, and paper. Recycling renewable resources or organic products via organic recycling route can be accomplished with low to zero energy, additional materials, and chemicals. Note, however, that some materials are composites of

organic and technical materials and hence would fall into technical class for purposes of recycling.

4. Reuse building components. Reusing intact building components from deconstructed buildings reduces the environmental impacts of building materials because these components require minimal resources for reprocessing. Progress in the techniques for deconstructing existing buildings, instead of demolishing them, means that used building components are becoming more widely available; likewise, businesses that specialize in the sale of components salvaged from deconstructed buildings are becoming more commonplace. One problem that remains to be solved, however, is how to recertify most used building products. That said, good progress has been made in developing visual regarding standards for some types of dimensional lumber, for example, western cedar and southern yellow pines.
5. Use recyclable and recycled content materials. To close the material loop in construction, all materials must have the capacity for recycling. Currently, this remains a very ambitious objective simply because few building materials are recyclable, and many others can be recycled only into a lower value application. For example, recycled concrete aggregates can be used as a subbase material but not – at least not readily, in the United Staes – as an aggregate in new concrete. Metals and plastics are perhaps the only materials that are fully recyclable without loss of their basic strength and durability properties. A wide range of recycled-content materials is available for the green building market. These generally contain either

postindustrial or postconsumer waste. Postindustrial waste refers to materials recycled within the manufacturing plant.

6. Use locally produced materials. Examining resources and emissions associated with transporting materials between various sites of extraction, materials production, product manufacture, and installation is one of the steps in an LCA evaluation. There is no doubt that minimizing transportation distances by using locally produced materials and locally manufactured products can greatly reduce the overall environmental impacts of materials. Defining what is meant by local can however, be a challenge. The LEED-NC building assessment standard sets 500 miles as the radius within which a product is considered local for the purpose of obtaining points. Another difficulty with assigning weight to locally produced products is that improved technologies may be passed over. A classic example is the introduction of Japanese automobiles to the US marketplace in the late 1970s, when the quality of workmanship of these cars compelled a rapid shift away from American products. The subsequent bailout by the US government of Chrysler Corporation in 1979-1980, couples with higher energy prices, forced US companies to rethink their products. Ultimately, fundamental changes took place in the design and production of American cars. Today, American cars are almost on a par with Japanese automobiles and exceed many European cars in terms of quality. In short, products not considered local may be far superior, result in lower life-cycle environmental

impacts, and encourage improvements in local products

The materials selection process may be summarized as follows: Rely on the three Rs - reduce, reuse, recycle.

LIFE-CYCLE ASSESSMENT

Life-cycle assessment (LCA) is a process of evaluating the effects that a product has on the environment over the entire period of its life thereby increasing resource-use efficiency and decreasing liabilities. LCA comprises several steps, which are defined in the International Organization for Standardization (ISO) 14000 series of standards that address environmental management systems. These steps include inventory analysis, impact assessment and interpretation of the impacts.

LCA can be applied as a whole building LCA in which the impacts of all the materials of the building are summed up to provide the total impact of a specific design. The information is often used to find the major sources of emissions and allows the project team to examine other design materials variations to lower these impacts. Energy consumption can also be included in the Whole Building LCA to demonstrate how energy conserving strategies, while possibly increasing embodied materials impacts, may result in a reduction in total impacts. LCAs can also be conducted for building products, most frequently in the form of EPDs. The EPDs data can be pulled into LCAs and aggregated to help produce Whole Building LCA.

Product / Manufacture Stage [A1-A3]			Construction Process Stage [A4-A5]		Use [B1-B7]							End-of-Life Stage [C1-C4]				Benefits & Loads Beyond System Boundaries [D]
					Building Fabric						Operation of the Building					
Raw Material Extract / Process / Supply	Transport	Manufacture	Transport to the Site	Assembly / Install in the building	Use / Application of Installed Products	Maintenance	Repair	Replacement	Refurbishment	Operational Energy Use	Operational Water Use	Deconstruction / Demolition	Transport to Waste Process	Reuse-Recovery-Recycle	Disposal	Reuse-Recovery-Recycle Potential
A1	A2	A3	A4	A5	B1	B2	B3	B4	B5	B6	B7	C1	C2	C3	C4	D
Cradle-to-Gate			Gate-to-Grave													
Cradle-to-Grave																
Cradle-to-Cradle																
System Boundries																

Table 11.3 Different stages of LCA Calculations. (Source: Koseci 2018)

As Table 11.3 shows, LCA is broken into five major stages A though D and substages, for example, stage A has substages A1 through A3. These stages are combined to produce four different types of LCAs, such as cradle-to-gate, gate-to-grave, cradle-to-grave, and cradle-to-cradle. For example, a cradle-to-gate LCA covers just the manufacturing substage (A1-A3) while cradle-to-cradle LCA includes all five stages and substage (A-D). This presents some challenges in practice because, in making comparisons, product LCAs must have addressed the same stages. For the materials cycle to be a truly closed loop, the preferred version is a cradle-to-cradle LCA.

LCAs report on the environmental performance of a product or building during each stage plus the total for all stages. This performance is measured in terms of a wide range of potential effects, for example:

- Fossil fuel depletion
- Other nonrenewable resource use
- Water use
- Global warming potential
- Stratospheric ozone depletion
- Ground-level ozone (smog) creation
- Nutrification/eutrophication of water bodies
- Acidification and acid deposition (dry and wet)
- Toxic releases to air, water, and land

Comparing these impacts takes careful analysis. For example, the total energy for a building's life cycle is composed of the embodied energy invested in the extraction, manufacture, transport, and installation of its products and materials, plus the operational energy needed to run the building over its lifetime, see Figure 11.1.

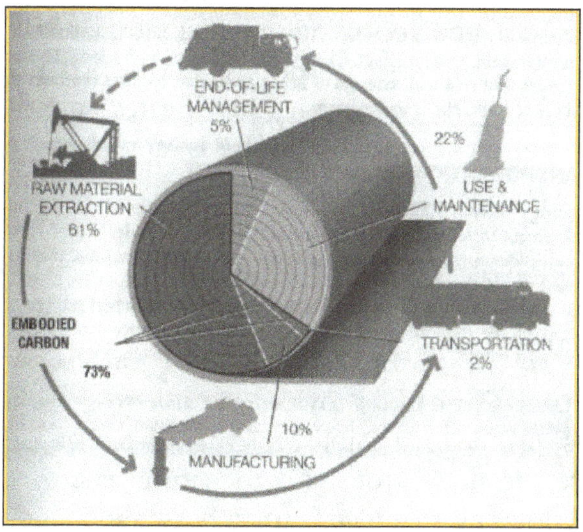

Figure 11.1 Life Cycle Assessment

For the typical building, the operating energy is far greater than the embodied energy, perhaps 5 to 10 times higher. Consequently, the operational stage has far more energy impacts than stages up through the construction stage. For other effects, however, the impacts of the stages up through construction can be far greater. Toxic releases during resource extraction and manufacturing process can be far greater than those occurring during building operation. The net result is that the designer using LCA tools must keep in mind the entire life cycle of the building, not just the stages leading to construction. In the following sections two major building LCA tools are covered: The Athena Impact Estimator for building (AIE) and Building for Environmental and Economic Sustainability (BEES), (Kibert, 2022).

ATHENA IMPACT ESTIMATOR FOR BUILDINGS

This whole-building tool is used by design teams to explore the environmental footprint of different material choices and core-and-shell system options. It was first released in 2002 and has undergone numerous updates since then. The Athena Impact Estimator (AIE) is applicable for new construction, renovations and additions in all North American building types. It can model over 1,200 structural and envelope assembly combinations. It allows for quick and easy comparison of multiple design options, (Athena Sustainable Materials Institute, 2024).

The Impact Estimator provides a cradle-to-grave life cycle inventory profile for a whole building. The inventory results comprise the flows from and to nature: energy and raw material flows plus emissions to air, water and land. How it Works? Users quickly describe building assemblies through dialogue boxes that request simple information like bay sizes and loads—the software does the rest, calculating a bill of materials and the associated environmental impacts. Users have flexibility to add materials as needed. Alternatively, users can import their own bill of materials from any CAD program.

The software reports footprint data for the following environmental impact measures consistent with the latest US EPA TRACI methodology: global warming potential, acidification potential, human health respiratory effects potential, ozone depletion potential, smog potential, and eutrophication potential. The Impact Estimator additionally reports fossil fuel consumption. The software is regionally customized—appropriate electricity grids, transportation modes and distances, and product manufacturing technologies are applied depending on the building location. Other user inputs such as building service life and type of building also affect the results.

Operating energy can be included in the building LCA if the user inputs an estimate for annual operating fuel consumption. The software will calculate total energy, including pre-combustion energy (the energy used to extract, refine and deliver energy) and the related emissions to air, water and land over the life cycle of the building. The software can subsequently compare the life cycle operating and embodied energy and other environmental effects of various design options, allowing the user to better understand the inherent trade-offs between (for example) adding more insulation and reduction in operating energy use.

The Impact Estimator takes into account the environmental impacts of the following life cycle stages: material manufacturing, including resource extraction and recycled content;

related transportation; on-site construction; maintenance and replacement effects; and demolition and disposal.

A typical array of information produced by AIE is shown in Figure 11.2. The information in this figure is not meaningful unless it is compared to alternative strategies.

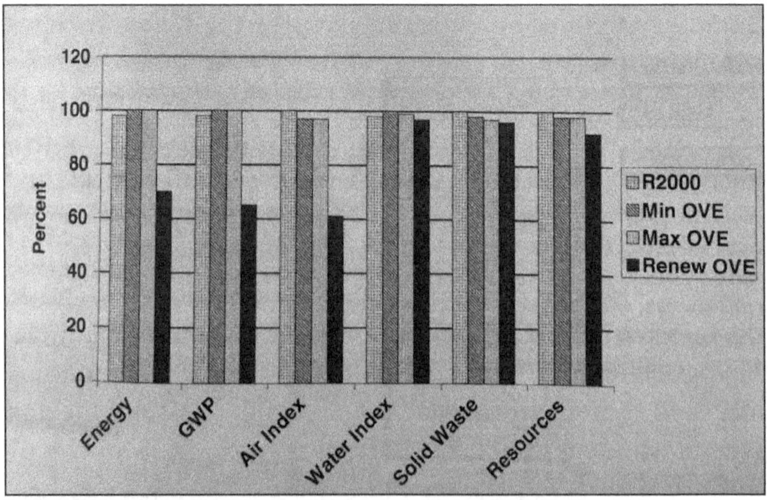

Figure 11.2 Sample output screen from the AIE showing energy use, various impacts, and other resource use for a comparison of four products (Source: Courtesy of the Athena Sustainable Materials, Institute)

BUILDING FOR ENVIRONMENTAL AND ECONOMIC SUSTAINABILITY

Building for Environmental and Economic Sustainability (BEES) is a tool developed by the National Institute of Standards and Technology (NIST) to allow designers, builders, product manufacturers and consumers to select cost-effective, environmentally preferable building products to LCA technology. Like AIE, BEES allows side-by-side comparison of building products for the purpose of selecting cost and energy-effective environmentally preferred products (EPPs) and includes both LCA and Life Cycle Costing (LCC) data, see Figure 11.3. The result is that the user obtains both environmental performance as well as economic comparisons.

In addition to the typical measures of performance, BEES provides data about air pollutants, indoor air quality, ecological toxicity, and human health for each material or product. BEES can compare building elements to determine where the greatest impacts are occurring and which building elements need the most improvement. The user assigns weights to categories, then combines the environmental and economic performance into a single performance score. For example, the user first decides how to weigh environmental versus economic performance, say, 50-50 or 40-60. The user then selects from among four different weighting schemes for the environmental performance measures. The latest version of BEES is available online and has a database of over 200 building products, including 80 brand-name products.

As an example, for floor coverings, there are 18 brand-name products and 17 distinct generic products. A sample output screen for a BEES LCA analysis is shown in Figure 11.4.

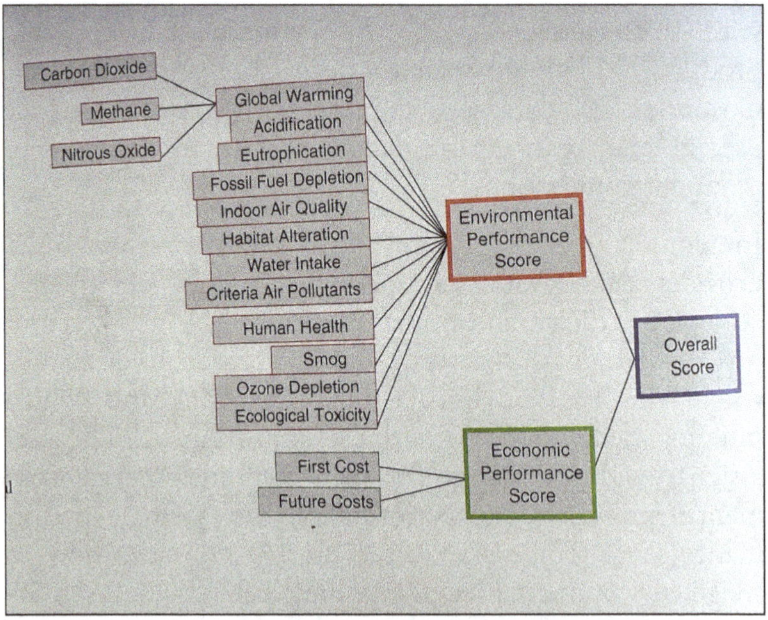

Figure 11.3 The BEES model combines environmental and economic performance into a single score for use in comparing product selection options. (Source: National Institute of Standards and Technology)

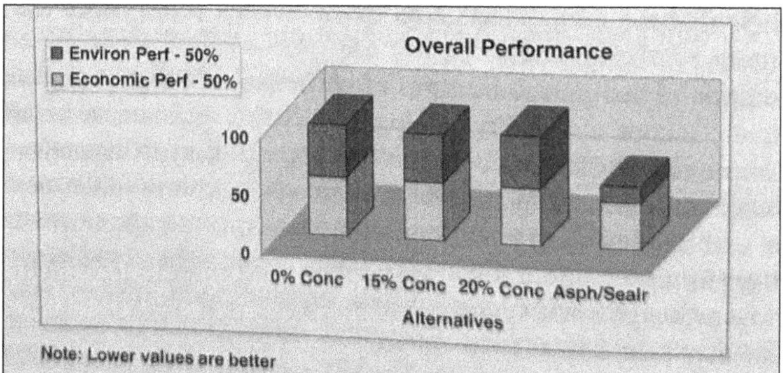

Category	0% Conc	15% Conc	20% Conc	Asph/Sealr
Economic Perf - 50%	50	49	48	39
Environ Perf - 50%	45	41	44	14
Sum	95	90	92	53

Figure 11.4 Sample output from BEES 3.0 showing comparative environmental performance concrete with various levels of fly ash and for sealed asphalt. BEES is a free LCA program available from the NIST (Source: National Institute of Standards and Technology)

MULTI-ATTRIBUTE STANDARDS (MASS)

MASs establishes the principles and procedures for developing multi-attribute sustainability standards for building products. Multi-attribute sustainability standard identifies and incorporates a broad range of objectives for social, economic and environmental criteria throughout all phases of a product life cycle and establishes minimum thresholds for conformance as relevant to a product category. The purpose of such a standard is to facilitate the reduction of social, economic and environmental stresses by setting forth product requirements which are effective, verifiable, industry-relevant, and consistent for all products within a category. A multi-attribute sustainability standard may be used for product certification, specification, or communication of environmental preferability, thereby stimulating the potential for market-driven continuous improvement. These standard endeavors to ensure that all multi-attribute sustainability standards for building products possess a common framework and consistently address sustainability indicators which are most relevant to the built environment. The principles of MASs are being developed by ASTM and it will be available to be followed by construction industry soon.

MOST COMMON BUILDING MATERIALS

This section addresses the current issues and status of major classes of construction products. Due to huge list of construction materials and products, we cannot cover them all, however, the materials discussed here are considered most important because of the scale of their application in construction which will include wood and wood products, concrete and concrete products, metals and plastics.

Wood and Wood Products

Wood and products made of wood are very important construction materials, merely because of their renewability. Enormous areas of United States are covered with trees, some 623 million acres or about one-third of the US landmass. Of this, 504 million acres are classified as timberland – that is, productive forest capable of growing at least 80-105 tons of commercial wood per acre per year. Approximately, 67 million acres are owned by the forest products industry, 291 million acres are held by 10 million individual private landowners, and another 49 million acres contained within the National Forest System are available for forest management (https://www.fia.fs.fed.us/).

A wide variety of wood products are used in construction, including dimensional lumber, engineered wood products, plywood, oriented strand board, and composite materials with wood fiber content. Wood products used in high-performance green buildings should originate in sustainably managed forests and should bear labels certifying this fact, Kibert, 2022. The major organization governing sustainable forestry internationally is the Forest Stewardship Council (FSC). The USGBC LEED building assessment standard provides a point related to certified wood products and for rapid renewable resources.

Sustainable Forestry Initiative (SFI) has been developed using certain principles by professional foresters, conservationists, and scientists that combine perpetual growing and harvesting of trees with long-term protection of wildlife, plants soils, and water quality in mind. On January 1, 2007, the SFI program became fully independent forest certification program. The multi-stakeholder board of directors of the SFI is now the sole governing body over the SFI standard.

It should be noted that there are several other third-party certification systems for sustainability harvested wood, including the American Tree Farm System and CSA Sustainable Forest Management. Green Globe building assessment system takes all of these into account in rewarding points, while the USGBC relies solely of FSC certification system.

Mass Timber/Structural Wood Products

The environmental impact of construction is significantly high with the embodied carbon of building materials being a major contributor to the emissions of greenhouse gases. Among the building materials, concrete and steel has significantly higher carbon footprints. By comparison wood products have an exceptionally low carbon footprint because wood products absorb carbon during their growth stage and, when harvested, the carbon is then sequestrated in the location they are utilized.

The CO_2 captured during the photosynthesis process gives timber an advantage over other building materials. To maintain this advantage, the sequestrated CO_2 needs to be managed at the end of the service life of the building to prevent its release into the atmosphere via decay and in the form of methane.

Mass timber products have a lower carbon footprint than most other traditional structural elements such as those made from steel or cementitious building materials. In addition to having a very low carbon footprint mass timber stores the carbon absorbed during their growth cycle, allowing buildings and cities to become carbon sinks.

Concrete and Concrete products

Concrete has an enormous and increasing number of roles in construction. It is composed of aggregate (rock), fine aggregate (sand), cement, water, and various additives. With respect to high-performance buildings, concrete has many positive qualities including the following:

- High strength
- Thermal mass
- Durability

- High reflection
- Usually locally available and can be used without interior and/or exterior finishes. In addition, it does not off-gas and affect indoor air quality, etc.

The key issue with concrete is the release of carbon dioxide into the atmosphere during the cement manufacturing process (European Commission, 2011). Cement comprises 9 to 14% of most concrete mixtures and it is the second to coal-fired utilities in carbon dioxide emissions. For each ton of powder cement produced, up to an equal mass of CO_2 is generated. However, during the life cycle of a concrete element, the cement reabsorbs about 20 percent of the CO_2 generated in the manufacturing process, at least partially mitigating the CO_2emitted during manufacturing. Minimizing the quantity of cement in a concrete mix is the strategy has a number of potential benefits. Fly ash and ground blast furnace slag, both of which have cementation properties, can be substituted at least partially for cement and result in increased in concrete performance. Fly ash can be readily substituted for over 30 percent of the cement volume; blast furnace slag, for more than 35 percent. These substitutions have the advantages of making beneficial use of otherwise industrial waste while simultaneously reducing the quantity of CO_2 associated with concrete production. Fly ash and blast furnace slag also can be blended with cement in the cement manufacturing process, resulting in reduced CO_2 emissions, reduced energy consumption, and expanded production capacity, Kibert, 2022.

The recycling properties of concrete is generally satisfactory. Crushed concrete can be used as subbase for roads, sidewalks, and parking lots. In the Netherlands, recycled concrete aggregate may substitute for one-third of the virgin concrete in concrete mixes. In general, recycled concrete aggregate is in high demand and has relatively high value.

Metals Use in Construction

Metals are used in construction industry due to their durability and strength to form structural components, pipework, cladding materials and other components. They are first choice material for structures, reinforcements, cladding, roofing, window frames, plumbing, heating equipment and many other applications. Metals can be found in old and historic buildings as well as in new and modern architecture. Most common metals used in construction include steel, aluminum, and copper.

For the variety of reasons, structural steel is often an ideal material when it comes to the construction of buildings, bridges, and other large structures. This high-quality metal is used for many purposes around the world, but there are some applications that rely on structural steel more often than others. Still beams, particularly I-beams, are commonly utilized as support trusses or primary framework of buildings. Steel beams are well known for their remarkable strength and support, which help ensure structural integrity. Because of their adaptability and utility, I-beams are the most often used form of steel beam.

Steel production today incorporates used steel products in the two production processes still being used. The basic oxygen furnace (BOF) uses 25 to 35 percent scrap steel for production that require drawability – for example, automobile fenders and cans – while the electric arc furnace (EAF) uses almost 100 percent scrap steel for products whose main requirement is strength – for example, structural steel and concrete reinforcement. Steel made from the BOF process generally has a total recycled content of 32 percent, which is composed of 22.6 percent postconsumer content and 8.4

percent postindustrial content. EAF-produced steel generally has a recycled content of about 96 percent, with a post-consumer content of 59 percent and postindustrial content of 37 percent (Statistics from 2002).

Recycled steel consumes a fraction of the resources and energy of steel produced from iron ore. Each ton of recycled steel saves 2,500 pounds of iron ore, 1,400 pounds of coal, and 120 pounds of limestone. Only one-fifth of energy needed to produce steel from iron ore is required to recycle scrap steel. Steel recycling systems in the United States are well established, so much so that recycling is dictated less by environmental concerns than by economists.

Aluminum recycling also has marked environmental benefits. Recycled aluminum requires only 5 percent of the energy needed to produce aluminum from bauxite ore, thus eliminating 95 percent of the greenhouse gases that would be generated by manufacturing aluminum from bauxite. Approximately 55 percent of the world's aluminum production is powered by hydropower, which, also controversial because of its environmental impacts, is a renewable resource. Recycling 1 pound of aluminum saves 8 pounds of bauxite and 6.4 kilowatt-hour of electricity. It is widely used in buildings because of its intrinsic properties of lightness and corrosion resistance. Aluminum is used in external facades, roofs and walls, in windows and doors, in staircases, railings, shelves, and other applications. Aluminum is an excellent material for architectural applications as it's environmentally friendly and ecologically responsible. It's a fully recyclable metal which is easily renewable to reduce waste metal. Aluminum is one of the best insulators for both temperature and acoustic insulation. Aluminum has a long lifespan and is maintenance friendly. The component of aluminum stops damage from UV rays and moisture and the metal doesn't rust or weather from harsh environments.

This impressive durability means that it's smooth finish doesn't require regular maintenance, unlike wood.

Use of Plastics in Construction

Plastic components used in construction include electric wiring, flooring, wall covering, waterproofing, fasteners, pipes, valves, fittings, hinges and decorative touches. Plastic hardware doesn't rust, so it can survive outside in any weather, indefinitely. The benefits of plastics: the production criteria are durable, low cost, water-resistant, lower energy and heavy chemicals and are lightweight. Disadvantages of using plastics are they pollute our environment and sometimes it is not biodegradable. Plastics are not used in home construction projects because it is simply not as strong as wood, metal, or brick. Also, plastic permanently deforms under stress (creeps), and is harder to nail, drill, and screw than wood. Many of these structural limitations can be overcome by mixing plastic with other materials to form composite building materials.

At present, however, there is little, if any, recycling of plastic building products into other end uses, which is a serious problem. Closed-loop behavior is, of course, desirable. But there are some success stories in plastic recycling. One is the development of processes that recycle high-density polyethylene (HDPE) into high-quality plastic lumber, a product with very high durability that is impervious to rot, insects, and saltwater damage and with a lifetime measured in hundreds of years.

Manufacturers of plastics derived from chlorine or that employ chlorine in their production are under severe pressure from environmental groups such as Greenpeace because of various impacts associated with their manufacture and disposal.

PVC, ubiquitous product in construction; it appears in piping, siding, flooring, and wiring, to name a few, is the main focus of these struggles. To date, the recycling rates of PVC are among the lowest for seven major classes plastics covered by the Society of Plastic Industry – less than one percent. And in the US, PVC is being defined by its industry based on its technical and economic merits, meaning that fundamental changes to the product or its manufacture are not anticipated soon. By contrast, the European PVC industry is exploring how to make fundamental changes in the production and disposal of its products, positioning PVC to be regarded as an environmentally responsible product. The major problems with PVC are the use of certain additives such as lead, cadmium, and phthalates and its disposal. According to green paper, only 3 percent of PVC waste is recycled; 17 percent is incinerated and remaining 80 percent is landfilled, with the total waste stream amounting to 3.6 million tons per year.

The risks associated with handling PVC, especially the loss of phthalate from soft PVC, were highlighted, along with problems caused by incineration – namely, the generation of dioxins, which are very hazardous chemicals. Unquestionably, PVC recycling must be improved, and reformulation of the basic product must be considered to remove barriers to its recycling. PVC product recycling faces many of the same problems associated with other plastics – the use of additives such as plasticizers, stabilizers, filters, flame retardants, lubricants, and colorants which are used to produce specific properties, Kibert, 2022.

END OF THE LIFE SCENARIOS FOR COMMON BUILDING MATERIALS

This section summarizes the fate of the major construction material such as concrete, timber and steel. As Figure 11.5 presents, a very low percentage of concrete is landfilled. The major portion of postconsumer concrete is downcycled and used as a subbase for roads, parking lots, and as a fill material in the various construction activities. On the other hand, the high percentage of postconsumer timber is landfilled (58 percent) 6 percent is incinerated, 13 percent is reused, and only 10 percent are downcycled. As discussed earlier, steel is highly recyclable building material. As shown, 93 percent of still is recyclable and 6 percent of steel is reused and only 1 percent is landfilled.

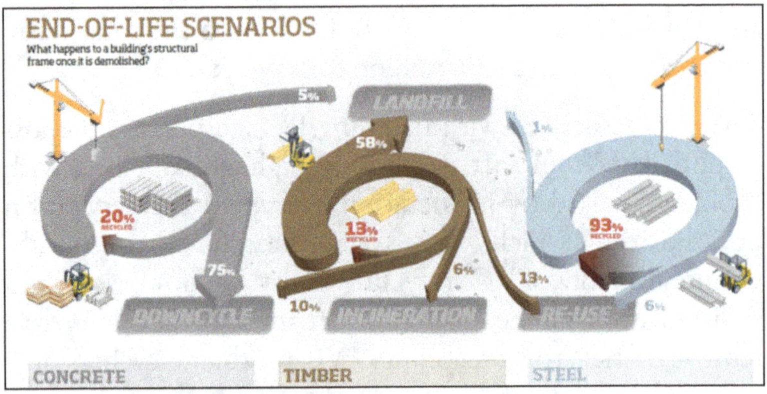

Figure 11.5 Schematic diagram of fate of major construction materials

REFERENCES

ASMI. 2014. Aetna Guide to Whole-Building, LCA in Green Building Programs (March) Aetna Sustainable Materials Institute.

European Commission. 2011. "Innovative ways to Reduce CO2 Emission from Cement Manufacturing." DG Environmental News (October 11). Available at http://ec.europa.eu,

Karl-Hendrick Robert, 2002, 'Natural Step Story: Seeding a Quiet Revolution' New Society Publishers.

Kibert J., Charles 2022. 'Sustainable Construction, Green Building Design and Delivery. Fifth Edition, John Wiley and Sons, Inc. Hoboken, NJ.

Koseci, F. C. 2018 'Integrated Life Cycle Assessment (LCA) to Building Information Modeling (BIM).' KTH. Master's thesis.

Kun – Mo Lee, 2004. "Life Cycle Assessment, Best Practices of 1404 Series" Committee on Trade and Investment. Anjou University of Korea.

Oregon Natural Step Construction Industry Group. 2004, 'Using Natural Step as a Framework Toward a Construction and Operation of Fully Sustainable Buildings. Formerly available at www.rtns.org.

Spiegel, Ross, Dru Meadows. 2010. Green Building Materials: A Guide to Product Selection and Specification, 3rd ed. Hoboken New Jersey, John Wiley and Sons.

Statistics from "2002: the inherent Recycled Content of Today's Steel," available at www.world-aluminium.org

TWELVE
CARBON FOOTPRINT AND CLIMATE CHANGE

BACKGROUND

Carbon footprint is the total measure of greenhouse gases released in the atmosphere from our built environment. Greenhouse gases (GHG) keep the earth warm and make it habitable for living organisms. An increase in carbon dioxide – a GHG, can lead to increase in global temperature. Carbon emissions can be natural or human made. For example, we produce GHG in our day-by-day activities from burning gasoline when we drive, burning oil or gas for home heating, or using electricity generated from coal, natural gas and oil. The optimal carbon footprint generated by an individual ranges between 6,000 and 16,000 pounds of GHG per year.

It may be surprising to learn that, regardless of income, the average American citizen cannot seem to drop below a certain threshold of CO_2 emission. The US is so large that daily commuting, business travel, and more daily necessities for the average American indirectly cause a release of CO_2, whether he/she uses buses, planes, cars, and boats for transportation, or simply hiring a necessary workforce to build new roads or public buildings.

One notable 2007 study conducted by the Massachusetts Institute of Technology assessed the average American's CO2 emissions. This study consisted of detailed interviews of the energy usage of 18 lifestyles, ranging from a 5-year-old child to a practicing Buddhist. The study found that the average annual carbon dioxide emissions per American was a shocking 20 metric tons, in comparison to a world average of 4 tons. The world average is less than half the amount of the lowest carbon emissions of anyone in the U.S., which is 8.5 tons, the source of this information can be found at:

https://www.inspirecleanenergy.com/blog/clean-energy-101/average-american-carbon-footprint

A carbon footprint is the total greenhouse gas (GHG) emissions caused directly and indirectly by individuals, events, products and organizations. It is calculated by adding together the emissions that result from every stage of a product or service's lifetime. For physical products, this timespan starts at the point of material production and manufacturing and goes right up to its use phase, and eventually end-of-life disposal. Throughout each product's lifetime, different greenhouse gases, like methane or nitrous oxide, may be emitted. By continuing to buy into these companies and products that emit greenhouse gases, we are contributing to the overall problem.

One study that took place in 2017 showed that China was the highest emitter of cumulative carbon dioxide. China's carbon footprint is concerningly high, and the air they breathe is of concerning quality. Unfortunately, China's use of fossil fuels has continued to increase as more factories are built as a way of keeping up with the global demand for manufacturing.

In this study, China was first in emitting the highest amount of carbon dioxide followed by:

- U.S. released 4.8 metric gigatons of CO2
- India, with 2.2 gigatons
- Russia with 1.5 gigatons
- Japan with 1.1 gigatons

It's worth noting that, after Japan, every country came in at below 1 annual gigaton, with most countries emitting less than 0.5 gigatons.

One simple but effective way of reducing one's carbon footprint is by eliminating meat from one's diet. Results from the 2013 National Household Food Acquisition and Purchase Survey found that industries that produce meat products (primarily chicken, fish, beef and pork) accounted for 30% of average weekly household GHGEs, the biggest share of any of the food industries.

In today's agricultural industry, many cows are also being bred via artificial insemination. This is done to save money, improve conception rates, and ensure certain behavioral traits are bred into the calves. Animal proteins have larger carbon footprints per calories than grains, legumes or vegetables due to the inefficient transformation of plant energy to animal energy. In addition, the methane that is released from bovine manure management has been linked to the release of methane in countless scientific studies. In short, not only is the artificial insemination and overpopulation of cows and subsequent levels of methane damaging to the ozone layer, but it's also about the way they are transported to us.

To put things into perspective, opting for a vegetarian meal one day a week could save the carbon equivalent of driving

1,160 miles per year. It also depends on which meat you cut down on, as some animal products are more greenhouse-gas intensive than others. Red meat is about 150% more GHG-intensive than fish or chicken, so if you feel that you really don't want to give up your animal protein, it is possible to choose wisely so that you reduce your footprint as much as possible by doing the following:

1. Switch to a clean energy company. Working with companies that use hydro, solar power or, more notably, wind energy is an effective way of helping to reduce your carbon footprint. You must make the decision to do this – you can't wait for your energy supplier to do this for you.
2. Eat low on the food chain. This has been touched upon earlier, but it's worth considering if you are concerned about your carbon footprint. Every day that you forgo meat and dairy, you are reducing your carbon footprint by 8 pounds, that's an impressive 2,920 pounds a year.
3. Don't buy fast fashion. This only seems to have come to the forefront of our conversations recently, but it's hardly a new problem. Clothing made poorly and quickly is usually disposed of quickly, too. These items of clothing fill up landfills, decompose and produce methane. There's also the transport of this clothing to consider; most fast fashion retailers have factories in countries like China, India, and Bangladesh, which requires significant burning of fossil fuels to ship to the U.S.
4. Avoid single-use plastic. This includes things like plastic cutlery, cosmetics packaging, plastic shopping bags, plastic drinking bottles, coffee cups and the packaging around fresh food produce. By

bringing your own reusable bags shopping, buying loose fruit and vegetables and drinking from a reusable water bottle instead of buying a new bottle of water every time, you will create notable changes in the way you consume plastic. These are all small and achievable changes, but they can really help to make an impact- plus, if everyone made small changes like these, think of the difference that could be made.

5. Drive less. If you can, consider cycling or walking to places where you would usually drive. Not only will this be an investment in your own health, but it will also save you money in the long run, reduce your use of cars and public transportation, and help you stay healthy, which will save a lot on anguish and medical bills in later life!

If you're interested in doing more for the environment, consider switching to clean energy. Unfortunately, the carbon footprint of the average American seems to rise every year, likely due to an increase in consumerism, demand for meat, transport, and population.

HOW IS A CARBON FOOTPRINT CALCULATED?

A carbon footprint is a calculation of an activity's impact on Earth's atmosphere, such as manufacturing a product, one's lifestyle, or running a business. A carbon footprint is typically computed by estimating not just the CO2 emissions caused by the activity, but also any additional greenhouse gases. It sums up the factors that affect the atmosphere, and subsequently, the climate.

All these factors are totaled up and expressed as a single value in terms of carbon dioxide equivalent (CO2e), which is the quantity of CO2 that would cause the same amount of warming. While this may seem complicated, everyone has a carbon footprint or an impact on the climate.

Many factors can cause carbon footprints, such as:

- Transportation
- Industrialization
- Energy production
- Personal homes
- Commercial buildings
- Agriculture
- Land use and forestry

As you follow the line down from each of these options, you find carbon footprints are essentially a given, considering the needs of the modern world. A carbon footprint includes both direct and indirect emissions, with direct emissions ascribed to a single person and indirect emissions attributed to a production process. Food, product consumption, transportation, and residential energy contribute to carbon footprints. Food is a significant contributor to carbon footprints, with meat being particularly problematic. Food transportation, pesticide use, and out-of-season purchases all add to carbon footprints. In addition, transportation, factory processing, and additional packaging all contribute to higher emissions in processed foods than in fresh foods. Energy inefficient homes squander considerable amounts of energy due to poor insulation, inefficient appliances, and excessive water use. Clothing, household items, and personal products all contribute significantly to an individual's carbon footprint due to emissions related to gathering resources, production, and transportation. Group transportation, such as trains or buses, emit less pollution per person, although they still emit pollutants. As you can see, almost everything in our energy-heavy world will leave behind a carbon footprint.

HOW DOES CARBON FOOTPRINT RELATE TO CLIMATE CHANGE?

Burning fossil fuels is responsible for roughly three-quarters of human-caused greenhouse gas emissions. According to the U.S. Energy Information Administration (EIA), we rely on fossil fuels to power our automobiles, heat our houses, and even consume electricity. The EIA has also concluded that greenhouse gas emissions have increased since industrialization, implying that humans are responsible for a significant portion of greenhouse gas production.

When fossil fuels like coal, natural gas, and oil are burned, carbon dioxide is discharged into the atmosphere. In addition, other biological materials, such as solid waste, produce carbon dioxide when they are burned. Carbon dioxide emitted in enormous quantities into the atmosphere impacts the earth, especially if it stays in the atmosphere for thousands of years. Carbon emissions directly impact humans, producing an increase in respiratory ailments due to smog and air pollution. Climate change is perhaps the most significant way that carbon emissions affect the environment. Our climate changes dramatically as the average global temperature rises, it heats. As a result, extreme weather phenomena such as tropical storms, wildfires, flooding, and severe droughts are caused by global warming. Because global warming can result in catastrophic weather events and disrupted ecosystems, everyone must recognize their effect on the future and seek to make a positive impact.

The objective of this chapter is as follows:

1. Describe the global carbon biochemical cycle and how it has been disrupted from the stable state during pre-industrial times
2. Articulate the connection between the human-caused, rapidly increasing atmospheric carbon concentrations and climate change
3. Discuss the carbon footprint of the built environment and how to quantify it, and
4. Describe how to use the knowledge gained by calculating the built environment carbon footprint to help rapidly and reverse the climate change.

HUMAN IMPACT ON THE BIOGEOCHEMICAL CARBON CYCLE

On Earth, there are several large-scale cyclic movements of certain chemicals that have profound effects on living organisms. Th pathways and rate of movement of these chemicals through both living (biotic) and nonliving (abiotic) compartments of the planet have enormous effects on life on Earth, Kibert 2022. These large-scale movements of key chemicals through the biosphere are referred to as biogeochemical cycles. The seven important biogeochemical cycles that have been identified are the cycle of carbon, water, oxygen, nitrogen, phosphorous, sulfur and rock. The carbon cycle is important because all known life-forms are composed of carbon and carbon compounds. CO_2, an important component of carbon cycle, regulates planetary temperature, making life on earth possible. Although CO_2 is a tiny portion of Earth's atmosphere – just 0.04 percent – it causes 63 percent of energy trapping that occurs in the atmosphere. The powerful ability of CO_2 to trap energy contributes to a climate change regime that supports life. Therefore, changes to atmospheric CO_2 levels affect average global temperature. Although CO_2 concentrations in the atmosphere have varied naturally for thousands of years, current levels exceed those measured in the ice cores from 420,000 years ago and are in fact reaching concentrations not experienced on Earth for several million years. These concentrations are continuing to increase at an accelerated pace and are upsetting the relative stability of Earth's climate. This upset is referred to as climate change and is root cause has been traced to human activities and technologies that are dumping enormous quantities of climate change gases, primarily CO_2, into the atmosphere.

A dramatic increase in biogeochemical carbon mass in atmosphere is caused by:

1. Elevated CO2 emission from buildings, powerplants, vehicles, cement plants and loss of biomass (deforestation).
2. Methane gas CH_4 which is far more toxic than CO_2 is also contributing to climate change. However, the magnitude of CH_4 emission rate is far less than CO_2 emission.
3. Upon deforestation processes, and reduction in biomass, enormous amounts of CO_2 and CH_4 are being released to environment.
4. As the Earth warms up ice cover in the northern Arctic boreal forests melts, exposes enormous quantities of biomass to air that were frozen for millions of years. The biomass then decomposes and releases CO_2 and CH_4. Boreal forest is the forest that grows in regions of the northern hemisphere with cold temperatures. Made up of mostly cold tolerant coniferous species such as spruce and fir.
5. As the ice cover disappears its reflective properties are lost. The darker biomass reveals in the process

absorbs the sun's energy, increasing the rate of decay.

6. Additional accelerator of climate change is the loss of snow and ice cover in the North Pole and Greenland. As such, the solar energy is being absorbed by the much darker ocean instead of being reflected, further increasing energy input to the atmosphere.
7. Fracking for oil and natural gas production is also having an effect because significant quantities of natural gas which is largely CH_4 are being released to atmosphere during fracking processes.
8. Additional sources of CH_4 released to the atmosphere include livestock and rice farming.

The result of confluence of all these human impacts is that the changes to the Earth's climate are occurring at an alarming rate. The nine out ten hottest years on record have occurred since 2000 with the warmest June in 2019, which tied with June 2020. In addition, 2024 summer was the hottest summer ever.

The carbon biogeochemical cycle exhibits two major behaviors: active or more rapid-cycling systems, and slow-cycling systems. The naturally active carbon cycle, comprising carbon movements among the atmosphere, ocean and the terrestrial and aquatic biosphere, contains about 41 billion tons of carbon. The human activities are the part of active carbon-cycles that are disrupting the otherwise relatively stable atmosphere, See Figure 12.1.

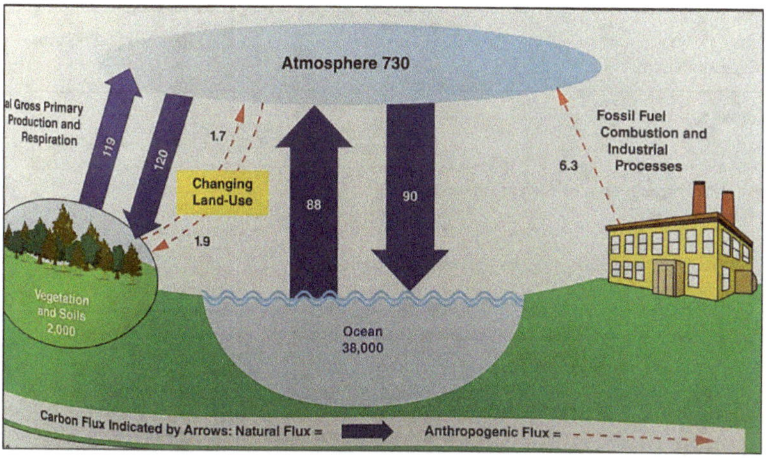

Figure 12.1 The global carbon cycle showing stocks and flows of carbon (Source: IPPC, 2001)

The imbalance resulting for human activities in illustrated in Table 12.1. Fossil fuel consumption, land use changes in which the biomass is removed, and cement production increase CO_2 emission to the atmosphere by about 0.7 gigatons per year. Natural oceanic and terrestrial systems absorb about 3.8 gigatons of this increase, leaving 3.2 gigatons of excess CO_2 per year to accumulate in the atmosphere. The 3.2 gigatons per year of addition to the 730-gigaton active carbon cycle, about 0.5 percent per year, are important because there is no natural mechanism capable of absorbing this excess at a rate that would prevent its accumulation. Indeed, the only systems capable of absorbing this excess are the natural mechanisms of erosion and sedimentation associated with Earth's lithosphere, see Figure 12.1. However, these mechanisms are associated with the slow-cycling portion of the carbon cycle, they function not on short time scales, such as years or decades, but on geological time scales of thousand years.

Human Activity	Gigatons (billion tons)
Fossil fuel burning	6.2
Cement production	0.1
Land-use change	0.7
Subtotal Human Caused Carbon Increase	***7.0***
Land removal	1.9
Ocean removal	1.9
Subtotal Human Caused Carbon removal from atmosphere	**3.8**
Net annual carbon accumulation in the atmosphere	**3.2**

Source: Intergovernmental Panel on Climate Change, 2013

Table 12.1 Human Activities contributing to Imbalances in the Global Carbon Cycle

The quickest path to slowing and reversing climate change is the is the reduction of human carbon emissions and the restoration of the biosphere. Because the built environment is the dominant cause of the excess CO_2 emissions, the solution lies in dramatically reducing energy consumption along with shifting to energy sources that have the lowest CO_2 emissions.

Shifting from coal to natural gas is an example of the latter because it would cut CO_2 emission in half for the same amount of energy produced, see table 12.2 (Kibert, 2022).

Fuel	Pounds of CO_2/million BTUs	Kilograms CO_2/kWh
Coal (anthracite)	228.6	0.354
Coal (bituminous)	205.7	0.318
Coal (lignite)	215.4	0.333
Diesel fuel and heating oil	161.3	0.250
Gasoline	157.2	0.243
Propane	139.0	0.215
Natural gas (methane)	117.0	0.181

Source: US Energy Information Agency 2013.

Table 12.2 CO_2 Emissions for Common Fossil Fuels

Due to extensive combustion of fossil fuels to support the built environment, transportation, and industry, atmospheric CO_2 concentrations have risen from pre-industrial levels of about 280 parts per million (ppm) to over 420 ppm today, an increase of 43 percent, see Table 12.3.

Because average atmospheric temperatures have been found to track the changes in CO_2 concentrations, it is reasonable to expect that dramatic upward changes in temperature will result from significant increase in these concentrations.

Gas	Chemical Formula	Preindustrial Level	Current Level	Increase since 1750
Carbon dioxide	CO_2	280 ppm	421 ppm	141 ppm
Methane	CH_4	700 ppb	1800 ppb	1100 ppb
Nitrous oxide	NO_2	270 ppb	337 ppb	67 ppb
CFC-12 refrigerant	CCl_2F_2	0	533 ppt	533 ppt
HCFC-22	$CHClF_2$	0	206 ppt	206 ppt
Source: 2013 & 2018 International Panel on Climate Change				

Table 12.3 Current and Preindustrial concentrations of Greenhouse gases

CLIMATE CHANGE AND GREENHOUSE GASES

The global warming that's causing our climate crisis is already having dire consequences. In just the past few decades:

- Rising temperatures have worsened extreme weather events.
- Chunks of ice in the Antarctic have broken apart.
- Wildfire seasons are months longer.
- Coral reefs have been bleached of their colors.
- Mosquitoes are expanding their territory, able to spread disease.

Who is causing this climate crisis? It's mainly us. Humans are the main cause of climate change — we burn fossil fuels and chop down forests, causing average temperatures to rise worldwide. That global warming trend is increasingly disrupting our climate — the average weather over many years. Earth has already warmed by about 1 degree Celsius, or 1.8 degrees Fahrenheit, since the 19th century, before industry started to boom. While we experience the effects, we're on our way toward 1.5 degrees C (2.7 F) by as early as 2030.

WHY IS A HALF-DEGREE MORE SUCH A BIG DEAL?

A warmer world — even by half-degree Celsius — has more evaporation, leading to more water in the atmosphere. Such changing conditions put our agriculture, health, water supply and more at risk. Picture a North Carolina cotton farm that's been around since 1960, with global average temperatures rising by roughly half a degree since it grew its first crop.

The increased evaporation and additional moisture to the atmosphere has led to 30% more intense rain during heavy downpours in that part of the U.S. Then a hurricane like 2018's Florence — already strengthened by warmer oceans and higher seas — dumps this excess rainfall on the farm. The crops get more flooded and damaged than they did half a century ago.

It's how you go from half-degree of warming to economic hardship.

Figure 12.2 Climate change is breeding storms with heavier rainfall, flooding farms — such as this one, which grows cotton

Whether it's a shift of 1.5 degrees or 2 degrees, these warming levels aren't magic thresholds. Every rise in warming is worse for the planet than the last. But they're not inevitable. It's not too late to slow the pace of climate change and avert the worst impacts of the climate crisis — if we act today. With every nation's help, we can attack this challenge.

Many scientists believe there is strong evidence that the average temperature of planet surface will increase from 9° to 13.5^{o}F in the 21st century. The likely result will be rapidly rising sea levels, substantially reduced crop yields, drought, and more energetic hurricanes cyclones, all threatening the very survival of human beings. Increasing concentrations of greenhouse gases produced by human activities, particularly CO_2, are the primary force pushing up global average temperature. CO_2 concentrations are now over 420 ppm compared to 280 ppm at the start of the Industrial

Revolution 230 years ago in the late eighteenth century. The geological record indicates that an average CO_2 level of 450 ppm defines the boundary between the ice-free planet. Climate scientists originally suggested that 350 ppm was the safe upper limit for CO_2, one we had already surpassed. The continuation of business as usual will likely cause CO_2 levels to increase from the current concentration of 420 ppm to over 450 ppm between 2030 and 2050. The UN is actively discussing trying to limit CO_2 increase to 550 ppm, a potentially catastrophic level, but representative of the as-yet-futile efforts to rein in these emissions. Clearly, strong and drastic action is needed to stop and reverse atmospheric CO_2 concentration to avoid the worst outcomes of climate change. It should be noted that CO_2 is the most common greenhouse gas by far, but more potent greenhouse gases, such as CH_4 that have a much higher impact per molecule.

The term global warming potential (GWP) in Table 12.4 warrants additional explanation. The GWP of greenhouse gas indicates the amount of warming a gas cause over a given period (normally 100 years). GWP is an index. CO_2 has an index value of 1, and GWP for all other greenhouse gases is the number of warming they cause compared to CO_2. Related to GWP is the greenhouse multiplier, which should be used to determine that impact of other gases in terms of CO_2, see Table 12.5.

Gas	Chemical Formula	Lifetime (years)*	GWP**
Carbon dioxide	CO_2	Variable	1
Methane	CH_4	12	25
Nitrous oxide	NO_2	114	298
CFC-12 refrigerant	CCl_2F_2	100	11,000
HCFC-22 refrigerant	$CHClF_2$	12	5,160

*) The lifetime of carbon dioxide is variable because about 50 percent is removed in a century while about 20 percent is resident in the atmosphere for thousands of years.
**) The GWP of carbon dioxide is defined as 1. The GWP of 72 for methane means that each molecule is 72 times more potent than a carbon molecule at trapping energy. Although methane is far more potent, its atmospheric concentration is relatively small. Source: International Panel on Climate Change 2007; International Panel on Climate Change 2013.

Table 12.4 Atmospheric Lifetime and Global Warming Potential of Greenhouse gases

Atmospheric Gas	Greenhouse Gas multiplier
CO_2 (carbon dioxide)	1
CH_4 (methane)	21
NO_2 (nitrous oxide)	310
CFC-11 (CCl_3F) refrigerant	1320
CFC-12 (CF_2Cl_2) refrigerant	6650
HCFC-22 ($CHClF_2$) refrigerant	1350
Surface ozone	100

Note: The multiplier indicates how many grams of CO_2 equivalent impact each gram of gas causes. Although some gases have large multipliers. The vast mass of $CO_2$is being emitted dwarf the mass of other gases and causes over 99 percent of the climate impact.

Table 12.5 Greenhouse Multiplier for Various Atmospheric Gases

The greenhouse multiplier is derived by combining GWP and other factors to calculate the climate change impact of climate change gases relative to CO_2. CO_2 is by far the significant greenhouse gas because of the vast quantities of CO_2 present in the various compartments of the biosphere and lithosphere. With respect to calculating the greenhouse gas multiplier, 1-kilogram CH_4 causes 21 times more energy trapping compared than 1-kilogram CO_2; thus, CH_4 has greenhouse multiplier of 21. The compact way of describing this is to state that, in terms of climate change, 1-kilogram CH_4 is the carbon dioxide equivalent (CO_2e) of 21 kg

CO_2e. using this methodology, the effects of all climate change gases can be expressed in terms of the CO_2e. t CO_2e often is used to refer to the tons of carbon equivalents; Mt CO_2e indicates megatons, Gt CO_2e is gigatons of CO_2 equivalent.

With respect to the scale of greenhouse gas emissions, CO_2 discharges are generally 10 times greater than the next most emitted greenhouse gas CH_4. While US CH_4 emissions were relatively flat in the period 1990 – 2018, CO_2 emissions climbed by almost 6 percent. With respect to sources of greenhouse gas emissions, the total US emissions have remained flat since 1990, see Figure 12.2.

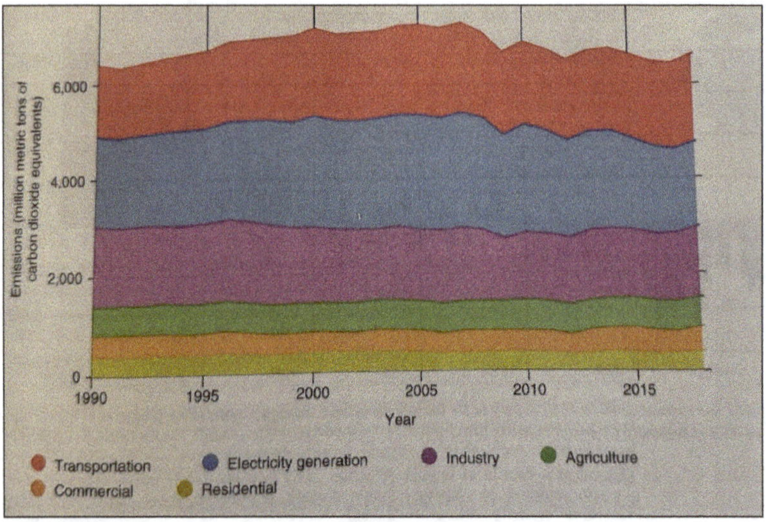

Figure 12.2 Total US greenhouse gas emissions have remained relatively flat over the past 28 years. Source: https://www.epa.gov/ghgemissions/inventry-us-greenhouse-gas-emissions-ans-sinks

It should be noted that increasing atmospheric CO_2 concentrations will lead to increasing acidification of ocean, with negative repercussions for all shell forming species and their ecosystems.

The models used by Intergovernmental Panel on Climate Change (IPCC) project that by the end of this century, the global average sea level will rise between 7 and 23 inches above the 1980 to 1999 average. Also, the IPCC stated that if the observed contributions from the Greenland and Antarctic ice sheets between 1992 and 2003 "were to grow linearly with global average temperature change, "the upper ranges of sea-level-rise would increase by 3.9 to 7.9 inches. In other words, in this example, the upper range for sea-level rise would be 31 inches, Kibert 2022.

Shitting temperatures, more violent storms, rising sea levels, melting glaciers, and other effects will displace people, affect food supplies, reduce biodiversity, and greatly reduce average quality of life. Creators of built environments, a major energy consumer, must dramatically reduce energy consumption, particularly reliance on fossil fuels.

The term Climate Change strategies refer to actions that attempt to limit the scale and rate of long-time climate change. Today, a wide range of measures are being attempted including:

- Shifting to low-carbon energy systems represented by Wind, Solar and Nuclear Technologies.
- Designing Compact Urban Areas
- High Efficiency Mass Transit Systems
- High-Efficient HVAC, Motors, and other building equipment

To Reduce CO2 Levels in the atmosphere, we need a vast and coordinated international efforts to maintain the current CO_2

levels in the atmosphere. The wide variety of possible remedies are associated with climate change mitigation ranges from non-technical to highly technical solutions, that remove carbon from the atmosphere and stores it in rock formations or caverns.

The next sections describe the carbon sequestration strategies that can help address climate change mitigation.

CARBON SEQUESTRATION

A natural or artificial process by which carbon dioxide is removed from the atmosphere and held in solid or liquid form is referred to as carbon sequestration. Carbon sequestration can be achieved using the following strategies:

- Terrestrial System Carbon Sequestration
- Geologic Carbon Sequestration
- Ocean Carbon Sequestration
- Carbon Capture and Sequestration/Storage

Terrestrial System Carbon Sequestration

Terrestrial carbon sequestration system includes vegetation, soils, and sediments that uptake and store CO_2. Terrestrial systems can store substantial portions of natural CO_2 emissions, but disturbances to these systems, such as land tilling and cultivation can reduce the ability of vegetation and soil to store carbon. Some clear trade-offs will have to be made if CO_2 absorption by vegetation and soil were to become a dominant climate change mitigation strategy. For example, converting greenfield/forests into agricultural lands should be stopped since tillage of lands under forest/greenfield would release significant amounts of CO_2 from soils and sediments. However, this would hurt food production if food production became a priority. Additionally, terrestrial carbon sequestration is not reliable because fire, disease, increasing temperature, and other disturbances can reduce vegetation and impact carbon storage. In short, because terrestrial carbon storage is uncertain and not well understood, it cannot realistically be the centerpiece of climate change mitigation strategies. However, it can be combined with other strategies to create an overall effective solution.

Vegetation can store enormous quantities of CO_2 and planting many billions of trees would ultimately have a significant impact on climate change. Ultimately, when wood is used as a construction material in place of materials with a carbon footprint, the result is a reduction in CO_2 because those materials will not have been manufactured (Building the Future 2020).

The building industry currently accounts for 39% of global energy-related CO2 emissions, 11% of which result from manufacturing building materials and products such as steel, cement, and glass. And yet, to date, emission reduction efforts have not really focused on decarbonizing the construction sector but the two other sectors responsible for the most emissions, electric power and transportation.

Understanding and reducing emissions from the building sector is vital if we want to limit global warming to 1.5 degrees Celsius and if we are serious about leaving a healthy flourishing planet for future generations.

When put into context, the need for and importance of creating more sustainable, low-carbon cities is indisputable. Between 2020 and 2050, 2.3 billion new urban dwellers will need housing and commercial buildings. Using conventional materials like steel and concrete would generate annual emissions of 0.53 billion tons of carbon dioxide equivalent (tCO2e) per year. Using wood for half of these buildings could reduce emissions by 0.15 billion tCO2e per year, with an additional 0.52 billion tCO2e per year stored in the buildings.

But the benefits of wooden construction go beyond sustainability. Wooden buildings have proven to have a positive effect on our health and well-being, fostering productivity and creativity, while connecting us with nature. From environmental to health benefits, the upside of building with wood is clear. Then, why don't we build more with wood? What is stopping us? What are these crucial elements of wooden construction that make the building process so beneficial? Do buildings with wood cause further deforestation? In Sweden, Ms. Frank explains, it does not. But how can we balance harvesting wood and natural forest regeneration and use the wood to store CO_2? (Building with Wood by Nyasha Harper-Michon, unece_info@un.org)

Geologic Carbon Sequestration

CO2 generated from combustion of fossil fuels can be stored in deep geologic formation to the depth of 2.5 miles below ground surface, where it could be injected into porous rock such as depleted mines and oil reservoirs.

For saving groundwater quality, EPA has come up with the series of requirements for designing the injection wells and a guideline for injecting CO2 into the subsurface. The rule also affords owners or operators injection depths flexibility to address injections in various geologic settings in the US in which geologic sequestration may occur, including very deep formations and oil gas fields that are transitioned for use as CO_2 storage sites.

The US Department of Energy (2015) estimates that in the United States, storage for between 900 and 3,400 gigatons of CO_2 is available in the deep rock formation that contain saline groundwater. Unmineable coal beds and oil and gas reservoirs provide other potential storage sites for CO_2. Per Kibert 2022, although the scale of these potential sequestration sites is enormous, there are significant uncertainties about the permanence, effectiveness, and cost of these strategies. Significant additional research will be required to gain a better understanding of the feasibility of geologic carbon sequestration.

Ocean Carbon Sequestration

The idea about intentional CO_2 storage in the ocean is to release a captured and compressed stream of CO_2 into the deep ocean at or above the sea floor. CO2 would dissolve into the sea water and become part of the ocean and its carbon cycle. The success of ocean storage depends on slow exchange of deep ocean water with the surface, thus isolating CO_2 from the atmosphere. The effectiveness of ocean storage will depend on the length of time that isolation from the atmosphere is maintained. The key factor in storing CO2in the deep ocean is the length of isolation time and is generally depends on the depth of ocean/sea where the CO_2 is injected/released. Ocean storage is still a work in progress and has not yet been implemented at large scales (IPCC 2005). According to the latest news(https://www.pnas.org), deep-sea sediments at high pressure and low temperature provide a virtually unlimited and permanent reservoir for carbon dioxide captured from fuel combustion. When injected below the ocean floor deeper than 3,000-meters, CO_2 will remain below a layer of more buoyant pore fluid. The ocean holds 60 times more carbon than the atmosphere and absorbs almost 30 percent of carbon dioxide emissions from human activities. This means that ocean is the key to understanding the global carbon cycle and thus our future climate, (https://www.csiro.au>articles>june)

CARBON CAPTURE AND SEQUESTRATION/STORAGE

Carbon capture and sequestration/storage (CCS) is the process of capturing carbon dioxide created during power generation and industrial processes and storing it so that it is not emitted into the atmosphere. CCS technologies have significant potential to reduce CO_2 emissions in energy systems and, in some cases, can remove up to 100 percent of the CO_2. The capture of CO_2 applies to point sources such as power plants and factories where the emissions are concentrated and can be processed to remove the CO_2. The connection between CCS and the built environment is that almost half of the fossil fueled power generation capacity is consumed by buildings. Significant contributors of materials with large carbon footprints are produced by industrial processes used to manufacture cement and steel, the second and fourth largest emitters of CO_2 per year. Consequently, there is a direct connection between the need for CCS and the construction and operation of built environment (IPCC, 2005).

MITIGATION MEASURES

To reduce the concentration of the greenhouse gases and at least prevent the upward concentration trend of GHG the following should be done by every nation and country on the face of Earth to stabilize the atmosphere:

1. Double the efficiency of 1 billion cars from 30 to 60 miles per gallon.
2. Decrease the number of car miles traveled by half
3. Use best efficiency practices in all residential and commercial buildings.

4. Produce current coal-based electricity with twice today's efficiency
5. Reduce 1,400 coal-fired power plants with gas-powered facilities.
6. Capture and store emissions from 800 coal and electric plants
7. Produce hydrogen from coals at six times today's rate and store the resulting CO_2.
8. Capture carbon from 180 coal-to-synfuels plants and store the CO_2.
9. Double the current global nuclear capacity to replace coal-based electricity
10. Increase wind electricity by 10 times compared to today for a total of 2 million large windmills.
11. Install 100 times the current capacity of solar electricity.
12. Use 15,400 square miles of solar panels to produce hydrogen for fuel cell cars.
13. Increase ethanol production 12 times by creating biomass
14. Eliminate tropical deforestation.
15. Adopt conservation tillage in all agricultural soils worldwide.

DEFINING THE CARBON FOOTPRINT OF THE BUILT ENVIRONMENT

The first step in evaluation of built environment's impact on climate change is the calculation of its carbon footprint. As discussed earlier, the carbon footprint is generally defined as the total amount of greenhouse gases produced to support human activities directly and indirectly, usually expressed in pounds, kilogram, or tons of CO_2 equivalent $t(CO_2e)$. Climate change, the long-term fluctuations in temperature, precipitation, wind, and all other aspects of the Earth's climate are the result of the total global carbon footprint. The built environment carbon footprint is by far the dominant fraction of the total global footprint. It results from carbon emissions connected to the construction, operation, and disposal of the built environment, plus the carbon emissions from transportation servicing buildings. For calculating the built environment carbon footprint, carbon emission can be classified as:

- Operational carbon
- Embodied carbon, and
- Commuting or transportation carbon.

Operational carbon (OC) is the carbon resulting from energy produced to operate the built environment for heating, cooling, ventilation, lighting, elevators, electronics loads, and other functions requiring electricity or thermal energy. This category of carbon emission is the most obvious and has received the most scrutiny over time. Another source of operational carbon that is not as well known or analyzed is the energy required to pump water and wastewater to and from the building.

Embodied carbon (EC) results from the front-end energy investment in the materials and products that comprise the built environment. It includes the extraction of resources, the manufacture of products, the installation of products and materials, building maintenance, and facility disposal.

Transportation/commuting carbon (TC) which accounts for 20 percent of total US carbon emissions, is greatly affected by distribution of buildings on the landscape and by the availability of mass transit systems with low carbon emissions per passenger mile. It makes sense to include the emission from transportation systems in accounting for the quantities of CO_2 attributable to buildings.

In summary, the carbon footprint of a building is the combination of these three components normalized to the building area.

$$\text{Carbon Footprint } (kgCO_2e/m^2) = [OC + EC + TC]/\text{Building Area}$$

The next section provides a method for calculating the carbon footprint for buildings. The approach described will provide a good estimate of each of the three sources of carbon associated with the built environment. The outcome of this effort will be to allow project teams to find the best combination of factors for minimizing a project's carbon footprint. This approach also can be used in urban planning efforts to support minimizing the transportation portion of the carbon footprint.

EMBODIED CARBON FOOTPRINT

The amount of energy and carbon associated with the materials and products that comprise the built environment is quite large and can be equivalent to from 5 to 20 years of operational energy, depending on the type of building and its energy profile. For high-performance green buildings with a low Energy Use Index, the years of operational energy that are equivalent to the embodied energy of the building may be much longer due to the lower annual energy consumption.

The entire EC footprint of a building can be assessed by adding the EC footprint of the building material used in construction of the building. For instance, EC footprint of the flooring can be assessed by reviewing the energy requirement of carpet manufacturing at different stages. Figure 12.3 shows the schematic diagram of carpet manufacturing, from material extraction, transportation, installation, maintenance and end of life disposal activities and percent of energy consumed during the entire life cycle of the carpet. The EC footprint of buildings can be greatly reduced by creating materials that are durable, low maintenance, and adaptable. Doubling the lifetime of a building from 50 to 100 years in effect cuts the EC footprint in half, a major effect. Clearly, good planning with a long-time horizon will ensure that frequent redesign of urban areas that requires removal of large numbers of buildings is unnecessary.

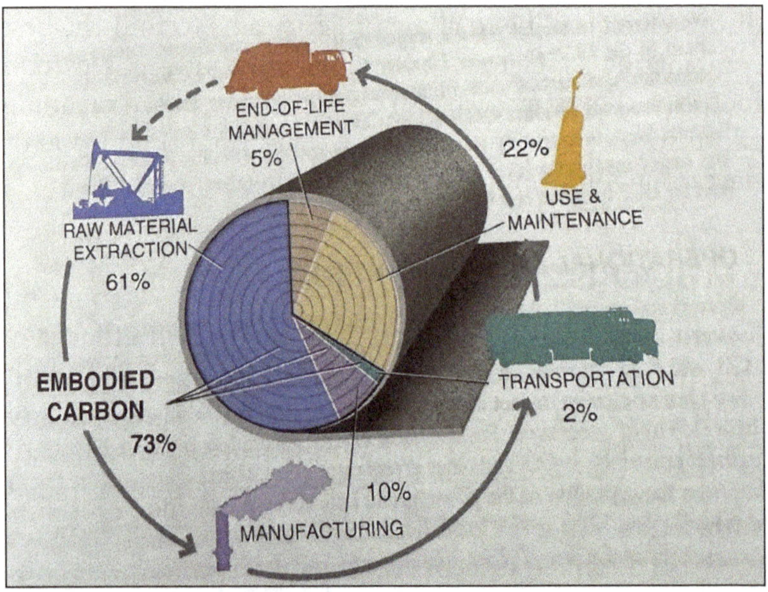

Figure 12.3 The embodied carbon of carpet is determined by examining the entire life-cycle of the materials from extraction through removal. Source: Peter Harris/BuildingGreen, Inc.

The life-cycle emissions of carbon are being determined for a wide range of materials, including carpet as presented in Figure 12.3. These data are being compiled into detailed databases, such as the Inventory of Carbon and Energy

(ICE) developed by researchers at the University of Bath in United Kingdom. The information displayed in Table 12.6 was extracted from ICE. Note that both the embodied energy and the EC of recycled metals are about one third to one-seventh the levels for virgin metals. This is true of all materials, whether they are concrete, plastic, glass, paper, or wood products. One approach to determining the carbon footprint of construction materials and products is to use an online embodied carbon calculator. For example, the embodied carbon in construction calculator (EC3) tool can be used for benchmarking, assessing, and reducing the carbon footprint of various materials during the raw material supply, transport, and product manufacturing stages of a product's life cycle. These are referred to as the A1 – A3 stages of life-cycle assessment, described in Chapter 11. The inputs to EC3 are building material quantities and carbon data from construction estimates and/or BIM models plus Environmental Products Declarations (EPD). The EC3 calculator assists a wide variety of stakeholders in assessing supply chain data and supports the process of establishing the target maximum level for a project's carbon footprint. Another significant attribute of EC3 is that it is a free, open access tool.

Material	Embodied Energy (MJ/kg)	Embodied Carbon (kg CO_2e/kg)
Aluminum – virgin	218	12.79
Aluminum – recycled	29	1.81
Asphalt 6% binder	3.93	0.076
Brick	3.00	0.24
Concrete 0% fly ash	0.55	0.076
Concrete 15% fly ash	0.52	0.069
Concrete 30% fly ash	0.47	0.061
Copper tube – virgin	57.0	3.81
Copper tube – recycled	16.5	0.84
Glass	15	0.91
Paint	70	2.91
Plastic (general)	80.5	3.30
Steel – virgin	35.4	2.89
Steel – recycled	9.4	0.47
Timber	10	0.72

Source: Hammond and Jones 2011

Table 12.6 Embodied Energy and Embodied Carbon of Common Materials Used in Construction As Indicated in the Inventory of Carbon and Energy Developed by the University of Bath in UK

Operational Carbon Footprint

Operational energy is the energy required to power the built environment. All industrial systems and electrical power systems that support them have carbon footprints. CO_2 and other greenhouse gases are emitted over the life cycle of power plants, and their climate change impact is expressed as kilograms of CO_2 equivalent per kilowatt-hour (kWh) of generation. For electrical energy generation, there are both direct and indirect emissions of CO_2 and other greenhouse gases. Direct emissions are those arising from the operation of power plant; indirect emissions arise from other phases of the life cycle, such as fuel extraction, transportation of fuel processing of the fuel, construction of the power plants, maintenance, and power plant decommissioning. An LCA for products of any kind, and it is based on the International Organization of Standardization 14000 series of standards. For any type of electrical generation plant, whether it is a fossil fuel plant, hydropower installation, nuclear power plant, or solar photovoltaic array, the exact same analysis is used to determine the carbon footprint as presented in Figure 12.4.

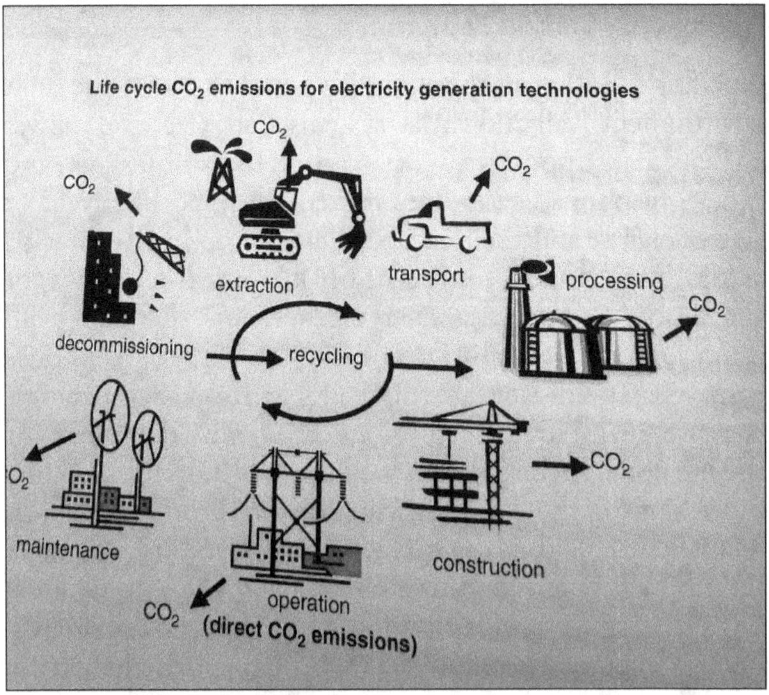

Figure 12.4 The carbon footprint of an electrical power generating station includes the emission from all phases of the life cycle in extracting and transporting the fuel required for the plant, the materials and products from which the plant is constructed, plus all emissions associated with the extraction, and transportation of these resources. The carbon footprint also includes the maintenance and decommissioning of the power plant at the end of its useful life. (Source: Parliamentary Office of Science and Technology 2006/2011)

Power plants that burn fossil fuel have by far the largest carbon footprint of all forms of power generation. This is due to the long chain, carbon containing molecules from which fossil fuels are derived. A typical coal-fired plant, for instance, will have emissions on the order of 1 kilogram of CO_2e/per kilowatt-hour of electricity generated (1 kg CO_2e/kWh). Oil-fired power plants also contribute a significant amount of carbon for each kilowatt-hour of electricity generated on the order of 0.6 kg CO_2e/kWh. Gas-fired power generation, which is becoming an increasing fraction of power generation, contributes on the order of 0.4 kg CO_2e/kWh.

Renewable forms of energy have significantly lower carbon footprints, as shown in Figure 12.5 and 12.6.

Biomass has a carbon footprint ranging from 25 to about 80 g CO_2e/kWh, while hydropower has a low carbon footprint of 10 g CO_2e/kWh or lower. Nuclear power has lowest carbon footprint of all forms of energy generation at 5 g CO_2e/kWh and lower. Photovoltaic power production has a lower but still surprisingly has carbon footprint compared to wind energy and hydropower, ranging from about 30 to 60 g CO_2e/kWh (Parliamentary Office of Science and Technology 2006/2011).

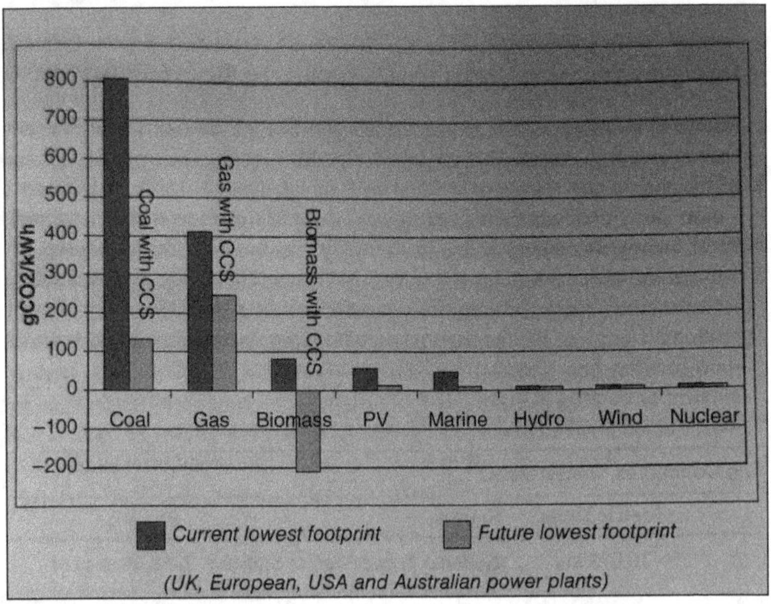

Figure 12.5 The carbon footprint of various electrical power generation technologies. carbon capture and storage technologies that remove carbon dioxide from combustion gases are being developed but have not been proven and tested on a large scale. Coal-fired power plants not only dominate the electrical power generation industry but also produce the greatest amount of carbon dioxide at about 1 kg CO_2e/kWh. Renewable energy systems range from about 50 grams of CO_2e/kWh for photovoltaics to about 5 g CO_2e/kWh for hydropower and wind power. Nuclear power plants also have a relatively small carbon footprint, at about 5 g CO_2e/kWh. (Source: Parliamentary Office of Science and Technology 2006/2011)

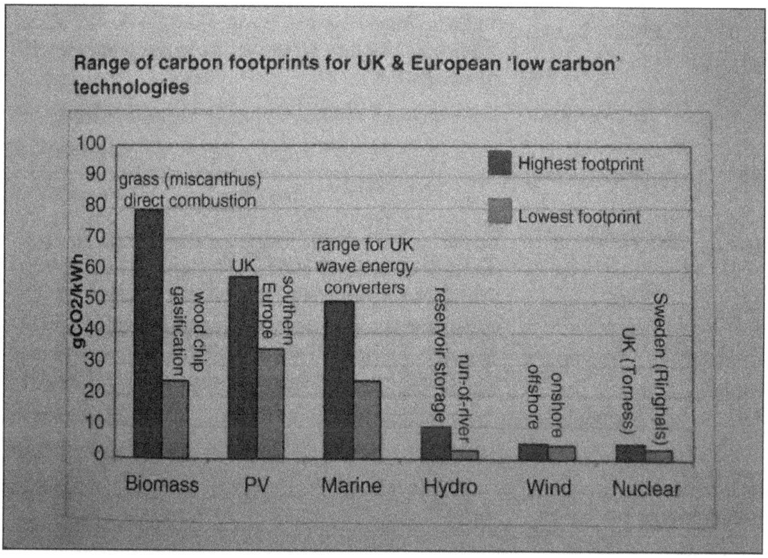

Figure 12.6 Contribution of renewable and nuclear power systems to climate change. Biomass at about 80 g CO_2e/kWh, has just one-tenth the climate change contribution of coal-fired power plants. At the lower end of the spectrum, wind energy and nuclear have just 5 g CO_2e/kWh. Source: Parliamentary Office of Science and Technology 2006/2011

TRANSPORTATION/COMMUTING CARBON FOOTPRINT

The location of buildings in the landscape is a function of decision made by owners and government. It is now well recognized that the distribution of the buildings causes some of the carbon and energy associated with transportation. The various investigations have revealed that the quantity of energy associated with just the daily commute of the workforce to and from commercial office buildings is significantly larger than operational energy of the facility. Average building energy was found to be about 92.9 kBTU/ft^2-yr, while transportation energy was 121 kBTU/ft^2-yr. The gap widens even further for newer code-compliant buildings, where the operational energy might be 51 kBTU/ft^2-yr, with transportation energy two times greater operational energy. An even larger gap occurs for high performance green buildings where the building energy may be 32 kBTU/ft^2-yr; the transportation energy would almost be four times greater than operational energy.

Each type of transportation has a different energy and carbon intensity as shown in Table 12.7 and 12.8.

Vehicle Type	Load factor (persons/vehicle)	Energy Use (BTU/vehicle-mile)	Energy Intensity (BTU/passenger-mile)
Cars	1.6	4,702	3,034
Personal trucks	1.8	6,156	3,345
Taxi and Van (demand response)	1.4	20,047	14,106
Vanpool	6.4	8,226	1,294
Bus – transit	9.1	36,760	4,025
Airline	113.2	263,971	2,332
Rail – Intercity (Amtrack)	20.5	30,972	1,187
Rail – light and heavy	25.8	20,022	776
Rail – commuter	31.6	51,888	1,643

Source: US Department of Transportation 2018

Table 12.7 Energy Intensity of Different Forms of Travel in the United States

Mode of Travel	CO_2 Generation
Average passenger car	0.571 (kg CO_2/vehicle-km)
Average air flight	0.26 – 0.40(kg CO_2/passenger-km)
Average rail	0.27 (kg CO_2/passenger-km)
Average light duty truck	0.780 (kg CO_2/vehicle-km)
Average motorcycle	0.37 (kg CO_2/vehicle-km)
Average bus	0.088 (kg CO_2/passenger-km)

Source: US Energy Information Agency 2016; EEA 2016: US Environmental Protection Agency 2018.

Table 12.8 Carbon Emission for Various Modes of Transportation

Each type of transportation has a different energy and carbon intensity, as shown in Tables 12.7 and 12.8. This information can be used to forecast the impacts of changes to the transportation system and the distances between various services on the carbon footprint. Figure 12.7 provides a snapshot of the relationship of average miles travelled to population density.

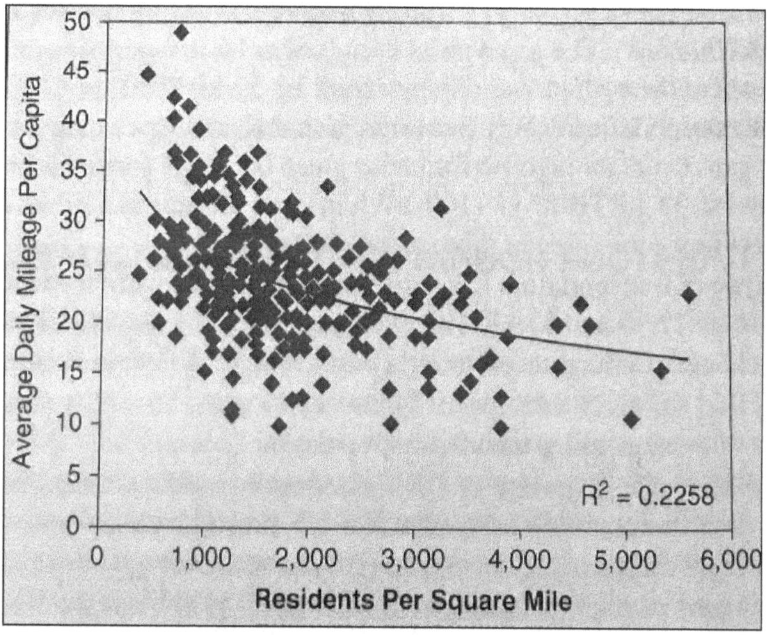

Figure 12.7 Relationship between population density and average daily miles travelled by automobile. Per capita vehicle travel tends to decrease with increase in density. Source: Todd Litman, Victoria Transport Policy Institute.

According to the Inventory of US Greenhouse Gas Emissions and Sinks 1990-2018 (the national inventory that the US prepares annually under the UN Framework Convention on Climate Change), transportation produced a major portion of total greenhouse gases in the US, about 28 percent (EPA 2020). Cars, trucks, commercial aircraft, and railroad contribute to transportation end-use sector emissions. The US EPA is addressing climate change by taking the following actions to reduce GHG emissions from the transportation sector:

1. Setting GHG emissions standards for cars and trucks.
2. Increase the use of renewable fuels
3. Taking the first steps to set a greenhouse gas standard for aircraft
4. Greening the federal fleet
5. Reducing the greenhouse gas emissions associated with moving goods
6. Informing customers' information on fuel economy and advanced technology cars
7. Foster the development of green racing protocols in cooperation with the Society of Automotive Engineers
8. Establishing a state and local transportation resource center.

REDUCING THE CARBON FOOTPRINT OF THE BUILT ENVIRONMENT

Significantly reducing built environment energy consumption and its corresponding carbon footprint is an especially important goal of sustainable construction. As noted in Chapter 9, the total energy associated with the built environment is probably on order of 65 percent of total US energy consumption, or about 65 quads. Although there are differences in energy sources of building, transportation and industry, the carbon footprint of the built environment is likely about the same percentage of the total human carbon footprint. Of the 100 quads of energy being consumed annually in US, 40 quads are consumed by building operations, and another 25 quads are consumed by transportation energy, the embodied energy of the materials and products of construction, and water pumping and processing. Consuming 100 quads of energy annually produces 6,600 million metric tons (MMT) of CO_2equivallent (MMT CO_2e), with the built environment contributing about 4,300 CO_2 CO_2e.

Climate change is the most serious issue of 21st century. To reverse course with respect to climate change, we must focus our activities on the built environment. Several strategies can be used to reduce the carbon footprint; among them are:

1. Dramatically reduce energy consumption
2. Shifting to renewable energy sources
3. Emphasizing compact form of development
4. Shifting to mass transportation
5. Designing buildings for durability and adaptability
6. Restoring natural systems
7. Designing low energy built environment hydrologic systems
8. Designing buildings for deconstruction and material reuse
9. Selecting materials for their recyclability
10. Including the carbon footprint of the buildings in building assessment systems.

Reducing atmospheric carbon concentration will require a concerted effort on the part of all stakeholders to the built environment. Building design must focus on long-term strategies that rebalance carbon emission into the atmosphere by greatly reducing the carbon associated with building construction and operations and with the distribution of buildings in communities. This later point addresses the problem of how buildings and their locations drive energy consumption and carbon emissions and transportation systems. Additionally, enormous effort must be made to restore the quantity of biomass on the planet to help in the reabsorption of carbon. Although many technical fixes have been proposed to reduce and absorb carbon in the atmosphere, thus far nothing has proved to remove the enormous quantities of carbon that would be required to stabilize the atmosphere.

CARBON NEUTRALITY AND ITS APPLICABILITY TO BUILDINGS

As notedthroughout this chapter, climate change is already causing enormous problems for ecosystems and human systems. The most obvious impacts are extreme weather conditions, including more frequent and stronger storms, drought, heat waves, landslides, wildfires, and floods. The IPCC Special Report (2018), "Global Warming of 1.5⁰C", warned its international audience about the impacts of global warming above pre-industrial levels: "Human activities are estimated to have caused approximately 1.0°C of global warming above pre-industrial levels, with a likely range of 0.8°C to 1.2°C. global warming is likely to reach 1.5°C between 2030 and 2052 if it continues to increase at the current rate."

To be clear, a 1.0°C temperature rise since the start of the industrial age has already occurred and the IPCC warns that the total rise must not exceed 1.5°C or there will be dire consequences. In short, a relatively small 0.5°C increase would result in even more serious impacts and alterations to every aspect of the planet. One possible solution that is increasingly being implemented is for all industries, including the construction industry, to determine the carbon footprint of their production systems and implement measures that offset their carbon footprint, making the net carbon footprint equal to zero. According to the European Parliament (2020), carbon neutrality means "having a balance between emitting carbon and absorbing carbon from the atmosphere in carbon sinks. Removing carbon dioxide from the atmosphere and then storing it is known as carbon sequestration. To achieve net zero emissions, all worldwide greenhouse gas emissions will have to be counterbalanced by carbon sequestration.

For buildings carbon neutrality can be interpreted in several ways. The most straightforward approach would be to determine the carbon footprint of its operational energy by converting the energy consumption into carbon generation based on the types of power plants serving the electrical grid. Similarly, for natural gas or other forms of energy, the number of kg-CO_2 eq per unit of energy is well known and the carbon footprint of this type of energy use is easily calculated. In the case of operational energy, the quantity and types of energy being consumed can be determined from energy models or from the actual utility bills for the facility.

The second approach would be to include the embodied carbon of the building in the calculation of the carbon footprint, see Figure 12.8. This information is available from EDPs or whole building LCAs as discussed in Chapter 11. The offset would apply to the combination of operational and embody carbon. Interestingly, many product manufacturers are implementing carbon neutrality strategies for their products, making the embodied carbon attributed to these products very low.

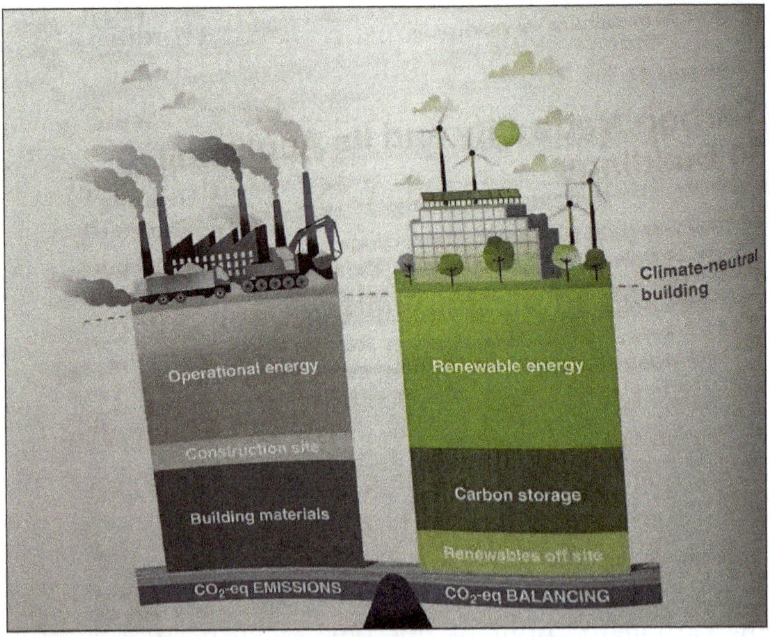

Figure 12.8 The carbon neutral building offsets carbon emissions from operational and embodied energy offset by strategies that generate renewable energy on or off the building site. Source: White Arkitekter

The third strategy is to also include the carbon footprint associated with the commuting energy of the building occupants and/or users. The location of buildings relative to each other resulting from urban planning and the available modes of transportation can result in exceptionally large commuting carbon footprints. However, including this carbon footprint in the total is much more difficult and involves a detailed survey of the building users to determine their modes of transportation, the energy performance of each mode of transportation, and the distance traveled.

REFERENCES

Average American footprint, Available at:

http://www.inspirecleanenergy.com/blog/clean-energy-101/average-american-carbon-footprint

Building the Future. 2020. Wood and Climate. ProHoltz Austria. Available at http://www.wooddays.eu/en/woodclimate/index.html.

Department of Transportation. 2018. "Driving to Green: The Transportation Energy Intensity of Buildings." Transportation Energy Data Book, 36th ed.

European Parliament. 2020. "What is carbon Neutrality and How Can it be Achieved by 2050?" European Parliament News, August 10. Available at http://www.europarl.europa.eu/news/en/headlines/society/20190926STO62270/what-is-carbon-neutrality-and-how-can-it-be-achieved-by-2050.

Hammond. G., and C. Jones. 2011. Inventory of Carbon and Energy, Version 2. Available at www.carbonsolutions.com.

Intergovernmental Panel on Climate Change (IPCC). 2001. "Climate Change 2001: The Scientific Basis. Intergovernmental Panel on Climate Change. Available at www.ipcc.ch/ipccreports/tar/wg1/.

Intergovernmental Panel on Climate Change (IPCC). 2005. Carbon Dioxide Capture and Storage. Available at www.ipcc.ch/site/assets/uploads/2018/03/srccs_wholrreport-1.pdf

Intergovernmental Panel on Climate Change (IPCC). 2007. "Climate Change 2007: Synthesis Report (AR4). Available at www.ipcc.ch/report/ar4/syr/.

Intergovernmental Panel on Climate Change (IPCC). 2013. "Climate Change 2013: The physical Science Basis. Intergovernmental Panel on Climate Change. Available at www.ipcc.ch/report/ar5/wg1/

Intergovernmental Panel on Climate Change (IPCC). 2018 "Global Warming of 1.5°C". An IPCC Special Report on the impacts of global warming of 1.5°C above the pre-industrial levels and related global greenhouse gas emission pathway, in the context of strengthening the global response to the threat of climate change, sustainable development, and efforts to eradicate poverty. Available at www.ipcc.ch/report/ar5/wg1/.

Kibert J., Charles 2022. 'Sustainable Construction, Green Building Design and Delivery. Fifth Edition, John Wiley and Sons, Inc. Hoboken, NJ.

Nyasha Harper-Michon, 2023. "Building with Wood", Publisher: Prestel; uneceinfo@un.org

National Oceanic and Atmospheric Administration (NOAA). 2019. "Land and Ocean Temperature Percentiles" (January 2018 to December 2018). NOAA National Climatic Data Center. Available at www.ncdc.noaa.gov/sotc/service/global/map-precentilemnp/201401-201412.gif.

Parliamentary Office of Science and Technology 2006/2011. Carbon footprint of Electricity Generation. Number 268. Parliamentary Office of Science and Technology.

US Energy Information Agency. 2013. "Carbon Dioxide Emissions Coefficients. Available at www.eia.gov/environment/emissions/co2_vol_mass.cfm

THIRTEEN
INDOOR ENVIRONMENTAL QUALITY

BACKGROUND

Indoor environmental quality (IEQ) is most simply described as the condition inside the built environment. It includes the following key components of high-performance green buildings:

- Indoor air quality (IAQ)
- Lighting, daylighting, and exterior view
- Acoustics
- Odor, noise and vibration control
- Thermal comfort and control
- Electromagnetic radiation

The good IEQ has shown to enhance productivity, decrease absenteeism, and improve building value. Levels of indoor air pollution can be as high, and sometimes higher than outdoor levels. Indoor air quality (IAQ) can affect anyone's health, but older adults, children, and people with health conditions like asthma and heart disease can be more vulnerable.

The following building elements affect IEQ:

1. Operation and maintenance of the building
2. Ventilation and performance standards, operational routines, etc.
3. Building occupants and their activities, personal hygiene, metabolism, etc.
4. Building contents include the equipment (heating, etc.), emissions of materials, etc.
5. Outdoor environment including climate, moisture, soil (dust particles, etc.), water, etc.
6. Building fabrics include building envelope (infiltration, water intrusion, etc.), floors, partitions, etc.

In addition, the factors that impact IEQ can be classified as:

- Physical,
- Chemical,
- Biological and
- Radioactive materials.

Controlling sound noise transmission in buildings is a major problem. Noise from air-handling systems, lights, transformers, and other sources can cause discomfort and health problems for building occupants. Sections of a building where low noise levels are required must be separated and insulated from noise-generating areas. Subtle background noises must be eliminated as much as possible. Otherwise, it can cause lower morale and decrease occupants' productivity.

Physical indoor environmental problems are traceable primarily to electrical and mechanical infrastructure of building. They include sound/noise transmission, lighting quality, thermal conditions and odors. Physical factors are generally non-toxic but are at least a nuisance to building occupants and can lead to health problems after exposure for extended periods.

A wide variety of chemicals can contaminate the indoor environment. Chemicals may be introduced into the indoor environment by painting, installation of carpets, or cleaning products. Chemical factors are classified according to the form they take at room temperature: vapor, gas, liquid, or particulates. Particulates include inorganic fiber, respirable particulates, such as dust and dirt; metals; and a variety of organic materials. Because small particulates can penetrate deep into the lungs, they are a serious concern. The size and density of particulates determine how deeply they can penetrate the respiratory system. Radon, a naturally occurring radioactive gas that has been connected to health problems, is a problem in many regions of the United States; thus, taking measures to mitigate it, is important for ensuring a good indoor environment.

Biological contaminants include bacteria, fungi, viruses, algae, insect parts, and dust, which may result in allergenic or pathogenic reactions. There are many sources for these pollutants: pollens from outdoors, viruses and bacteria, from humans, hair and skin flakes from household pets to name but a few.

Covid-19, the novel coronavirus that became widespread in 2020 and caused a pandemic, prompted an emergency investigation into the role of buildings in facilitating the spread of virus and potential role of HVAC systems in removing the airborne virus. Preliminary research indicates that the very small virus particles, about 0.1 microns in size, may be active for up to three hours after they are emitted. The World Health Organization (WHO) issued guidelines that stated that the coronavirus could spread indoors, especially when people were in poorly ventilated, crowded rooms for extended period. Because HIVAC systems can potentially spread the virus if high-speed airflows are present, new and existing buildings could be retrofitted to decrease the probability of transmission.

SOUND AND NOISE TRANSMISSION

Control of sound and noise transmission in buildings is a major undertaking. Noise from air-handling systems, lights, transformers and other sources can cause discomfort and even health problems for building occupants. Building designers and engineers often are intimidated by the challenge of dealing with sound and noise transmission because it is a somewhat intangible concept in a world of mostly tangibles; meta, size, color, and so on.

The basic premise in creating an acoustically acceptable indoor environment is to ensure and sound levels in particular areas of a building are at or below an acceptable range for the specific application. For instance, it would be mistaken to locate a helicopter pad just outside of a library. Clearly, sections of a building where low noise levels are required must be separated and insulated from noise-generating areas. When it comes to acoustics, designers can easily prevent obvious problems – for example, taking care not to locate a conference room next to a chiller plant. But more subtle problems may be overlooked, such as neglecting to insulate a wall that separates a restroom from a private office.

High noise levels in commercial buildings can lead to morale problems and loss of productivity when occupants become irritated and annoyed and thus distracted from work. The other major noise-related problem for building occupants is caused by exposure to unhealthy noise levels generated by air handlers, transformers, lighting, elevators, machinery, and motors.

LIGHTING QUALITY

Problems associated with lighting quality are like those associated with noise in that the cause is a poorly understood building support system. As a requirement for a high-quality indoor environment, lighting is probably better understood than sound, but nevertheless it often is overlooked in building design.

It is widely acknowledged that natural sunlight is the best light source for the eye. Unfortunately, these days most people spend an inordinate amount of time indoors and away from the natural light. Thus, the ideal healthy indoor light environment is the one that allows natural light indoors or whose lighting system replicates natural lights as closely as possible.

Type of Light	Color Characteristics
Incandescent (argon-surrounded filament)	White with yellow tint
Incandescent (halogen-surrounded filament)	White
Fluorescent	White with blue tint
Light-emitting diode	White or white with blue tint
Mercury vapor	White with blue tint
Metal halide	White with blue-green tint
Sodium vapor (high pressure)	Amber white
Sodium vapor (low pressure)	Yellow

Source: Kibert, 2022.

Table 13.1 General Color Characteristics of Typical Building Lighting Systems

Natural sunlight has an equal spectral distribution of the visible light frequencies combined to appear as white light. By contrast, artificial light sources are bound by the laws of physics; hence they are limited in the frequencies of visible light that they emit. A list of common artificial light sources and their general color characteristics is shown in Figure 13.1. Incandescent lights, particularly the halogen type, give the best color rendition as natural light. Fluorescent and mercury vapor lights emit white light with a distinct preponderance of blue frequencies. Fluorescents can be made to offer warm color ranges, but the color of fluorescents lighting is not natural and typically tends to produce a too-bright, sterile atmosphere. Sodium-based lights produce a yellow light and are often used for outdoor applications.

In commercial buildings, the primary source of artificial lights is incandescent, fluorescent, and LED lighting fixtures. Mercury vapors and metal halide sources are also used in large rooms or high-bay areas. In a typical building, general lighting in office areas is either LED or fluorescent. LED lights are used for more direct applications where a fluorescent tube is not applicable – for example, accent lighting.

In August 2023, the sales of incandescent light bulbs were mostly stopped. It was originally stopped in 2007, however, Trump administration rolled back and finally instated by the current Biden administration on August 1, 2023.

Figure 13.1 Incandescent lights are very popular

The rule passed by President Joe Biden's Department of Energy in April 2022 stated that light bulbs must emit a minimum of 45 lumens per watt. A lumen is a measure of brightness. That effectively outlaws the manufacture and sale of common incandescent bulbs, the kind you screw into most light sockets in your home. That's because traditional incandescent bulbs provide just 15 lumens per watt, according to light bulb manufacturer Philips. By contrast, most LED bulbs will get you 75 lumens per watt, or more. Trump once famously complained about the quality of the light coming from LED bulbs, telling House Republicans "I always look orange" in the energy-efficient lighting.

In December 2022, the Department of Energy proposed a rule that would more than double the current minimum light bulb efficiency level, to over 120 lumens per watt for the most common bulbs. That would go into effect by the end of 2024 and effectively ban CFL bulbs. The Department of Energy estimates the rules will save US consumers close to $3 billion on their utility bills, and it will also cut planet-warming carbon emissions by 222 million metric tons over the next 30 years.

Flickering lights can also cause irritation and health problems. Ballasted lights - for example, fluorescent, mercury vapor, metal halide, and sodium lights are subject to flickering when the ballas malfunctions. This can easily lead to sore eyes and headaches and, ultimately, lower productivity. Glare is also a problem; however, unlike the other problems described here, it is not a consequence of artificial light but rather involves the light source and reflector positioning. Windows, desktops, and computer screens, even shiny paper, are all reflectors that can cause uncomfortable glare. Depending on the intensity of the light, glare can quickly lead to discomfort and headaches, especially when reading, typing, or looking at a computer screen.

THERMAL CONDITIONS

The climatic conditions in which people are working have a profound impact on how they behave and how well they work. But because everyone is different, what is perhaps comfortable for one person in an office may be profoundly uncomfortable to his or her neighbor. In general, the indoor comfort range is in the center of the psychometric chart. Generally accepted ranges for comfort are as follows: in the winter temperatures between 68°F and 75°F and relative humidity between 30 and 60 percent; in summer, temperatures between 72°F and 80°F and relative humidity between 30 and 60 percent. Relative humidity is the amount of moisture in the air relative to the amount of moisture air can hold when it is completely saturated.

Air velocity is another variable in the indoor climate that is not a fundamental property of the air. Air velocity varies generally, depending on where a person is in relation to vents, doors, windows, and fans. It is an integral aspect of air conditioning that indoor air be circulated; hence, it must have a certain velocity. The goal of HVAC designer is to introduce the highest velocity air where it has little or no effect on the building occupants, usually along ceilings or walls, so that by the time it comes in contact with people, it has slowed to an undetectable rate. High velocity air is more likely to cause discomfort in cool indoor climates and, conversely, to be welcome in warm indoor climates.

ODORS

One of the most common and annoying environmental problems. Including tobacco smoke, human body odors, cleaning, and personal glooming products. Even simple odors may consist of many substances.

Solving these problems is not easy, because the human olfactory system is highly complex and not well understood; moreover, the chemical sources that create many of these odors also are poorly understood. Even simple odors in office settings are complex, consisting of many substances. Typical sources of odor in the indoor environment include tobacco smoke, human body odors, and cleaning and personal grooming products. Off-gassing of building materials is another common source of smells. Complicating this issue is the pronounced difference in individual sensitivity to odors. Visitors to an office are generally far more sensitive to odors than their long-term occupants, for example. Because human reactions to odors are so varied, it is nearly impossible to predict how anyone or group of people will react.

VOLATILE ORGANIC COMPOUNDS

Volatile organic compounds (VOCs) are carbon containing compounds that evaporate readily at room temperature and are found in many housekeeping, maintenance, and building products made with organic (carbon-based) chemicals. Paints, glues, paint strippers, solvents, wood preservatives, aerosol sprays, cleaners and disinfectants, air fresheners, stored fuels, automotive products, and even dry-cleaned clothing and perfumes are all sources of VOCs. In any indoor environment, there can be up to 100 different VOCs in varying concentrations. Carbon filters can be used to absorb VOCs, but they must be replaced regularly, as the odors deplete carbon. There are six major classes of VOCs:

1. Aldehydes (formaldehydes)
2. Alcohol (ethanol, methanol)
3. Aliphatic hydrocarbons (propane, butane, hexane)
4. Aromatic hydrocarbons (benzene, toluene, xylenes)
5. Ketones (acetone)
6. Halogenated hydrocarbons (perchloroethylene, trichloroethylene, methylene chloride)

Formaldehyde is highly reactive and may be in all three states of matter. It is highly soluble in water and can irritate body surfaces that normally contain moisture – for example, eyes and upper respiratory tract. Formaldehyde gas is pungent and easily detectable by its odor at concentrations well below one part per million. It is perhaps the most commonly occurring VOC in construction, found in many common products, such as paints, wood products, and floor finishes. When combined with other chemicals, it can be used as glues and binders in numerous products. Urea formaldehyde (UF) foam insulation, particleboard, interior-grade plywood, wallboard, some paper products, fertilizers, chemicals, glass and packing materials contain significant amounts of formaldehydes.

RADON

Radon is a radioactive gas released from the normal decay of the uranium, thorium, and radium in bedrocks. It is an invisible, odorless, tasteless gas that seeps up through the ground and diffuses into the air. In a few areas, depending on local geology, radon dissolves into ground water and can be released into the air when the water is used. Radon gas usually exists at very low levels outdoors. However, in areas without adequate ventilation, such as underground mines, tunnels and homes with poor ventilation radon can accumulate to levels that substantially increase the risk of lung cancer.

Radon is present in nearly all air. Everyone breathes in radon every day, usually at very low levels. However, people who inhale high levels of radon are at an increased risk of developing lung cancer. Radon can enter homes through cracks in floors, walls, or foundations, and accumulate indoor due to poor ventilation issues. It can also be released from building materials, or from water obtained from wells that contain radon. Radon levels can be higher in homes that are well insulated, tightly sealed, and/or built on soil rich in the elemental uranium, thorium, and radium. Basement and first floors typically have the highest radon levels because of their closeness to the ground. Figure 13.2 shows the mechanisms through which radon gas can enter the homes/buildings.

Radon decays quickly, giving off tiny radioactive particles. When inhaled, these radioactive particles can damage the cells that line the lung. Long-term exposure to radon can lead to lung cancer, the only cancer proven to be associated with inhaling radon. There has been a suggestion of increased risk of leukemia associated with radon exposure in adults and children; however, the evidence is not conclusive.

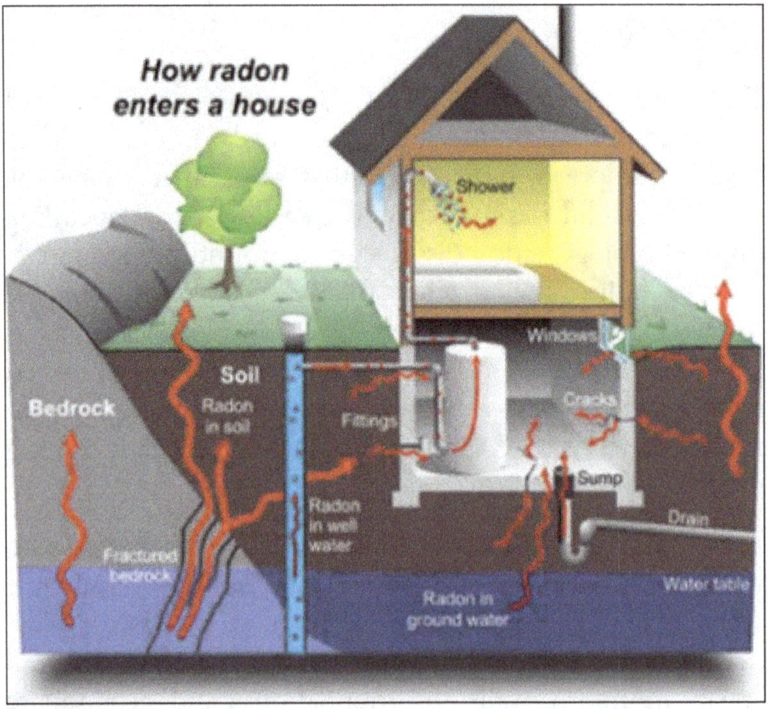

Figure 13.2 Schematic diagram of radon intrusion into the buildings

Cigarette smoking is the most common cause of lung cancer. Radon represents a far smaller risk for this disease, but it is the second leading cause of lung cancer in the United States. Scientists estimate that 15,000 to 22,000 lung cancer deaths in the United States each year are related to radon.

Exposure to the combination of radon gas and cigarette smoke creates a greater risk of lung cancer than exposure to either factor alone. Most radon-related cancer deaths occur among smokers. However, it is estimated that more than 10 percent of radon-related cancer deaths occur among nonsmokers. Radon was identified as a health problem when scientists noted that underground uranium miners who were exposed to it died of lung cancer at high rates. The results of miner studies have been confirmed by experimental animal studies, which show higher rates of lung tumors among rodents exposed to high radon levels.

Scientists agree that radon causes lung cancer in humans. Recent research has focused on specifying the effect of residential radon on lung cancer risk. In these studies, scientists measure radon levels in the homes of people who have lung cancer and compare them to the levels of radon in the homes of people who have not developed lung cancer.

Researchers have combined and analyzed data from all radon studies conducted in Canada and the United States. By combining the data from these studies, scientists were able to analyze data from thousands of people. The results of this analysis demonstrated a slightly increased risk of lung cancer for individuals with elevated exposure to household radon. This increased risk was consistent with the estimated level of risk based on studies of underground miners. Techniques to measure a person's exposure to radon over time have become more precise, thanks to several studies carried out in the 1990s and early 2000s, (Bas, Ed. 1993). The question is how the presence of radon gas in homes and buildings can be measured or identified?

Testing is the only way to know if a person's home has elevated radon levels. Indoor radon levels are affected by the soil composition under and around the house, and the ease with which radon enters the house. Homes that are next door to each other can have different indoor radon levels, making a neighbor's test result a poor predictor of radon risk. In addition, rain or snow, barometric pressure, and other influences can cause radon levels to vary from month to month or day to day, which is why both short- and long-term tests are available.

Short-term detectors measure radon levels for 2 to 90 days, depending on the device. Long-term tests determine the average concentration for more than 90 days. Because radon levels can vary from day to day and month to month, a long-term test is a better indicator of the average radon level. Both tests are relatively easy to use and inexpensive. A state or local radon official can explain the differences between testing devices and recommend the most appropriate test for a person's needs and conditions.

The U.S. Environmental Protection Agency (EPA, 1991) recommends taking action to reduce radon in homes that have a radon level at or above 4 picocuries per liter (pCi/L) of air. About 1 in 15 U.S. homes is estimated to have radon levels at or above this EPA action level. Scientists estimate that lung cancer deaths could be reduced by 2 to 4 percent, or about 5,000 deaths, by lowering radon levels in homes exceeding the EPA's action level.

The EPA has more information about residential radon exposure and what people can do about it in the Consumer's Guide to Radon Reduction. As part of the Phase I Environmental Site Assessment study which was discussed in Chapter 1 is to reveal if at the given site/location the presence of radon gas above the EPA action level is possible, for further info please see the link below:

http://www.cancer.gov/cancertopics/factsheet/Risk/radon

ASBESTOS

Asbestos is another potentially deadly IAQ problem. Unlike radon, however, the health implications of asbestos have been documented in detail, for it has been a major environmental problem for many years. When it was discovered that threadlike particles in asbestos could lodge in human lungs, its use began to phase out. Exposure to asbestos has been linked to stomach and lung cancers.

The term asbestos refers to a group of silica-based minerals in fibrous bundles. Introduced in 1930s and widely used in the United States from 1940 to 1973, asbestos comprises many naturally occurring materials that are processed to produce a manageable form for use in construction, insulation, and fire retardation materials. Indoor building materials containing asbestos include thermal insulation on ceilings and walls; insulating materials walls; insulation on pipes, ducts, boilers and tanks; and finishing materials, such as ceiling and floor tiles and wall boards. Table 13.2 shows several building materials where asbestos may be present.

Cement Pipes	Elevator Brake Shoes
Cement Wallboard	HVAC Duct Insulation
Cement Siding	Boiler Insulation
Asphalt Floor Tile	Breaching Insulation
Vinyl Floor Tile	Ductwork Flexible Fabric Connections
Vinyl Sheet Flooring	Cooling Towers
Flooring Backing	Pipe Insulation
Construction Mastics (Floor Tile, Carpet, Ceiling Tile, etc.)	Heating and Electrical Ducts
Acoustical Plaster	Electrical Cloth
Textured Paints/Coatings	Electrical Wiring Insulations
Decorative Plaster	Electrical Cloth
Ceiling Tiles & Lay-in Panels	Chalkboards
Packing Materials (for wall/floor penetrations)	Fire Doors
Spray-Applied Insulation	Roofing Shingles
Blown-in Insulation	Roofing Felt
Fireproofing Materials	Base Flashing
Taping Compounds (thermal)	Thermal Paper Products
Packing Materials (for wall/floor penetrations)	Fire Doors
High Temperature Gaskets	Caulking/Putties
Laboratory Hoods/Tabletops	Adhesives
Laboratory Gloves	Wallboard
Fire Blankets	Joint Compounds
Fire Curtains	Vinyl Wall Covering
Elevator equipment Panels	Spackling Compounds
Note: The above list does not include every product/material that may contain asbestos. It is intended as a general guide to show which types of materials may contain asbestos.	

Table 13.2 a summary table showing the list of typical suspect asbestos-containing materials

High quantity release of asbestos into the airstream usually occurs during maintenance, renovation, and other construction activities, and this is when the asbestos becomes dangerous. There is very little danger to human health if the material is left undisturbed; asbestos becomes a health hazard only when its fibers are released into the air. Most experts agree that if asbestos surfaces are not deteriorating or being abraded, thus not releasing asbestos fibers are best left alone. Removal of asbestos is very costly and can be done safely only by professionals. An unsafe removal process can do more harm than good by releasing more particles into the air, where they can continue to contaminate a building for years, Kibert, 2022.

LEAD-BASED PAINTS

Lead-based paint (LBP) or other surface coatings (paint, varnish, stain, or other applied coating) are considered material if they contain an amount of lead equal to, or more than, one milligram per square centimeter (1.0 mg/cm2), more than 5,000 ppm, or more than 0.5 percent by weight. Section 1017 of the Housing and Urban Development (HUD) Guidelines, Residential Lead-Based Paint Hazard Reduction Act of 1992, otherwise known as "Title X, defines LBP hazard as "any condition that causes exposure to lead that would result in adverse human health effects" resulting from lead-contaminated dust, bare, lead-contaminated soil, and/or lead-contaminated paint that is deteriorated or present on accessible, friction, or impact surfaces. Therefore, under Title X, intact lead-based paint on most walls and ceilings would not be considered a "hazard", although the paint should be maintained in good condition and monitored to ensure that it does not deteriorate and become a hazard. It is important to note that LBP may create a lead hazard when being removed.

The condition of these materials must be monitored when they are being disturbed. In the event LBP is subject to abrading, sanding, torching and/or cutting during demolition or renovation activities, there may be regulatory issues that must be addressed.

Engineering controls and air monitoring shall be implemented during disturbance of chipping or deteriorated LBP to eliminate the risk of public and worker exposure. Prior to any demolition work, a licensed asbestos and lead inspector should conduct a site inspection to determine a necessity for any abatement work.

Abatement projects producing hazardous lead waste must meet land disposal restrictions under the Resource Conservation and Recovery Act (RCRA). Cal/OSHA requires that worker safety protection be implemented to prevent lead exposure above the Permissible Exposure Level (PEL) during construction activities. The PEL for lead is 50 $\mu g/m^3$.

COMBUSTION OF BY-PRODUCTS

Combustion by-products are created under conditions of incomplete combustion. The primary sources of combustion by-products that contribute to the contamination of indoor air are gas, wood, and coal stoves; unvented kerosene space heaters; fireplaces under downdraft conditions; and tobacco smoke. The major by-products include carbon dioxide (CO_2), carbon monoxide (CO), nitrogen dioxide (NO_2), sulfur dioxide (SO_2), and particulates. Their health effects can vary, depending on the type of by-product produced (Bas 1993). Each of these by-products is described more fully next.

Carbon dioxide: Carbon dioxide is colorless, odorless, and tasteless gas. Although it is a by-product of combustion, it is relatively harmless; it is, after all, also a natural product of respiration. That said, and even though it is nontoxic, if the concentration of CO_2 is too high, the result can be unpleasant and perhaps unhealthy for a building's inhabitants. And since it is a natural product of respiration, it can also be an indicator of the quality of ventilation and IAQ.

Carbon monoxide: CO is another colorless, odorless, and tasteless gas, but it must not be confused with CO_2. The effects of high-level CO exposure can range from nausea and vomiting to headaches and dizziness to coma and death. The health effects of low-level CO exposure are not clearly defined, but its toxicity is unquestionable. The symptoms of CO poisoning, which include nausea, dizziness, confusion, and weakness, may be confused with those of the flu. People with anemia or a history of heart disease can be specifically sensitive to CO exposure.

Nitrogen dioxide: Concentrated NO_2 is a dark brown gas with a strong odor. Exposure can cause irritation of the skin and eyes and other mucous members. Controlled human exposure studies and epidemiological studies in homes with gas stoves illustrate that, depending on the level of exposure, NO_2 can alter lung function and cause acute respiratory symptoms. Because of its ability to oxidize, NO_2 has been shown to damage the lungs directly. Symptoms of exposure may include shortness of breath, chest pain, and a burning sensation or irritation in the chest. People with chronic respiratory illnesses, such as asthma and emphysema may be especially sensitive to NO_2.

Sulfur dioxide: SO_2 is a colorless gas with a suffocating odor. It is highly soluble in water and thus is readily absorbed by mucous membranes. Once it is inhaled, SO_2 is dissolved and forms sulfuric acid (H_2SO_4), sulfurous acid (H_2SO_3) and bisulfate ions. During normal nasal respiration, SO_2 is absorbed primarily by the nasal tissues; only 1 to 5 percent reaches the lower respiratory tract. However, when a person breathes through the mouth – for example, during heavy exercise – significant quantities of SO_2 can penetrate the lower respiratory tract even at low concentrations. The primary physical effect of SO_2 exposure is broncho constriction, which begins at considerably lower levels for people with asthma than for healthy individuals. The constriction develops almost immediately upon exposure, but it also subsides just as quickly when exposure ends. The intensity of constriction is directly related to the amount of SO_2 per unit of time that reaches the lower respiratory tract, not necessarily the level of exposure. Also, the effect of SO_2 does not increase with time.

Construction particulates: Particulates produced by combustion can affect lung function directly. The smaller particulates, the more deeply they penetrate the lungs and thus the more dangerous they become. The particles can serve as carriers for other contaminants or as mechanical irritants that interact with chemical contaminations.

MOLD AND MILDEW

Mold is a fungus that roots deep into softer organic materials that are wet. Mildew is a surface-level fungus that grows on top of wet and hard surfaces. In general, mildew grows on visible surfaces, whereas mold may develop anywhere. Mold is also more prevalent than mildew, creating more health risks.

Moisture can act as a breeding ground for molds, bacteria, and mites. Mites are the most prominent cause of house dust allergies. They are found in beds and pillows, especially humidity levels are high. An indoor moisture level of 30 to 50 percent relative humidity is recommended to maintain good health as well as comfort.

Biological contaminants also may multiply in standing water, in cooling towers, in water-damaged ceilings, and on surfaces where moisture in the air condenses on cold walls. Additionally, damp organic materials, such as leather, cotton, furniture stuffing, and carpets, can be contaminated with fungi. Airflow rates also have an important role in the concentration of airborne biological pollutants. Reduced flow rates tend to provide a favorable medium for molds, dust, and fungi. The HVAC equipment in a building plays a very important role in maintaining proper air flow rates.

SICK BUILDING SYNDROME AND BUILDING RELATED ILLNESSES

Sick building syndrome or SBS is a temporary medical condition caused by being in a building or type of enclosed space where the indoor air quality is poor. There are many factors that can contribute to poor indoor air quality. In 1984, a report by the World Health Organization suggested that up to 30% of new or remodeled buildings have low air quality, which can lead to SBS.

Diagnosing SBS can be tricky because the causes are many and often emulate cold or flu symptoms. Among the possible symptoms are:

- Throat irritation
- Breathing difficulties
- Tightness in the chest
- Runny nose
- Allergy-like symptoms, such as sneezing
- Burning sensations in the nose

- Dry, itchy skin rashes
- Headaches
- Chills
- Dizziness
- Difficulty concentrating
- Forgetfulness
- Fatigue
- Irritability
- Nausea
- Body aches
- Fever, and more

The key to diagnosing SBS is that a person's symptoms improve after leaving a building in question and only when they return, do the symptoms return as well. If a building is deemed to have poor indoor air quality, there are a few steps people can take to treat SBS:

- Use low fume and fragrance-free cleaning products.
- Dust and vacuum regularly to keep the environment clean.
- Test for possible indoor mold or fungus.
- Update or purchase air purifiers for buildings or room.

Air cleaning can be a useful addition to indoor air quality control alongside ventilation to fight against SBS. For more information visit www.MatrixAir.com or e-mail at info@matrixair.com.

The building-related illnesses (BRI) refer to symptoms of diagnosable illness that can be attributed directly to a defined IAQ problem. This is contrary to SBS where there is no specific cause that can be identified. Typically, SBS is diagnosed when the affected employees' symptoms disappear almost immediately upon leaving the building.

MULTIPLE CHEMICAL SENSITIVITY

While some chemicals are known to cause problems, some people are sensitive to a wide range of substances. This is known as multiple chemical sensitivity. It is also called idiopathic environmental intolerance (IEI).

Multiple chemical sensitivity, or IEI, is under debate in the medical field currently. Some healthcare providers question whether the underlying illness is not medical but psychiatric. They suggest that the symptoms are caused by anxiety. Others say it is a medical disorder triggered by exposure to chemicals, electromagnetic forces, or other substances in the environment. Symptoms often begin with short-term, severe chemical exposure, such as a chemical spill. Or they may happen after exposure over time.

After this, low levels of chemicals found in everyday materials can trigger physical symptoms in people with multiple chemical sensitivity. This includes chemicals in soaps, detergents, cosmetics, and newspaper inks.

What are the symptoms of multiple chemical sensitivity?

Below are the most common symptoms. But each person's symptoms can vary. Symptoms may include headaches, rashes, asthma, muscle and joint aches, fatigue, memory loss, and confusion.

So far, multiple chemical sensitivity, or IEI, is a diagnosis under debate. Some in the medical field lean toward the symptoms being physical signs of psychiatric illness. This means they think it's not primary medical illness. Others in the medical field and some organizations agree these are physical reactions to substances. There is ongoing review about this condition as a clinical diagnosis. Multiple chemical sensitivity shares symptoms with other chronic conditions that are poorly understood, including fibromyalgia and chronic fatigue syndrome. For more information refer to:

https://www.hopkinsmedicine.org/health/conditions-and-diseases/multiple-chemical-sensitivity

LEGIONELLOSIS

The term legionellosis refers to two important bacterial diseases: Legionnaire's disease and Pontiac fever, both caused by the bacterium Legionella pneumophila. The diseases do not spread via person-to-person contact but rather through soil-air and water-air links both indoors and outdoors. The bacteria can survive in water for up to a year under certain conditions. Legionella prefers stagnant water, which is found in the drain pans of HVAC units and cooling towers. Fans then can transfer the bacteria, to be inhaled by unsuspecting victims. Sources of legionella in residences and other buildings also may include hot tubs, vaporizers, humidifiers, and contaminated forced-air heating systems. Algae and other aquatic life-forms can promote the growth of legionella by providing the bacteria with food.

PONTIAC FEVER

Pontiac fever is an acute, nonfatal respiratory disease caused by various species of Gram-negative bacteria in the genus legionella. It causes a mild upper respiratory infection that resembles acute influenza. Pontiac fever resolves spontaneously and often goes undiagnosed. Both Pontiac fever and the more severe legionnaires disease may be caused by the same bacteria, but Pontiac fever does not include pneumonia. See "en.m.wikipedia.org".

In July 1968, 95 out of 100 people employed in – ironically – a public health building in Pontiac, Michigan, became ill with a flulike ailment. In fact, if the number of cases had not comprised such a high proportion of the employees, the disease probably would have been diagnosed with the flu. The employees all claimed to suffer from headaches, fever, and muscle aches and pains. Called Pontiac fever, the disease was eventually traced back to a faulty HVAC system. However, it was not until the discovery of legionnaire's disease, nearly 10 years later, that bacterium which caused Pontiac fever was finally identified.

LEGIONNAIRE'S DISEASE

Legionnaire's disease is a type of pneumonia caused by legionella. Both the disease and the bacterium were discovered following an outbreak traced to a 1976 American Legion convention in Philadelphia, Pennsylvania. This disease develops within 2 to 10 days after exposure to legionella, and early symptoms may include loss of energy, headache, nausea, aching muscle, high fever [often exceeding 104°F, and chest pains. Later, many bodily systems, as well as mind, may be affected. The disease eventually cause death if high fever and antibiotics cannot defeat it. Victims who survive may suffer permanent physical or mental impairment. The Center for Disease Control has estimated that the disease infects 10,000 to 15,000 persons annually in the United States; others have estimated as many as 100,000 annual US cases.

Legionnaire disease is a sever multisystem illness that can affect the lungs, gastrointestinal tract, central nervous system, and kidneys. It is characterized by a low attack rate (2 – 3 percent), along incubation period (2 – 10 days), and sever pneumonia. Unlike Pontiac fever, hospitalization is required. Nost victims are men in their 50s and 60s who are smokers and/or have underlying respiratory problems. Alcohol consumption, diabetes, and recent surgery can also be contributing factors.

ALLERGIC REACTIONS

Allergies are certain reactions in which immune system reacts abnormally to a foreign substance. The foreign substance could be simply indoor air pollution. The body will mimic the effects of a real illness by stimulating the production of white blood cells to combat the allergen. The individual usually does not experience an allergic reaction until after the second exposure to a specific allergen. The first exposure results in the manifestation of the allergy. Allergies that cause an allergic response include viable and nonviable agents. Viable agents include bacteria, fungi, and algae. Common nonviable agents include house dust, insect and arachnid body parts, animal dander, mite fecal pellets, remains of molds and their spores, pollens, and dried animal excretions.

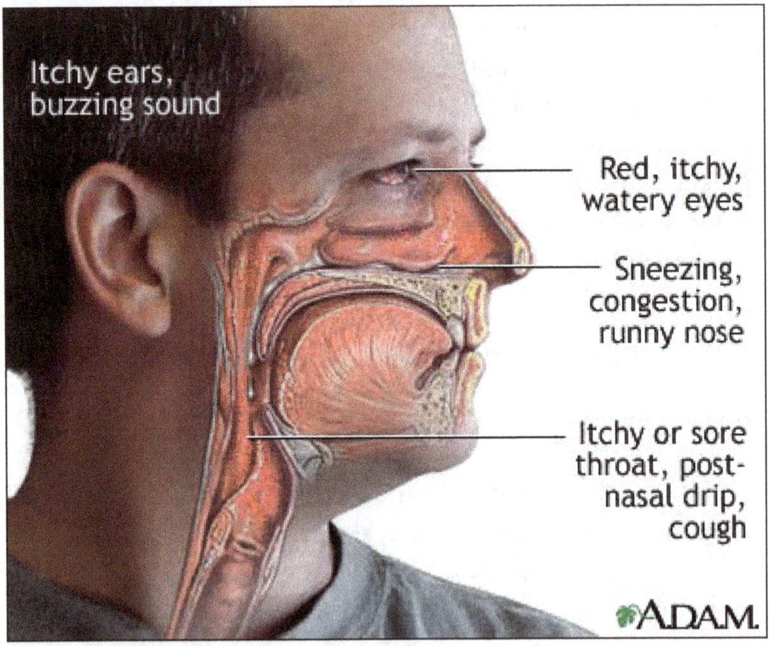

Figure 13.3 The sensitive respiratory tract that may exhibit allergic reactions and irritation

Regular cleaning to remove dust, the use of high efficiency filters, and regular filter changing can help to reduce or eliminate biological contaminants.

INTEGRATED IEQ DESIGN

Clearly, the complex range of IEQ issues warrants an integrated approach to the design of buildings to maximize the quality of built environment. "The Design Guide (www.wbdg.org) provides a good overview of Integrated IEQ design and suggests that these measures for attaining good IEQ in buildings be undertaken:

1. Facilitate quality IEQ through good design,

construction, commissioning, and operation and maintenance practices

2. Value natural aesthetic and wellness concerns such as the importance of views or the integration of natural and man-made elements
3. Provide thermal comfort with maximum degree of personal control over the temperature and airflow
4. Supply adequate quantity and quality of ventilation and intake of outside air to ensure acceptable IAQ and to reduce the probability of the transmission of COVID-19 and other viruses.
5. Prevent airborne bacteria, mold, and other fungi as well as radon, through building envelope design that properly manages moisture sources from outside and inside the building, and with heating, ventilating, air conditioning (HVAC) system designs that are effective at controlling indoor humidity.
6. Use low emission materials that do not emit pollutants.
7. Assure acoustic privacy and comfort by employing sound-absorbing material and equipment isolation.
8. Control disturbing odors through contaminant isolation and removal, and by careful selection of cleaning products. Pursuing energy efficient strategies to remove harmful odors and recover energy used in conditioning the interior environment.
9. Create a high-performance luminous environment through the careful integration of natural and artificial light sources.
10. Provide high quality potable water.

We will discuss the above-mentioned items in the following sections:

FACILITATE QUALITY IEQ THROUGH GOOD DESIGN, CONSTRUCTION, AND OPERATION/MAINTENANCE

When planning a structure, it is crucial to consider factors such as weather condition, terrain, soil type, temperature, precipitation, wind and wind direction. By combining these factors, then the project team should take advantage of these natural characteristics of the area to come up with some designs parameters which will take full advantage of these natural conditions to maintain a good IEQ within the built environment. For example, the feeling of moisture in the air, the indoor air temperature, strange smells, and exposure to sunlight from a window, tremendously contribute to IEQ.

In addition, the project team should make every effort in choosing the building material which do not contribute to undesirable indoor air quality conditioning inside the built environment. As discussed in Chapter 11, choosing every piece of building material and building product should be based on the realistic assessment of LCA and EPD in which LCA analysis has been followed by the reliable third-party entities. In another words, specifying materials that contain zero or low VOCs and using grilles and grates at building entrances to prevent the entrance of dirt, toxics, and other materials into the building (see Chapter 5) are examples of materials and product specifications that can contribute to good IEQ. Choosing the right building location and its orientation and designing an integrated daylighting system using IBM and choosing appropriate types of windows and glazing can both facilitate good daylighting and minimize solar thermal heat gains in the building. During construction, operation phase good house-keeping practices including the right air filters and cleaning products can be effective in maintaining a good IEQ condition inside the built environment.

VALUE AESTHETIC DECISIONS

The word "Biophilia" originates from Greek, 'philia' meaning 'love of'. It literally means a love of life and human connection to nature which is an intuitive and natural drive imprinted into our DNA. The project team should keep in mind this intuitive human love of nature and good views in designing the built environment which will fulfil the human desire and expectations.

PROVIDE THERMAL COMFORT

So many factors are involved in maintaining thermal comfort in a built environment. Using passive design ideas (discussed in Chapter 9) for heating, cooling and adequate ventilation is not only sustainable, but also provides a good IEQ for the building occupants. The thermal comfort for the building occupants is the function of air speed, humidity and temperature. A good HVAC system provides an adequate level of these three conditions within the built environment. Off course these conditions should also be in control of building occupants for having desirable levels of comfort based on their preference.

SUPPLY ADEQUATE LEVELS OF VENTILATION AND OUTDOOR AIR

Depending upon the number of building occupants and level and type of contaminants in the indoor air, concentrations of air pollutants may increase over time and become one or two orders of magnitude more than the outdoor air. The rule of thumb tells us that the level carbon dioxide in indoor air increases, so the level of other chemicals such as VOCs also increases. Therefore, monitoring the CO_2 levels of indoor air usually reveals a good indication of the other contaminants level within the built environment. By using such information, the ventilation rate of the building can be adjusted to achieve the desired levels of IAQ.

The global appearance of COVID-19 in 2020 is causing major reconsideration of ventilation rates to reduce the probability of the spread of this virus as well as other potential viruses.

PREVENT MOLD AND MILDEW

As discussed earlier, mold has become a major issue of IEQ system design. The building envelope is the first barrier to prevent intrusion of moisture and exotic objects such as mold and mildew into the built environment. Controlling humidity with HVAC system is also essential to preventing the growth of mold in the building, especially at extreme load conditions when outside humidity is either very high or very low. Mold is measured by the number of spores per cubic meter of air, and it is important to ensure that the level of mold in the indoor air is less than that of the outside air and in no case more than 700 spores per cubic meter of air.

USE LOW EMITTING MATERIALS

As their name implies VOCs are volatile and depend upon their Henry's Law coefficient, the volatilization rates of these chemicals are different. Many, like formaldehyde, are additives into the building materials such as paint to enhance its properties (e.g., making paint more durable and more rapidly drying). Toluene, xylenes, and benzene are other examples of synthetic VOCs that are toxic and harmful to human and ecosystem health. Despite benefits they may provide; VOCs pose a threat to health of building occupants and are being eliminated in the high-performance green buildings. Chemicals that are used in the building – for example, in cleaning supplies and copy machines – should be specially stored in spaces that prevent their migration into the surrounding building environment. In addition, these chemicals should be stored in insulated locations that minimize human exposure. Radon control should also be considered in areas where it has been identified as present in the local soil. For renovation projects, the removal of asbestos and lead-based paint as explained in Chapter 11, should be accomplished in a manner that prevents exposure to workers and future exposure to building occupants.

ENSURE ACOUSTIC COMFORT AND PRIVACY

Transmission of noise and sounds through buildings can affect both the health and comfort of building occupants, and significant effort should be made to minimize noise generation and transmission by the use of sound-absorbing materials; sound and noise attenuating walls, floor and ceilings; by isolating air handlers and other rotating machinery from the building; and by designing HVAC systems that are quiet and do not transmit conservation between spaces.

CONTROL DISTURBING ODORS THROUGH CONTAMINANT ISOLATION AND PRODUCT SELECTION

Some building spaces, such as copy rooms, janitors' closet, storage rooms, and designated smoking areas should be regularly depressurized and isolated from the other spaces in the building. This strategy prevents the migration of chemicals and odors typical of these spaces to the occupied areas of the building.

CREATING A HIGH-PERFORMANCE LUMINOUS ENVIRONMENT

Daylighting has enormous benefits because it contributes directly to human health and can provide significant energy savings when part of a well-designed, integrated lighting system. A wide range of high-performance lighting systems are available that can provide high-quality, high-efficiency lights.

PROVIDE QUALITY WATER

The building water system should be designed to provide the appropriate quality of water for all purposes. Potable water is needed for drinking, kitchen sinks, water fountains, lavatories, and dishwashers, and its quality should be monitored to ensure that it does not contain inappropriate levels of various metals and bacteria. The type of water that does not meet potable water standards can be incorporated into the building for usc such as flushing toilets and urinals and for landscape irrigation purposes. As the case with all aspects of high-performance building, periodic maintenance helps ensure that the building's water system provides the quality of water required for the activities in the building.

ADDRESSING THE MAIN COMPONENTS OF INTEGRATED DESIGN

In the following sections, we address the design of major subsystems that affect IEQ, including integrated lighting, daylighting, and views; thermal comfort and comfort control; acoustic comfort; electromagnetic radiation; and design of building HVAC systems.

INTEGRATED LIGHTING, DAYLIGHTING, AND VIEWS

Designing lighting and daylighting components of high-performance building is an important part of built environment. The lighting in the building not only should promote and enhance the health of its employees but at the same time should save energy as discussed in Chapter 9. Approximately, over 30 percent of US energy is consumed in buildings, reducing the energy consumptions in buildings will reduce CO2 generation and help to combat climate change. Therefore, substituting artificial light with natural light by choosing an effective daylighting strategy is very crucial in creating a healthy living environment to the building occupants and meanwhile reducing the energy consumption which is mainly coming from fossil fuel consumption.

In addition to using an integrated lighting system for buildings, the provision of views to the outside for building occupants is also important. Using sky-lighting and clerestory windows may bring ample sunlight to the building but fails to provide outside views to the building inhabitants. The degree of visual comfort and satisfaction to the building inhabitants is a function of both daylight and artificial lighting levels. Generally, these two forms of lighting are complementary to each other. The artificial lighting should be utilized at night-time or on rainy/cloudy days in absence of sunlight. However, there is a transition point in modern green buildings where artificial lighting and daylighting are tradeoffs, depending on the availability of daylighting. Modern daylighting control systems have the capability of throttling artificial light levels in response to the availability of daylighting, thus optimizing the use of electrical energy in the building for lighting purposes (Kibert, 2022). It should be noted, in the era prior to electricity, building illumination was provided largely by openings, windows, skylights and glazing, and significant design effort was invested in using natural light to the maximum extent possible.

The Whole Building Design Guide (WBDG), is a gateway to up-to-date information on integrated whole building design techniques and technologies. WBDG is one of the largest web-based portals providing government and industry practitioners with one-stop access to up-to-date information on a wide range of building-related guidance, criteria and technology from a 'whole buildings' perspective. According to the WBDG, the daylight design consists of systems, technologies, and architecture. The following list indicates some of the components of typical daylighting system design, although all of them may not be present at the same time:

- Daylight-optimized building footprint
- Climate-responsive window-to-wall-area ratio
- High-performance glazing
- Daylight-optimized fenestration design
- Skylights (passive or active)
- Tubular daylight devices

- Daylight redirection devices
- Solar shading devices
- Daylight-responsive electrical lighting controls
- Daylight-optimized interior design (such as furniture design, planning, and room service finishes).

According to Kibert, 2022, as is the case with most aspects of passive design, the design of the daylighting systems begins with the building footprint. In general, for good daylighting, buildings should be oriented in the east – west axis that maximizes north and south exposures, see Chapter 9. It is important that width of the building footprint in the north – south direction be minimized and in no case be less than 60 ft wide. German regulations stipulate that, in an office building, the occupants must be within 15 meters of an outside window to support both daylighting and views, see Figure 13.4.

Figure 13.4 The outside is brought in at Hillside Middle School in Salt lake City, Utah. Floor-to-ceiling windows throughout the design provide not only transparancy and daylighting but also excellent views from common areas. Source: GSBS Archtects and Benjamin Lowry, Photographer

The following are some of the design considerations in designing daylighting systems:

- Increase the number of perimeter daylight zones.
- Promote daylight penetration high in the space by locating widows high on the wall or by providing roof monitors and clerestories.
- Reflect daylight by using light colors to increase room brightness
- Slope ceilings to direct more light into space. Avoid direct-beam sunlight on critical visual tasks
- Filter daylight with vegetation, curtains, and louvers to help distribute the light
- Be aware that different building orientation requires different daylighting strategies. For example, light shelves, while effective on the south side of a building, would not be effective on the west and east sides.

In general, it is a good idea to use either a computer model or a physical model to assist in the design of an integrated lighting and daylighting system.

THERMAL COMFORT DESIGN AND CONTROL

Per ANSI/ASHRAE Standard 55-2017, thermal comfort is the condition of mind that expresses subjective satisfaction with the thermal environment. The human body can be viewed as a heat engine where food is the energy input. The human body will release excess heat into the environment, so the body can continue to operate. The heat transfer is proportional to temperature difference. In cold environments, the body loses more heat to the environment and in hot environments the body does not release enough heat.

Both the hot and cold scenarios lead to discomfort. Maintaining this standard of thermal comfort for occupants of buildings or other enclosures is one of the important goals of HVAC (heating, ventilation, and air conditioning) design engineers. Thermal neutrality is maintained when the heat generated by human metabolism is allowed to dissipate, thus maintaining thermal equilibrium with the surroundings. The main factors that influence thermal comfort can be grouped into:

- Environmental factors and
- Personal factors

Environmental factors include:

- Air temperature
- Thermal radiation
- Air speed and
- Relative humidity

Personal factors include:

- Metabolic rate, and
- Clothing

Three of the four environmental factors – temperature, humidity, and air speed are familiar. Thermal radiation is the effect of direct solar or other radiation on the skin, and it can affect people in the building spaces where there is direct sunlight, without shading or glare protection, into the space. Thermal comfort can be achieved through a wide variety of combinations of those factors. For example, it is well known that air speed can compensate for higher temperatures. Thus, ceiling fans used in rooms provide airflow that makes the higher temperatures more tolerable.

Similarly, lower humidity can make higher temperatures more acceptable. In addition to the four environmental factors noted, other factors, such as clothing, activity levels, and personal factors such as individual health, affect thermal comfort. Thermal comfort control is the ability of occupants to adjust at least one of the four environmental factors to their liking. Giving building users at least some degree of control over thermal comfort is recognized as contributing to the health and productivity of the occupants and is considered a significant measure in the design of high-performance green buildings. As the range of values deviates progressively from the ideal, fewer people are satisfied. This observation could be expressed statistically as the percentage of individuals who expressed satisfaction by comfort conditions and the predicted mean vote (PMV). The PMV predicts the mean response of a larger group of people who vote according to ASHRAE thermal satisfaction scale where:

In general, the rule is that if 80 percent of a population agrees that the thermal conditions are comfortable, then the combination of environmental factors is providing comfortable conditions. This is the basis for the recommendations in ASHRAE 55-2017. Thermal environment conditions for human occupancy, for acceptable combinations of environmental factors.

Inclusion of clothing levels and metabolic rates in determination of thermal comfort is also included in ASHRAE 55. The clothing level (CLO) is numerical value describing thermal insulation provided by clothing and ranges from 0.5 to 1.5. The CLO valuation assumes that a person is standing. If an individual spends most of the day sitting, CLO value may need to be increased, depending on the type of chair. CLO values are determined for the average occupant for each season on a space-by-space basis.

Psychological parameters, such as individual expectations, also affect thermal comfort. The thermal comfort temperature may vary greatly between individuals and depending on factors such as activity level, clothing, and humidity. People are highly sensitive to even small differences in environmental temperature. It was found that many people would be satisfied by an ideal set of values.

The Predicted Mean Vote (PMV) model stands among the most recognized thermal comfort models. It was developed using principles of heat balance and experimental data collected in a controlled climate chamber under steady state conditions (2) The adaptive model, on the other hand, was developed based on hundreds of field studies with the idea that occupants dynamically interact with their environment. Occupants control their thermal environment by means of clothing, operable windows, fans, personal heaters, and sunshades. The PMV model can be applied to air-conditioned buildings, while the adaptive model can be applied only to buildings where no mechanical systems have been installed. There is no consensus about which comfort model should be applied for buildings that are partially air-conditioned spatially or temporally.

The MPV index predicts the mean response of a larger group of people who vote according to the ASHRAE thermal sensation scale where "In general, the rule is that of 80 percent of population agrees that thermal conditions are comfortable, the combination of environmental factors is providing comfortable condition. This is the basis for the recommendations in ASHRAE 55-2017, thermal environment conditions for human occupancy, for acceptable combinations of environmental factors, Kibert, 2022.

The metabolic rate (MET) estimated the typical level of activity of the occupants within a given space. MET is expressed on a decimal scale and ranges from 0.7 to 8.7. The 0.7 level represents sleeping or resting, while above 1.0 is light activity; greater than 2.0 represents moderate activity and perspiration. When values rise above 1.0, evaporation of perspiration becomes a factor in an individual's level of comfort. An estimate of the average metabolic rate of the occupants in a given space is determined as an input to assess thermal comfort.

ASHRAE Standard 55-2017 addresses the use of various new technologies to deliver thermal comfort using lower-energy approaches. Air movement in general, is becoming a more popular strategy for cooling occupants as opposed to lower operational temperature because the energy requirements are lower. ASHRAE 55-2010 includes a new method for determining cooling effect of air movement above 30 feet per minute. This method allows ceiling fans or other means of elevating air speed to provide comfort at higher summer temperatures than per previously permissible. New provisions based on field-study research allow elevated air speed to broadly offset the need to cool air in warm conditions. ASHRAE 55-2010 allows modest increases in operative temperature beyond the predicted PMV limits as a function of air speed and air turbulence, both of which increase the cooling sensation by using convection to remove heat from the skin.

ACOUSTIC COMFORT

Acoustic comfort is the state of well-being experienced in an appropriate sound environment, where noise is controlled and managed. This comfort goes far beyond a simple decibel measurement: it compasses the quality of sound, its impact on our concentration, communication, rest and health. Acoustic discomfort can be caused by noise that is too loud, by continuous noise of less intensity, or by irregular noise, even if it is infrequent during the day, for example, passing an aircraft.

Noise is a major disruptor of the body it has direct effect on the nervous system. It is a significant stress factor that affects our physical and mental well-being. Noise levels that do not directly affect hearing can, especially if they are frequent and prolonged, lead to fatigue, sleep disturbance and reduced cognitive performance. In workshops and outdoors, workers and craftsmen wear hearing protection devices.

Consideration of acoustic comfort in buildings generally falls far down the list of priorities, both in sustainable construction and in conventional building design. Acoustics are, in fact, very important to the health, well-being, and productivity of people in the offices, schools, and virtually every other type of facility. Therefore, a good acoustic environment for building occupants helps increase their performance and reduces the incidents of illness and lost workdays. Acoustic comfort is part of a bigger picture of overall space comfort and includes not only acoustics but also other issues, such as thermal comfort, lighting quality, the availability of daylight, and other similar factors.

Noise is prevalent in buildings and can come from outside traffic noise, voices within the building, mechanical equipment in adjacent spaces, copiers, phones, and numerous other sources. To produce a good acoustic environment, several problems must be addressed: (1) noise outside the building, (2) noise from adjacent spaces, and (3) lack of self-control in the spaces of the building. Although the noise in the spaces may not be harmful to hearing, the presence of distracting noise reduces concentration on work or study and decreases the productivity of individuals.

SOUND AND NOISE CONTROL TERMINOLOGY

The unit of sound intensity is called decibel. In the US noise reduction is measured by Sound Transmission Class (STC), which is a number that represents the noise reduction, in decibels (dBA) of a building element such as a wall or window. Note that the decibel scale is logarithmic; thus, a 10-dBC reduction in sound between two spaces corresponds to about a 50 percent reduction in sound volume. For example, a 40 dBA sound is about half as loud as a 50-dBA sound.

A wall with an STC rating of 20 provides a 20-dBA sound reduction. A wall with an STC rating of 20 and with a 60 dB sound level on one side would reduce the noise level to 40 dB on the other side. A typical home interior wall constructed of one sheet of 1/2–inch drywall on either side of a wood frame has an STC of about 33. Typical STC values are presented below:

STC	Sound Level
25	Normal speeches can be understood quite easily and distinctly through the wall
30	Loud speeches can be understood fairly well, normal speech heard nut not understood
35	Loud speeches audible but not intelligible
40	Onset of Privacy
42	Loud speeches audible as a murmur
45	Loud speeches not audible; 90% of statistical population not annoyed.
50	Very loud sound such as musical instruments or stereo can be faintly heard; 99% of population not annoyed
60	Superior soundproofing; most sounds inaudible

Source: North American Insulation Manufacturer's association 1997.

Table 13.3 Effects of STC Rating on Sound Transmission Through a Building Element

EXTERIOR NOISE ISSUES AND CONTROL

The problem of the exterior noise issue had to be dealt with during the selection process of the site for construction with the idea of the intended use of the built environment. For instance, if the site will be used as a school facility, or a quiet residential dwelling, then during selecting the land a careful review of traffic volume, and type of activities in the immediate neighboring properties should be very crucial in eliminating the future acoustic problems.

In general, sites that are in high noise areas such as industrial areas, and highways, should be avoided. If the site has already been established, and it is in proximity of a high noise area, building spaces can be re-organized such as storage areas, rest rooms, janitorial closet, mechanical rooms to be relocated on the side facing the source of the noise. More sensitive areas, such as classrooms, are on the quiet side of the building. Earth berms or other structural solutions, such as concrete barriers, may be required if there is more than one direction from which significant noise is generated. Window also play an important role in controlling the exterior noise; therefore, they should be selected with special consideration for controlling noise. Double- and triple-pane glass with inert gas infill may be the best solution for situations where there is significant exterior noise problem.

INTERIOR SPACE ACOUSTIC REQUIREMENTS

Each type of interior space has different considerations and requirements, depending upon the type of activities that are happening in the space. Private offices, for example, require a space where private conversation can occur without being heard in adjacent spaces and where the acoustic conditions support workers health and productivity. These types of spaces have generally problems of noise transmission through partitions, excessive noise levels in the room, and noises from the building's air handling systems. Some of the solutions recommended by Whole Building Design Guide are:

- To extend walls from floor to structural deck above
- Insulate partitions to achieve the required STC value to reduce noise transmission from adjacent spaces
- Locate offices and conference rooms so that they are not adjacent to mechanical equipment rooms.

Classrooms are spaces designed for learning, and modern classrooms generally have multimedia communication environments. Good acoustics are needed for effective verbal communication, which means that there must be relatively low noise levels. Some types of noises that interfere with the learning process are noises from the outside of the school, such as the nearby traffic and aircraft flying over, hallway noise, and noise from the other adjacent classrooms, mechanical equipment and duck work, and noises within the classroom itself. The recognition of the need to have a high-quality acoustical learning environment in classrooms resulted in the publication of American National Standards Institute, Inc. For more information, please refer to ANSI/ASA S12.60, which provides acoustic performance criteria, design requirements, and design guidelines for new school classrooms and other learning spaces. It aims to ensure optical learning conditions by minimizing noise reverberation within classrooms and other learning spaces.

HVAC SYSTEMS AND IEQ

Proper design of a building's HVAC system is perhaps the most important approach for providing a healthy indoor environment. Conversely, a poorly designed HVAC system can be harbinger of trouble. The HVAC system provides a means for moving, exchanging, filtering, and conditioning all the air in a building. Because the HVAC system plays such an important role in IEQ, it is imperative that it be understood and maintained properly. The following describes the advantages offered by an effective HVAC system and the problems caused by a poorly designed system.

HVAC SYSTEM DESIGN

High-performance homes require better HVAC design strategies than conventional homes, and builders frequently find that the high-performance HVAC learning curve can be challenging. HVAC contractors who work with high-performance builders can't use the same strategies that work for mainstream builders. Here are a few design concepts and strategies to help builders better understand HVAC design for high-performance homes.

Historically, many homes in northern latitudes have had only heating systems, but this is changing. In these heating-dominant climates, HVAC system designs would naturally first need to meet the heating needs of a building. So, the system should be designed to provide enough heat and secondarily meet the air conditioning needs of a building. In parts of the country with a cooling dominant-climate, the cooling and heating load calculation are approached from the other direction. Now split-ductless systems provide both cooling and heating in one energy efficient system that is appropriate for both heating and cooling climates.

High-performance homes have significantly lower cooling and heating requirements than conventional homes. A 1,500-square-foot home in Climate Zone 5 (See Chapter 9, Figure 9-17). For example, might have a design heating load of approximately 35,000 Btu/h if it is built to current codes. That same home built with a high-performance envelope, might have a heating load of approximately 26,000 Btu/h. If this same home were built to the Passive House level, the heating load might be below 15,000 Btu/h.

Split-ductless systems are a great solution for high-performance homes with low cooling and heating needs. The systems are energy-efficient and come in a range of capacities to match any size home. Split-ductless systems offer a variety of options to serve any space. They can connect multiple ducted or ductless indoor units to one outdoor compressor system. These options allow split-ductless systems to fit many different circumstances and visual requirements. For example, Mitsubishi Electric Cooling & Heating (Mitsubishi Electric) now offers a one-way cassette that fits between standard joists and is flush mounted on the ceiling when homeowners/builders want to blend the HVAC design aesthetic into a room and conceal the indoor unit.

HOW TO DESIGN A HVAC SYSTEM

HVAC system design includes three basic steps: calculating heat loss and gain, selecting equipment and designing duct runs. For each step in designing an HVAC system, there is an associated technical manual provided by the Air Conditioning Contractors of America (ACCA). The ACCA technical manuals are so widely adopted that some design steps are commonly referred to by the name of the manual. Step one in designing an HVAC system is conducting a room-by-room cooling and heating load calculation. This Manual J®calculation is generally done by the HVAC contractor or mechanical engineer after the builder has provided the envelope details and a set of plans using software tools such as Wrightsoft®, Adtek, or SpecPro.

Step two in the HVAC design process is selecting the right size system using Manual S®. Improper sized systems such as undersized systems will not provide enough heating to provide comfort on cold winter days. Oversized systems increase costs needlessly and will not perform properly due to short cycling when the system is too large for the house and rapidly reaches the desired temperature. An appropriately sized system will run longer, which puts less strain on the equipment

If you are using compact ducted indoor units, there will be a third step: proper duct design. Duct design is accomplished by following the procedures in Manual D®for static pressure and CFM flow and then Manual T ® (for termination) to select register grilles. Duct design can be challenging, and poor design often leads to comfort problems.

UNIQUE HVAC DEMANDS FOR HIGH-PERFORMANCE HOMES

In high-performance homes, all ductworks should be inside the conditioned space. Any time ductwork is situated outside of conditioned space, there is an energy penalty and the risk of condensation in/around the ductwork increases. International Energy Conservation Code® (IECC) 2018 may allow ductwork to be buried in insulation, but this is still considered outside the conditioned space. Ducted mini-split systems are designed with these space constraints in mind and should always be installed within the conditioned space. Ductless systems also require room-by-room considerations. Bedrooms may have very low heating or cooling demands, yet some builders and HVAC contractors put a wall-hung head in every bedroom without considering the actual needs of each bedroom. This is likely to increase system costs and lower system performance.

THE INS AND OUTS OF LOAD CALCULATION

Load calculations are based on the coldest and hottest days of the year and are done to ensure that comfort will be provided during the extremes. However, most of the year, temperatures fall somewhere in the middle. This middle territory is referred to as the "shoulder season" for HVAC design. The ability to go from maximum capacity to a low-capacity output is called the "turndown ratio." Value-priced split-ductless systems frequently do not ramp down to lower capacities and therefore have poor turndown ratios. If a system has a poor turndown ratio, it is more likely to turn off and then back on frequently. Called short cycling, this situation degrades comfort, energy efficiency, and equipment life.

Oversizing systems is frequently seen in the field. It indicates a poor job of doing room-by-room load calculations or the existence of shortcomings in the distribution design. Cooling and heating loads are getting lower as building codes become more stringent. In high-performance homes, very low cooling and heating loads require new thinking to ensure the HVAC system performs just as well as the building envelope.

Earlier, split-ductless systems (including unitary air source heat pump systems) had poor cold climate performance. In the last 10-15 years, Mitsubishi Electric and other manufacturers have created air source heat pump technology that is designed to perform efficiently in cold climates. In cold climate areas with design temperatures below 30°F, it is important to select a system which will perform efficiently without supplemental heat. Mitsubishi Electric Hyper Heat systems use technology that will provide energy efficient heating in the winter down to -13°F. The Northeast Energy Efficiency Partnership (NEEP) has a cold climate air-source heat pump qualified product list that is a great resource for selecting a system which will perform well in these areas.

For more information visit:

https://zeroenergyproject.com/2018/07/12/hvac-design-strategies-for-high-performance-homes/

HVAC DESIGN MEASURES FOR PREVENTING COVID-19 AND ANOTHER VIRUS TRANSMISSION

Unlike mold and mildew which generally are caused by poor moisture control design and/or moisture barrier installation, viruses are brought into a building by people working in or using the building. COVID-19 is highly contagious and is spread mostly by air movement. The virus spreads when infected people talk, sneeze, or cough, resulting in droplets that can land on or near others. Touching objects that have been contaminated by infected people is another potential pathway for transmission. The top priorities must be enforced before considering other measures: wearing masking and physical distancing of at least 6 ft. The very small virus particles are about 0.1 microns and are carried in large droplets of 5 to10 microns. The larger droplets evaporate, leaving behind an aerosol of virus particles, which are thought to be active for up to 3 hours. Problematic conditions include crowded offices and spaces where people are present for extended periods of time, and in spaces where the HVAC systems move air at high speed, spreading the virus from infected people to others. Some measures that have been recommended for existing building retrofits and new building design include:

- Variable-speed fan motors in the air-handling system to allow lower speed airflows
- More sophisticated, pressure sensitive airflow-control systems that manage airflow based on number of people in space, usually detectable by measuring CO_2 concentrations
- Use of high-performance air-purification systems

Guidance issued by ASHRAE outlines HVAC strategies that can be employed to affect infectious aerosol transport air distribution patterns, differential room pressurization, personalized ventilation, source capture ventilation, filtration (central or local), controlling temperature and relative humidity and proven technologies such as ultraviolet germicidal radiation (UVGI) (ASHRAE 2020). In terms of potentially removing or killing off viruses, including the coronavirus, there are four general categories of intervention that can be applied: filtration, irradiation, heating, and miscellaneous technologies see Table 13.4.

Filtration	Irradiation	Thermal	Other
Air is forced through fiber-based material or membrane	Uses electromagnetic radiation, most commonly ultraviolet-C light, to deactivate pathogens	High-intensity targeted heating	Includes ionized purifiers and ozone generators; mechanism of action varies by technology
Efficacy varies, depending on fiber type	Cannot be used in occupied spaces, since light may be harmful to skin and eyes	Can be deployed in stand-alone units or through HVAC systems	
	Only practical when light can reach surfaces; many technical issues must be considered when installed	Required air to be exposed to heat for significant periods and thus may be difficult to apply when there is a continuous flow	
Virus still active	Virus is made inactive via Sterilization Process		
Which can be challenging for a continuous stream of air or aerosol			

Table 13.4 Methods That Can be Used to Purify Contaminated Air in Building HVAC Systems

The strategies that can be used to address viruses vary depending on the building types. ASHRAE already addresses ventilation system design for healthcare facilities but there are not yet any specific requirements for public facilities, schools, homes, and commercial/institutional buildings. In general, higher ventilation rates reduce the risk of transmission but may result in the intake of outdoor pollutants. High efficiency air filtration is a well-known strategy that can be effective but will not eliminate all risk, Ultraviolet (UV) light, especially in UV-C spectrum has good germicidal effect and can be used with other measures such as filtration and good ventilation control to lower the overall risk to building occupants and users. ASHRAE has made it clear that it is now important to consider infectious aerosol dissemination in all facilities and that risk assessment should be conducted to support the design of buildings. ASHRAE has also stated that it is important to go beyond the minimum requirements stated in existing standards and include in current research reports, papers, and studies that advance the state of knowledge regarding mitigation measures.

EMISSIONS FROM BUILDING MATERIALS

Traditional construction materials like cement or steel come with a large environmental price tag, adding to the building sector's climate performance. Currently, most decarbonization efforts are focused on reducing emissions from heating, cooling and lighting. This is set to change as Europe's buildings become more energy efficient, gas and coal heating is replaced, and electricity generation is made cleaner.

In fact, looking at buildings' carbon footprints throughout their entire lifecycle – from the manufacture of their materials to their demolition – it is expected that the share of emissions embodied in construction materials to be just as high as operational emissions by 2050. In the most extreme cases, embodied emissions already make up more than 90 percent of a building's carbon footprint.

What materials are chosen to build with is critical towards determining – and limiting – the full climate impact of offices, shops and houses.

This is because the production of today's most widely used construction materials often is very polluting. "Just three materials – concrete, steel and aluminum – are responsible for 23 percent of overall global emissions today," a 2023 UN Environment Program (UNEP) report states. In Europe, construction accounts for 250 million tons of CO2 emissions every year.

More sustainable construction materials exist, but they tend to be costlier, and builders often distrust them. Still, developers and architects are increasingly using nature-based alternatives, such as timber, clay and even hemp, in their projects. At the same time, many manufacturers are finding ways to green up their products. What materials can home, schools and restaurants be made of in a net zero future?

BRINGING SUFFICIENCY AND EFFICIENCY INTO THE EQUATION

Avoiding new construction altogether is generally the most sustainable option: Extending the lifetime of existing buildings, making better use of them and re-using their components and materials when they must be deconstructed is the single option likely to save the most emissions, limit resource extraction and cut waste.

"We need to think about growth and consumption in general," Kim Gundlach, architect and researcher at the Natural Building Lab (NBL), told Clean Energy Wire (CLEW). In many places, including Germany, it is often easier to demolish and re-build rather than adapt or refurbish an existing building. At the same time, researchers agree that

Europe's building stock is not used to its full potential. "It's really important that we continue to use the spaces that we already have and make them fit for further use.

Dirty construction habits block path to truly climate-friendly buildings. Using materials that have already been manufactured is one of the lowest-hanging fruits to decarbonize new construction and renovation projects. There is tremendous potential to re-use materials already present in our built environment, especially in developed economies with a large and ageing building stock.

For new buildings, researchers say that both designing efficiently with fewer materials (which can be achieved while retaining the same structural strength) and taking a closer look at existing building standards that might conflict with climate neutrality targets should be more prominent in the discussion to make buildings more sustainable.

Reused and recycled buildings challenge wasteful construction status quo shifting to bio-based materials. Constructing with bio-based materials would allow the building sector to go from significant polluter to carbon heroes. Materials such as timber, bamboo, straw or hemp all capture carbon as they grow. Schools, homes and offices built using these materials could store more carbon than their construction emitted in the first place and thus make them sink over the buildings' entire lifespans, which usually cover at least several decades. If stored CO_2 in a 100- year-old-tree stays in a building for longer than 100 years, then we have managed to delay its release and help the atmosphere breathe a little.

The shift to (properly managed) bio-based materials could lead to emissions savings of up to 40 percent in many of the world's regions by 2050, even compared to savings from low-carbon concrete and steel, according to UNEP's 2023 report. In general, the construction sector could reduce emissions by 18 percent by 2030, if 50 percent of new residential construction shifted to bio-based materials.

However, "The economic incentive is just not strong enough for a lot of people to actually work with these materials, this is due to very complex building certification and permitting processes in US, as well as issues of liability and warranty, especially for re-used materials and non-standardized products.

It is important to make construction processes much more flexible and adaptable, since we do not know what materials will be available in future climatic conditions. Therefore, we need to strengthen regional value chains for more regional resilience: smaller structures may be able to adapt much more organically.

TIMBER

Mass timber (engineered wood, such as cross-laminated timber) can compete with steel and reinforced concrete in building structures. It holds great potential as a sustainable construction material and already is widely used in projects today. Timber could help society reach climate targets: a cubic meter of wood contains around one ton of carbon, depending on the tree species. Among its many benefits, mass timber is highly durable, fireproof, and building components can be prefabricated off-site in a modular or serial way, reducing construction waste and speeding up building processes. Wood is also light, making it ideal for adding stories to existing buildings, which create living areas without sealing up more ground. Moreover, it is an effective insulator, reducing the need for heating and cooling.

However, just how climate friendly wood products depend on the extent of sustainable forest strategies and responsible felling, as well as on broader social and ecological contexts. Trees take a long time to grow and the resilience of forests to climate change is an important factor for their productivity. "There's a lot of research into how the construction sector can deal with different types of wood, such as hard wood, because we simply don't know which trees will grow in future and whether forests will continue to grow as vigorously as they do now or whether they will have problems with droughts

To focus on timber as a building material is important. However, instead of new buildings we need to focus on renovations and extensions and work with the existing building stock as well as hardwood materials are important for a climate-just construction sector."

CLAY

Already a popular construction material in many parts of the world until not too long ago, clay is experiencing a renaissance as a sustainable alternative to conventional building materials. Clay buildings are environmentally friendly, sound-insulated, non-combustible, and the material generally is locally available.

Clay is used in ceilings, walls, roofs and floors, as well as for panels in drywall construction, and clay plasters are also common. Pressed or molded into bricks, clay (or rammed earth) can also have immense load-bearing capacities. Clay has great temperature and moisture-regulating properties and also has a positive effect on room acoustics and room climate. However, the material is sensitive to weather.

Due to increasing awareness of issues related to IAQ, both public and private industry are promoting the use of low-emission building materials. Communicating IEQ requirements to subcontractors and suppliers is an important step in the process of creating a healthy building, but there is still debate about how to include materials emissions requirements in specifications. I additions, the EPDs should include the type of chemicals and their Reference Exposure Levels (RELs) in building materials. A chronic REL is an airborne concentration level that would pose no significant health risk for individuals indefinitely exposed to that level. Chronic RELs have been developed for over 80 hazardous substances; another 60 chemicals are under review. The measured maximum indoor air concentrations of the chemicals must fall to less than 50 percent of its associated REL, see Table 13.5 adopted from DTSC's Office of OEHHRA (2009), Public Health Goals.

Chemical Name	Chemicals Abstract Service Number	Inhalation REL (μ/m^3)
Benzene	71-43-2	27
Chloroform	67-66-3	150
Ethyl glycol	75-00-3	400
Formaldehyde	50-00-0	55
Naphthalene	91-20-3	9
Phenol	108-95-2	5,800
Styrene	100-42-5	21,000
Toluene	100-88-3	5,000
Trichloroethylene	79-01-6	600
Xylenes	Several	700 – 22,000

$\mu g/m^3$= microgram per cubic meter
Source: California Department of Toxic Substances Control (DTSC) Office of Ecological and Human Health Assessment (OEHHA), 2015

Table 13.5 Chronic RELs for Selected Organic Chemicals associated with IAQ

The impact of the following chemicals which are used as an additive in processing of the major building materials are discussed next.

ADHESIVES, SEALANTS, AND FINISHES

Adhesives, sealants, caulks, coatings, and finishes are placed in the building when wet and are expected to dry, or cure, on the premises. The release of VOCs is inherent part of this process. The solvents used in formulating these materials are the source of most VOCs emitted during drying and later during building occupation.

Adhesives are materials or substances that bind one surface to another. They affect a wide range of construction materials; adhesives can be applied with flooring and wall coverings, or they may be a component of a material, such as plywood, particleboard, moveable wall panels, and office workstations. Adhesives are applied in the liquid or viscous state, then cure to a solid or more solid state to achieve bonding. Most adhesives release VOCs and pose the greatest thread during their application and curing. When applied, adhesives should be used with areas with increased ventilation for 48 to 72 hours to avoid accumulation of VOCs. The packaging label or other installation information for adhesives should always be consulted for additional product-specific precautions.

One method of characterizing adhesives in terms of their influence on IAQ is to identify the resin used in the base. Resins can be natural or synthetic. Natural resins usually have low emission potential, but in synthetic resins, emission potential can vary dramatically. Currently advances are being made in the development of adhesives with low or no emissions.

Sealants are applied in joints, gaps, or cavities to eliminate penetration of liquids, air and gases. Note: although the construction industry differentiates between indoor and outdoor sealants, (the former being referred to as caulks, and the latter as sealants). Sealants usually are selected based on their flexibility and resin base. Like adhesives, sealants can be hazardous during application and curing. Their emission potential is directly related to the percentage of base resins and solids. Fortunately, sealants, which raise concern with regard to their VOC emission potential, are used indoors in small quantities. Alternate water-based sealants manufactured using nontoxic components are available. One such product for interior use is a vinyl adhesive sealant. An acrylic latex exterior sealant for building joints is also on the market.

FINISHES

Finishes encompass a wide range of products, including paints, varnishes, stains, and sealers. Finishes are the major component of building materials and furnishing whose primary purpose is to provide protection against corrosion, weathering, and damage. Secondarily, they may also add aesthetic value to building materials. All finishes have similar characteristics. They require resins and oils to form a film and to aid adhesion by promoting penetration into the substrate. All coatings require carriers (water or organic solvents) that provide viscosity for application. Carriers also improve adhesion through evaporation.

Paints and stains require solids, including pigments, to provide various colors. The number of solids is a good indicator of the VOC emission potential of the finish. The sanding or burning of finishes generates potential IEQ hazards such as dust from talc, silica, mica, and especially lead.

Water-based finishes typically are emitting low; however, organic solvent-based finishes are more likely to be high emitting. The current trend is to replace conventional finishes with water-based alternatives, although paints have been targeted primarily in this effort. Very few stains, sellers, and vanishes have been successfully adapted for low VOC emissions, because, to date, alternative finishes generally do not perform as well as their traditional counterparts. The new products often require more application to achieve results similar to those of traditional products. The color selection of alternative paints is limited as well. And although hypoallergenic, preservative-free paints are available, their shelf life and color selection are limited. It is also important to point out that water-based products may have low VOCs but contain other hazardous materials. Unlike organic-based paints, water-based paints require preservatives and fungicides, such as arsenic disulfide, phenol, copper, and formaldehyde. These additives are considered chemical hazards by the National Institute of Occupational Safety and Health (NIOSH).

USGBC LEED and Green Globes building assessment standards provide credit for the use of low-emission paints and coating if their VOC emission does not exceed the VOC and chemical component limits required by Green Seal's GS-11, issued in 2015.

PARTICLEBOARD AND PLYWOOD

The major components of plywood and particleboard are Urea Formaldehyde (UF). Both plywood and particleboard emit UF upon manufacturing and being installed. The emission rate of UF is affected by the temperature and humidity of the ambient air.

Particleboard is a composite material made from wood chips or residues, bonded together with adhesives under heat and pressure. Particleboard is relatively expensive and is available in sheets that measure 4 by 8 ft. The major IAQ concern with particleboard is off gassing of formaldehyde. Most particleboards contain UF, and a very small percentage then contain phenol formaldehyde (PF). Particleboard containing PF emits far less formaldehyde than board made by UF. PF is used in particleboard where a high-moisture environment is anticipated, especially restrooms and kitchens.

The most common construction application of particle-board is as a core material for doors, cabinets, and a wide variety of finishing, such as tables and prefabricated wall system. Particleboard also is used in wood-framed housing, primarily for non-structural floor underlayment. Usually, a finished floor is installed over particleboard. Particleboard is also used as a backing for paneling. Once the board is covered, the VOC content is inconsequential because formaldehyde emissions are delayed for as long as the board remains covered.

Although particle board can be manufactured with lower formaldehyde emissions, it is still of great concern because of the possibility of large, exposed surface area in relation to the volume of the given space. Emissions of trace amounts of formaldehyde can continue for several months or even years. These emissions do decrease over time, but rates increase as temperature and/or humidity rises. It is estimated that emission rates double with every increase of 12°F.

Plywood

Plywood is composed of several thin wood layers oriented at alternating 90° angles that are permanently bonded by an adhesive. Exterior plies are referred to as faces, and interior plies are known as the core. Plywood generally is classified as hard-wood or softwood. Approximately 80 percent of all softwood plywood is used as wall and roof sheathing, siding, concrete framework, roof decking, and subflooring. Hardwood plywood is used for building furniture, cabinets, shelving, and interior paneling. The type of adhesives used to bond the plies plays a major role in assessing the effects of the plywood on IAQ. The surface area of the plywood in relation to the volume of the space is another determining factor for proper IAQ. Interior-grade plywood is generally bonded with UF resins. The off gassing of UF in plywood can be compounded by finishes or sealants used in conjunction with the plywood. Size, temperature, humidity of the space, surface area, and finish of the plywood all can affect the concentration of formaldehyde emissions, Kibert, 2022.

FLOOR AND WALL COVERINGS

Of all building materials, carpet has generated the most debate, which is ironic, considering that emissions from carpet system are relatively low compared with emissions from other building materials. Most of emissions associated with carpeting are due to the adhesives used to secure it. Thus, when selecting carpet, the entire system and the emission of each constituent must be evaluated. The components of a carpet system are the carpet fiber, carpet backing, adhesives, and carpet pad (generally used in residential applications only). Carpet backing is used to hold the fibers in place. Often two backings are used: one keeps the fibers in place, and the other adds strength and stability. The secondary backing is made from fabric, jute, or polypropylene bonded with either styrene-butadiene rubber (SBR) latex or a polymer coating, such as synthetic latex. SBR latex contains the chemicals styrene and butadiene, which are known as irritants of mucous membranes and skin. SBR latex adhesives are found in primary and secondary backings and emit low but steady amounts of the by-product 4-phenylcyclohexene, the chemical that is responsible for the "new carpet" smell and is suspected of being a possible source of building occupant illness complaints.

Adhesives may be used twice in common carpet systems: to glue the backing to the fiber and/or to glue the carpet system to the substrate. Carpet pads are optional part of the carpet system. They generally do not contribute to IAQ problems. There are five basic types of pads: bonded urethane, prime polyurethane, sponge rubber, synthetic fiber, rubberized jute.

There are five basic carpet fiber materials. Wool is the only natural fiber, and accounts for less than 1 percent of the carpet market. The remaining four – nylon, olefin, polyester, and polyethylene terephthalate – are synthetic fibers. Derived from petrochemicals, synthetic fibers are stronger, more durable, and usually less expensive than wool; they are also less likely than wool to release small fibers into air. Both USGBC LEED-NC and Green Globes building assessment standards provide credit for using low-emission carpeting system if the system meets or exceeds the requirements of the Carpet and Rug Institute's Green Label Plus program.

RESILIENT FLOORING

Resilient flooring is a pliable or flexible flooring. Tiles and sheets are the two basic forms, both of which are attached to a substrate using adhesives. Resilient flooring can be composed of vinyl, rubber, or linoleum. Vinyl flooring is made primarily of PVC resins, with plasticizers, to provide flexibility; filters; and pigments for color. Bubber flooring comes in two basic forms – smooth-surface or molded – and is made from a combination of synthetic rubber (styrene butadiene), nonfading organic pigments, extenders, oil plasticizers, and mineral filters. Linoleum is a natural, organic, and biodegradable product. Its main components are linseed oil, pine rosin, wood flour, cork powder, pigments, driers, and natural mildew inhibitors. Linoleum tiles are durable, greaseproof, and water- and fire-resistant. They also are easily maintained and long-lasting.

Typically, no individual compound in resilient flooring has high VOC emissions. Plasticizers are the main source of emissions. Using more rigid, less plastic tile is recommended to avoid potential hazards. Note, however, that low-emitting tiles may be glued with high emitting adhesives.

WALL COVERINGS

Wall coverings are a popular alternative to paints. Many of the available coverings pose little or no threat to IAQ. The three basic types of wall coverings are paper, fabric and vinyl. Paper itself has no impact on IAQ, but the adhesives used to apply it may contain formaldehyde. However, most paper adhesives are purchased as powder and mixed with water; thus, they emit little or no VOCs. Fabric wall coverings may contain formaldehyde, which sometimes is used to keep the material from fading and to improve resistance to water. Fabric coverings also can act as a sink by absorbing extraneous VOCs in a building and re-emitting them into space. Two major concerns with vinyl wall coverings are the environmental conditions of the project location and the construction of the walls receiving the finish. In temperate climates, when moisture may not be readily evaporated, vinyl-covered walls can become moldy.

INSULATION AND CEILING TILES

Insulation and acoustic ceiling tiles can contribute VOC and particulate contaminants from a variety of sources. Depending on their composition, these materials may incorporate a variety of adhesives and fibrous materials that can combine to complicate the IAQ issue.

Insulation

Most insulation is made of fiberglass, mineral wool, and cellulose (made from recycled wood). Asbestos was also used frequently until the late 1970s. Fiberglass and mineral wool have raised IAQ concerns because of the small fibers that are produced when the material is disturbed. Fiberglass is listed by the International Agency for Research on Cancer as a possible carcinogen. Cellulose insulation is generally spray-applied and is considered a nontoxic material. In this materials category, foam insulation has received most of the attention because of its impacts on the environment rather than on IAQ. That said, VOCs are emitted from synthetic foam during manufacturing or while spray foam is used.

Acoustic Ceiling Tile

The suspended acoustic ceiling is one of the most common structures found in commercial buildings today. Most acoustic ceiling tilesare made from mineral or wood fibers, which are wet and compressed to the desired thickness, size, and pattern. The tiles usually are coated with latex paint at the factory. The primary concern regarding the effects of acoustic ceiling tiles on IAQ is the occurrence of microbial growth on either mineral fiber or fiberglass tile exposed to moisture. Another concern is that porous tiles can absorb VOCs and then re-emit them.

REFERENCES

ASHRAE .2020. "ASHRAE Position Document on Infectious Aerosols" (April 14). ASHRAE Board of Directors.

Bas, Ed. 1993. Indoor Air Quality in the Building Environment, Troy, MI: Business News Publishing.

DTSC: 2015. Inhalation RELs, documentation. Available at: oehha.ca.gov/air/general-info/oehha-acute-8-hour-and-chronic-reference-exposure-level-rel-summary

HVAC Design Strategy for High-Performance Homes. 2018. Available at:

https://zeroenergyproject.com/2018/07/12/hvac-design-strategies-for-high-performance-homes/

Kibert J., Charles 2022. 'Sustainable Construction, Green Building Design and Delivery. Fifth Edition, John Wiley and Sons, Inc. Hoboken, NJ.

Kim Gundlach of Natural Building Laboratory (www.nbl.berlin) Architect and Urban Design.

US Environmental Protection Agency (EPA). 1991. Building Air Quality: A Guide for Building Owners and Facility Managers. EPA/400/1-91/033 (September). Washington, DC: US Environmental protection Agency.

Sundell, J; Lindval, T; Berndt, S. 1994. "Association Between Type of Ventilation and Airflow Rates in Office Buildings and Risk of SBS among Occupants. Environ. Int. 20 (2):239-251.

UNEP's Annual Report. 2023. Keeping the Promise, an Official UN Site.

World Health Organization (WHO). 1984. Report on Sick Building Syndrome (SBS) and Building Related Illness (BRI).

World Building Design Guide (WBDG) is available at www.wbdg.org.

www.ingramcontent.com/pod-product-compliance
Lightning Source LLC
LaVergne TN
LVHW052015200926
848481LV00005B/1500